Food Security in the United States

To Mom,
who has a special
fondness for food.
Love
Lawrence
6/11/84

Westview Replica Editions

The concept of Westview Replica Editions is a response to the continuing crisis in academic and informational publishing. Library budgets for books have been severely curtailed. Ever larger portions of general library budgets are being diverted from the purchase of books and used for data banks, computers, micromedia, and other methods of information retrieval. Interlibrary loan structures further reduce the edition sizes required to satisfy the needs of the scholarly community. Economic pressures on the university presses and the few private scholarly publishing companies have severely limited the capacity of the industry to properly serve the academic and research communities. As a result, many manuscripts dealing with important subjects, often representing the highest level of scholarship, are no longer economically viable publishing projects--or, if accepted for publication, are typically subject to lead times ranging from one to three years.

Westview Replica Editions are our practical solution to the problem. We accept a manuscript in camera-ready form, typed according to our specifications, and move it immediately into the production process. As always, the selection criteria include the importance of the subject, the work's contribution to scholarship, and its insight, originality of thought, and excellence of exposition. The responsibility for editing and proofreading lies with the author or sponsoring institution. We prepare chapter headings and display pages, file for copyright, and obtain Library of Congress Cataloging in Publication Data. A detailed manual contains simple instructions for preparing the final typescript, and our editorial staff is always available to answer questions.

The end result is a book printed on acid-free paper and bound in sturdy library-quality soft covers. We manufacture these books ourselves using equipment that does not require a lengthy make-ready process and that allows us to publish first editions of 300 to 600 copies and to reprint even smaller quantities as needed. Thus, we can produce Replica Editions quickly and can keep even very specialized books in print as long as there is a demand for them.

About the Book and Editors

Food Security in the United States

edited by Lawrence Busch and William B. Lacy

Despite the fact that every year it produces a larger surplus of agricultural products than any other country in the world, the U.S. still must contend with a number of important but often unaddressed issues related to food security, including problems of soil erosion, water supply, energy availability, nutrition, farmworker health and safety, and product distribution. This book, containing contributions from authorities in both the natural and social sciences, expands the range of issues pertinent to the security of the U.S. food system, taking into account the adequacy and sustainability of the food supply, equity in access to food by the entire population, the nutritional quality of food, and the costs and benefits (social, economic, and health) of the food system as it is presently organized. Each of the authors considers an aspect of U.S. food security from the point of view of a specific discipline, as well as in terms of broader policy implications.

Drs. Busch and Lacy, associate professors of sociology at the University of Kentucky with appointments in both the College of Agriculture and the College of Arts and Sciences, are the authors of *Science, Agriculture, and the Politics of Research* (Westview, 1983).

Food Security in the United States

edited by Lawrence Busch
and William B. Lacy

Westview Press / Boulder and London

A Westview Replica Edition

Published in 1984 in the United States of America by
Westview Press, Inc.
5500 Central Avenue
Boulder, Colorado 80301
Frederick A. Praeger, Publisher

Library of Congress Cataloging in Publication Data
Main entry under title:
Food security in the United States.
(A Westview replica edition)
1. Agriculture and state--United States. 2. Soil conservation--Government policy--United States. 3. Agriculture--Water-supply--Government policy--United States. 4. Food supply--Government policy--United States. 5. Nutrition policy--United States. I. Busch, Lawrence. II. Lacy, William B., 1942- . III. Series.
HD1761.F596 1984 363.8'0973 83-21604
ISBN 0-86531-810-7

Printed and bound in the United States of America

10 9 8 7 6 5 4 3 2 1

Contents

Tables and Figures

Tables

Figures

Preface

This volume originated in discussions we held with Dr. Reed Hertford, then of the Ford Foundation, in the Spring of 1981. The discussions focused on the lack of concern in the United States with the problem of food security. Eventually the generous support of the foundation combined with that of the Kentucky Agricultural Experiment Station made possible a number of activities.

Among these was the sponsoring of a workshop, under the auspices of the Committee for Agricultural Research Policy, on the subject of food security. The workshop was held in September 1982 at Carnahan House, the conference center of the University of Kentucky. Participants were carefully chosen to represent diverse disciplines and viewpoints, and for their willingness to communicate within an interdisciplinary setting. Over a period of several days the participants each had an opportunity formally and informally to present their ideas on U.S. food security from the perspective of their disciplines and experiences, to engage in interdisciplinary dialogue and exchange, and to consider the broader policy implications. In addition to the authors of this volume, Wendell Berry, Kay Kaiser, Mark Lancelle, C. Oran Little, and Kenneth Pigg participated in the often heated discussion.

After the workshop the presenters were asked to complete their draft papers. When the papers were received, they were sent out to several other participants for a written review. The reviews were then returned to the authors and revised papers were written. As a result, we believe that the papers in this volume represent more than merely several unrelated chapters; though they do not cover all aspects of the problem of food security, they nevertheless form a coherent whole and the basis for both additional research and informed policy decisions.

Lawrence Busch
William B. Lacy

Acknowledgments

Any printed volume necessarily represents more than the work of the author. An interdisciplinary edited collection presents special problems: Not only must each of the chapter authors contribute, but the various conventions, vocabulary, and grammatical styles of each discipline should be consistent. This requires a true group effort and a commitment of all to the task.

Our conference coordinator, Nancy Welt, played a pivotal role setting up the initial workshop, arranging meals, slide projectors, travel, and more. Following the workshop, she wrote innumerable letters to participants, helped to clarify technical details, and generally coordinated the entire project.

Carol Calenberg typed the camera-ready manuscript with speed and accuracy. Deborah Witham, editor of the Kentucky Agricultural Experiment Station, edited the manuscript and helped to ensure consistency of style and references and minimal duplication. Ann Stockham compiled the index (on rather short notice). Janice Taylor provided the necessary accurate accounting and bookkeeping for this project.

Laura Lacy provided editorial assistance. In addition, a number of graduate students provided assistance, including Paul Marcotte, Jack Thigpen, and DeeAnn Wenk.

Finally, we must thank both the Ford Foundation and the Kentucky Agricultural Experiment Station for providing both the funds and physical facilities, without which this work could never have been attempted.

Of course the usual disclaimer applies: Any errors or omissions remain the responsibilities of the authors and editors.

Lawrence Busch
William B. Lacy

January 1984

Contributors

C. Phillip Baumel is a Distinguished Professor of Economics at Iowa State University in Ames. He has served on the Rural Transportation Advisory Task Force and the Iowa Farm Bureau Field Crops Advisory Committee. Dr. Baumel has published in the field of transportation and agricultural economics.

Marty Bender is a Research Associate at the Land Institute in Salina, Kansas. His work there includes the study of resource use (energy and water) and agriculture, and the development of perennial grain crops and their impact on U.S. agriculture.

Lawrence Busch is a Professor of Sociology at the University of Kentucky in Lexington with joint appointments in the College of Agriculture, the College of Arts and Sciences, and the College of Medicine. Dr. Busch is also the Co-chair of the Committee for Agricultural Research Policy. He has published on agricultural research policy.

L. J. (Bees) Butler is a Program Associate with the Food Systems Research Group at the University of Wisconsin in Madison. Dr. Butler has done extensive research on the Plant Variety Protection Act and the U.S. seed industry. He currently serves as a member of the Council for Agricultural Science and Technology task force on Germ Plasm Preservation and Utilization in U.S. Agriculture.

James A. Christenson is Chair of the Department of Sociology, at the University of Kentucky in Lexington. Dr. Christenson is also the editor of the journal, Rural Sociology. He has conducted research and published in the areas of evaluation research, public services, and American values and beliefs.

Katherine L. Clancy is Associate Professor of Human Nutrition at Syracuse University in New York. Dr. Clancy is engaged in research on the relationships of food and nutrition to regional food systems and on other issues related to food fortification. She is also working in the area of public health.

Molly Joel Coye is the Chief of the Occupational Health Clinic at San Francisco General Hospital. She is also an Assistant Professor of Clinical Medicine at the University of California at San Francisco's Medical Center. Dr. Coye has served as a consultant to the World Health Organization in Nicaragua.

Donald N. Duvick is the Director of the Plant Breeding Division at Pioneer Hi-bred International Corporation in Johnston, Iowa. Dr. Duvick is responsible for his company's plant breeding research for the United States and Canada and is a member of the company's Board of Directors. He has conducted research on genetic diversity and germ plasm development and use, and serves on national and international committees and boards concerned with germ plasm utilization and conservation.

Burton C. English is a Staff Economist at the Center for Agricultural and Rural Development at Iowa State University in Ames. Dr. English focuses on national agricultural resource use and the evaluation of agricultural policy as pertaining to natural resources.

C. Dean Freudenberger is a Professor of International Development Studies and Economics at The School of Theology, Claremont, California. Dr. Freudenberger has degrees in agronomy and social ethics. During a period of 26 years, he has served in 121 foreign countries as an advisor on food policy and food production. His new book, Food for Tomorrow, a critique of American agriculture, is forthcoming.

William H. Friedland is Professor of Community Studies and Sociology at the University of California, Santa Cruz. Among his recent publications is Manufacturing Green Gold: Capital, Labor, and Technology in the Lettuce Industry.

Joan Dye Gussow is Chair of the Department of Nutrition Education at Teachers College, Columbia University in New York. Dr. Gussow is past president of the Society for Nutrition Education and is on the boards of several organizations including the American Farm Foundation and the Land Institute. She is the author of numerous publications including The Feeding Web--Issues in Nutritional Ecology.

Marvin Hayenga is Associate Professor of Economics at Iowa State University in Ames and has done research on agricultural markets and food industry economics. He teaches graduate and undergraduate courses in agricultural marketing, and the structure and organization of the food and agricultural sector.

Wes Jackson is the Director of the Land Institute in Salina, Kansas. The Land Institute is a small non-profit organization concerned with the development of perennial grain crops and sustainable alternatives for agriculture. Among Dr. Jackson's publications is New Roots for Agriculture.

Dietrich Knorr is a Professor of Food Science and Human Nutrition at the University of Delaware in Newark. Dr. Knorr teaches food processing and works in biotechnology and food quality. Among other publications he recently edited Sustainable Food Systems and Alterations in Food Production.

Bruce Koppel is Research Associate in the Food Systems Program at the East-West Center in Honolulu, Hawaii. Dr. Koppel has written on technology assessment and agricultural development.

William B. Lacy is the Associate Chair of the Department of Sociology at the University of Kentucky in Lexington with joint appointments in the College of Arts and Sciences and the College of Agriculture. He is also the Co-chair of the Committee for Agricultural Research Policy and co-author of the recently published book, Science, Agriculture, and the Politics of Research.

Sally J. Lawrence is a Research Assistant in the Department of Sociology at the University of Kentucky in Lexington.

E. Phillip LeVeen is an economist teaching agricultural policy at the University of California at Berkeley and also directs a small non-profit organization called Public Interest Economics West. Dr. LeVeen has done research on the reclamation act and rural development in California.

James M. Meyers is the Associate Director of Extension Programs at the University of California at Berkeley. He has served as the Executive Secretary of the National Agricultural Research and Extension Users Advisory Board (1979-81). Dr. Meyers was a member of the evaluation staff for the recently completed National Evaluation of Extension.

David Pimentel is a Professor of Entomology and Agricultural Sciences at Cornell University in Ithaca, New York. Dr. Pimentel has written a number of scientific papers and books including Food, Energy, and Society.

Neil Sampson is the Executive Vice-President of the National Association of Conservation Districts in Washington, D.C. He has served as a conservation consultant in West Africa designing a soil and water conservation program. His most recent book is Farmland or Wasteland: A Time to Choose.

Paul D. Warner is a Professor of Sociology at the University of Kentucky in Lexington, Kentucky. He is also an Extension Specialist with the Kentucky Cooperative Extension Service. Dr. Warner has published in the areas of program evaluation, program planning, and organizational development. His recent book (with James A. Christenson) is The Cooperative Extension Service: A National Assessment.

Sylvan H. Wittwer is the Director Emeritus of the Michigan Agricultural Experiment Station and Associate Dean of the College of Agriculture and Natural Resources at Michigan State University in East Lansing. Dr. Wittwer has published on the state of U.S. agricultural research.

Introduction: What Does Food Security Mean?

Lawrence Busch and William B. Lacy

Food security is a necessary though not sufficient condition for the success of civilization. Ancient civilizations rose and fell based upon their ability to maintain a secure, stable food supply. The river valley civilizations of ancient Egypt were dependent upon their ability to forecast the annual flood. The irrigation civilizations of Mesopotamia existed only so long as labor to clear the canals of silt was readily available.

Conceptually, the problem of food security has been with us at least since Malthus. The frightening possibility that food supplies might outstrip population growth returns to haunt us at least once every decade. Indeed, less than a decade ago the shortfalls in food production in several areas of the world raised the prospect of a "world food crisis." Since then, a voluminous body of literature has dealt with the complex issue of maintaining an adequate food supply for the world as a whole.

In contrast, in the United States agricultural production has outstripped population growth for at least a century. As myriad scientific and popular publications of the Department of Agriculture inform us, American agriculture produces a larger surplus every year than that of any other country in the world. Thus, one might ask, why should the issue of food security even be relevant to the United States? The scientifically produced agricultural cornucopia appears to be sufficient to relegate the problem of food security to others. Yet the extraordinarily complex set of institutions that together constitute the agricultural sector of our economy may serve to mask fundamental problems of U.S. food security.

Certainly, we need not worry about where our food will come from tomorrow. Next year and probably even the next decade are fairly secure. However, the very complexity of the situation demands that we look beyond the short term. Over the last several years a number of books and articles have focused upon particular issues relating to food security--soil erosion, water supply, nutritional status, health and safety, declining growth of crop and animal productivity. Each of these documents raises a number of questions about the future of our food supply. However, few have attempted to put the pieces of this complex jig-saw puzzle together. While we have _some_ understanding of _some_ of the pieces, we have had difficulty in comprehending the totality.

Food security has at least three dimensions: the first of these is availability, having enough food available for the entire population at all times to sustain human life. To accomplish this we must have a production system that (1) produces enough in the short run, (2) is sustainable in the long run, (3) does not place undue risks on agricultural producers, and (4) responds rapidly to disruptions in the food supply due to natural disasters, civil disturbances, environmental imbalances, or other causes.

A second dimension of food security is accessibility. The food supply must not be limited by what economists call "effective demand." Low income populations and inner city residents must have equal access to the food supply. Simply making food available is not enough; one must also be able to purchase it.

A third dimension of food security is adequacy. An adequate food supply will provide for the differing nutritional needs of the various segments of the population. Adequacy can be conceptualized in terms of balanced diets, offering the necessary variety of foods throughout the year. At the same time, an adequate food supply will provide food that is free from disease and toxic substances.

Moreover, each dimension of food security must also consider the social, economic, and health costs and benefits of the food system as it is presently organized. A secure food system should not impose undue social, economic, or health costs on any special segment of the population.

In light of this multidimensional definition of food security, many related issues emerge:

Land. During the 1930s soil erosion in the United States rose to record levels, resulting in the Dust Bowl. Recent studies indicate that soil erosion is once again at 1930s levels. Perhaps as much as 6 billion tons of soil are eroded each year. Most of this damage is subtle, not forming visible gullies on hillsides but rather removing the soil in thin sheets. Hence, relatively flat states such as Iowa suffer from the most serious soil erosion. Part of this problem is due to a failure to practice proper crop rotation techniques. In other instances, it is due to heavy cropping of marginal lands. These croplands are experiencing additional degradation in terms of depletion of soil nutrients and soil compaction. Moreover, prime farmland is being removed from crop production at a rather significant rate and converted to other uses.

The financial crisis facing farmers over the last several years has served to exacerbate erosion problems. Heavily indebted farmers have turned to short-run, cost-cutting measures. Planting an extra corn crop, grazing a few more cattle, or harvesting a few more trees pays this year's bills, but it creates serious erosion problems if continued on a regular basis. The massive number of farm bankruptcies in recent years, in spite of record farm program costs, reflects the seriousness of the problem.

Water. In North America rainfall is very unevenly distributed across the continent. In most years the Eastern portion of the United States receives more than its share of rainfall, while arid lands of the far Midwest and West receive little precipitation. In the East, agriculture may be threatened by increasingly acidic rain. The consequences and extent of acid rain remain to be adequately documented, although some evidence of lowered crop yields exists.

In the far Midwest, the situation is clearer: in order to irrigate relatively marginal land, major aquifers, such as the Ogalala, are being depleted at an alarming rate. Unlike the quickly replenished aquifers of the East, these aquifers contain a substantial proportion of fossil water--that is, water deposited hundreds, even thousands of years ago and not replenished on an annual basis. The depletion of these aquifers will result in serious economic losses in the areas affected.

Finally, the large, irrigated and highly productive agricultural valleys of the West are fed by an expensive and wasteful system of canals. Much of the water evaporates before reaching its destination. Overuse of this federally subsidized water source is creating increasing problems of salinization in the Western valleys. Such salinization is expensive to reverse. Moreover, less valuable crops must often be grown on salted lands.

Energy and environment. Although agricultural production requires a relatively small share of the total energy budget of the United States, it is extraordinarily energy inefficient. Far more energy goes into producing crops than the crops themselves contain. Moreover, much of this energy is fossil energy which will become increasingly scarce in the future. Forecasts suggest that fossil energy use will peak just after the turn of the century, less than a score of years away.

At the same time, the supply of arable land is decreasing. Highways, urban sprawl, and other nonagricultural activities are encroaching on agricultural land. Yet, increasing production on the remaining land requires a proportionately greater amount of energy. As Pimentel notes (Ch. 5): "to reduce the land area by half and maintain total corn yield, about three times more fossil energy is required." A similar case may be made for water. As easily accessible sources are depleted, the energy (as well as economic cost) of maintaining an adequate water supply increases.

High energy use is also a potential environmental problem with pollution in the form of fertilizer and pesticide runoff. The concentration of animals in feedlots has also created potential pollution problems. Since the mid-'70s and the dramatic rise in oil prices occasioned by the formation of OPEC, energy use in agriculture has leveled off. As a result, pollution from farm chemicals has been somewhat reduced. Nevertheless, problems of energy efficiency and environmental pollution persist.

Seeds. Today, only a handful of companies provide the bulk of the seed used for producing major cereals in the United States. Much of this seed originates from the same genetic pool. Food security demands that the genetic pool utilized in agricultural production be as diverse as possible. Only a diverse genetic pool can ensure some measure of protection against the constantly changing population of plant pests. When uniformity is the rule, then mutant insects or pathogens can cause significant damage to crops. For example, the southern corn leaf blight that swept through the U.S. corn belt in 1969 and 1970 reduced yields by as much as 50 percent.

At the same time, however, farmers' demands for high-yielding, uniform crops that lend themselves easily to mechanical harvesting decreases significantly the diversity of seeds produced. Not surprisingly, farmers desire seed that will maximize their profits, and rapidly adopt a particular variety or hybrid with these features. The result is genetic uniformity. For example, Butler reports (Ch. 7) that nearly 37 percent of all corn grown in the United States is of one variety. Unfortunately, even the determination of what are "safe" levels of uniformity is highly problematic. Yet, despite the amibiguity surrounding what constitutes diversity, we cannot afford to ignore the problem.

Seeds not utilized in production are stored in government-sponsored germ plasm banks. These storage facilities consist of large refrigeration units and fields for growing stored materials to produce new seeds at regular intervals. However, they suffer from a number of severe limitations. First, seeds in these banks are collected from all over the world. Yet, when new seeds are produced, they are grown under conditions often quite different from those of their native environment. As a result, the seeds are gradually transformed through a selection process each time they are reproduced in the field. Second, the thousands of stored varieties must be catalogued in order to be accessible to breeders. Given the virtually infinite number of possible classifications, this presents an extraordinarily difficult problem. Omitting categories later deemed relevant may restrict access; on the other hand, too many categories may make the system unworkable. Third, germ plasm banks may actually contribute to increased uniformity in that they make the "best" varieties available to breeders very rapidly. Thus, when desirable plant material is identified, it is rapidly incorporated into breeding programs around the world (Busch and Lacy, 1984). Finally, germ plasm banks are often poorly funded and poorly maintained. Unfortunately, such work is a low status activity. Hence, in the event of a serious pest outbreak, germ plasm banks may be unable to respond rapidly.

Labor. Over the last century, the labor force employed in American agriculture has declined dramatically. Unlike the labor force in virtually every industry in the United States, agricultural labor is often employed irregularly, works in relatively dangerous and sometimes unhealthy conditions, and is excluded from most of the social welfare legislation pertaining to the nonagricultural labor force. Moreover, the problem is exacerbated by the inadequate statistics collected on the agricultural labor force. In fact, Friedland (Ch. 8) argues that the data is so poor that an adequate policy with regard to labor cannot be developed. This deficiency is in part due to the failure to see agricultural labor as a problem at all.

In the past, agricultural labor was evenly distributed across the nation. In fact, much of it was, and still is, family labor. Few farmers hired more than one or two "hands" to help with the work. Such farmers and farmworkers knew each other well. However, with increased scale in farming and the rise of highly specialized single-commodity farms, both capital and labor have become more concentrated both regionally and in the production of a given commodity. In the future this concentration may bring to agriculture more supervisory and management personnel, as well as the industry-labor relations problems that have long been a feature of the manufacturing industries. Subsequent financial gains will probably accrue to the more highly capitalized operations and to supervisory personnel, rather than to those who are lowest paid.

Marketing and transportation. Over the last century, and particularly since the end of the Second World War, agricultural markets in the United States have become increasingly regionalized. Tomatoes are grown in California; oranges are grown in Florida; corn is grown in Iowa; and dairy cows are raised in New York and Wisconsin. As a result of this increasing regional specialization of agricultural production, food is often shipped thousands of miles before it reaches consumers. Even supermarkets in producing regions frequently obtain their produce from great distances due to the lack of direct farmer to supermarket marketing arrangements. In addition, only a small number of rail links tie the Eastern to the Western portion of the United States. Disruptions of these rail links, for whatever reason, seriously impair our ability to move food from one coast to the other.

In addition, the U.S. highway system is in need of major repair. In particular, the off federal-aid system of rural county roads is deteriorating. An estimated 61 percent of bridges in this system--and particulary those in some agricultural states--are deficient (Baumel and Hayenga, Ch. 12).

Still another transport-related aspect of food security is the increasing percentage of nitrogen fertilizer that is produced overseas. Foreign production now accounts for about one-fourth of the total and is directly tied to rising natural gas prices in the U.S. Disruption of these supplies could significantly reduce domestic production of all crops.

Wholesaling and retailing. The food retailing industry in the United States has grown increasingly concentrated since the end of World War II. Supermarket chains have replaced individual stores and "convenience" stores have virtually eliminated the "mom and pop" grocery store. In addition, food stores of all types are less numerous in inner cities than they were just a few years ago. Most stores currently stock two to three days worth of food. Therefore, in the event of a prolonged disruption of the highly concentrated food supply chain, shortages would quickly develop.

Consumers are not prepared to cope with this problem. Recent studies reveal that consumers have a relatively small supply of food in their homes, most of it quite perishable in nature (e.g., Christenson, et al., Ch. 13). Particularly at risk are single people, the elderly, and poor inner city residents.

Nutrition. Currently, many Americans receive more calories per day than they need, leading them to be overweight. For most, reduction of total calorie intake below acceptable levels is unlikely to occur (Gussow, Ch. 10). For some consumers, however, increased economic concentration and/or unemployment will place an adequate food supply beyond their financial resources. The very existence of the Food Stamp Program as well as reports of hunger illustrate that this is a reality for some Americans.

Paralleling this problem is the difficulty the average American has in obtaining a balanced diet by eating the proper amounts of foods from each of the various food groups everyday. The number of items on supermarket shelves now numbers over 12,000. Despite enrichment of foods with vitamins and minerals, and the publication of nutritional data on food packages, maintaining a balanced diet has become complex and difficult in this context. The consumer is required to be an expert in nutrition in order to receive an adequate supply of vitamins and minerals each day. Moreover, those aspects of food of which we are aware are assumed to be the only relevant ones to maintaining a balanced diet. While our understanding of nutrition has improved markedly over the years, complete knowledge of the enormous complexities of nutrition is both unavailable and probably beyond human faculties. Finally, the impact of so-called "fast foods" and the increasing role that animal products play in our diet must be considered. It is possible, of course, that these changes will be accompanied by no ill effects. However, we have no way of easily discerning those outcomes.

Food processing. In the late 19th century the most serious problems associated with food involved either the transmission of disease through food products or the deliberate adulteration of food. In the early part of this century, federal laws were enacted restricting the processing, packaging, and contents of foods. As a result of these laws, for the most part, problems of adulteration and disease are no longer of major consequence. However, in an effort to differentiate among food products and to produce ever more unusual and exotic foods to attract consumers, the nature of food available to the general population has changed. New and unusual ways of preparing and preserving food have been invented. These involve subjecting the food to unusually high or low temperatures, or to other processes not utilized in the previous thousands of years of humankind's existence. In addition, 2,500 food additives are now available and approved for use in processed foods. Virtually no studies exist of the interaction of these thousands of food additives. Only a few studies exist of the long-term effects of such food additives. Finally, we know little of what the change in texture, accomplished through elaborate processing procedures, does to the nutritive value or effects on our alimentary canals of the food itself. Knorr and Clancy (Ch. 11) suggest that perhaps the whole question of food safety needs to be recast as one of food quality and that the goal should be minimal processing of food.

Research. The abundant food we have today is partially a product of more than one hundred years of agricultural research. However, during the period since World War II and particularly since 1960, agricultural research has become increasingly narrow and fragmented. Legislators' and farmers' demands for quick, immediately applicable results have forced researchers away from fundamental issues relating to food security (Lacy and Busch, Ch. 14). The various agricultural disciplinary associations have tended to encourage this fragmentation and often have discouraged concern for interdisciplinary issues. In addition, the publish or perish approach to research in our large agricultural research universities has encouraged those scientists who wish to obtain tenure and promotion to focus on small, easily definable topics in which research output will be forthcoming on a regular basis. In addition, the general interest in increasing agricultural productivity has focused only on short-term, low risk research and paid scant attention to the long-run sustainability of the food system. Moreover, the agricultural research system has become isolated from the larger scientific community, has had its basic science capacity reduced severely, and has seen its staff grow increasingly older. Finally, woefully inadequate funding, particularly at the federal level, has eroded the institutional bases of agricultural research at both the federal and state levels. The research and development requirements for a secure food system will be diverse and complex, and will require a major refocusing of the research priorities. The agricultural research system is at a crucial point in its history which requires those involved to reexamine ways of meeting these new requirements and enlisting the necessary resources to do so.

Extension. Since the passage of the Smith-Lever Act in 1914, each state has had a Cooperative Extension Service. Today, each county of the United States has one or more agents dispensing knowledge, advice, and know-how relating to agriculture, home economics, 4-H, and in some cases, community development. Until recently, virtually all extension activities were informed by a communications model that clearly separated research and development from the diffusion of innovations (Meyers, Ch. 15). In fact the very existence of extension as a separate organization has reinforced the view among many researchers that client contact is someone else's job. As a result, extension has tended to focus more and more on dissemination of information, while only rarely providing researchers with information about client needs and demands.

In addition, much research and extension effort is focused on production issues. Relatively little time is utilized for issues of health and safety or natural resource conservation. This is in part due to the enormous range of clients that extension serves and their varying influence on the research and extension system.

Farmer and farm worker health and safety. One of the three most dangerous occupations in the United States, along with mining and construction, is farming. Each year a significant percentage of farmers are killed or injured while on the job. Although safety has been improved in recent years as a result of better equipment design, occupational disease rates remain the same (Coye, Ch. 9). Acute pesticide poisoning, as well as illness resulting from chronic low-level exposure, remains a serious concern for farmworkers.

Unfortunately, poor record keeping, minimal training of physicians, and similarity of symptoms to those of other diseases, leads us to vastly understate the magnitude of the problem. In addition, few workers are adequately covered by workers' compensation laws. Even among those covered, lack of knowledge by both physicians and farmworkers limits the number of claims filed. Of particular note is the extraordinary accident rate of aircraft utilized in spraying agricultural chemicals. Approximately 20 percent of such aircraft report unscheduled landings each year. Finally, we should note that those regulations that do exist are poorly enforced and sometimes inadequate. Moreover, enforcement is fragmented across numerous agencies.

International implications. Unlike most other nations, the United States produces far more food than it can consume. This surplus food is exported both to Europe and to various nations of the Third World. European countries, of course, pay world market prices for virtually all that they purchase, and are careful to avoid any importations that would seriously undermine their own farm sectors. In contrast, Third World countries have been and continue to be the recipients of food aid which is either free or priced substantially below the world market. While no one would deny the propriety of the United States providing food in situations of famine, the provision of significant food stocks to Third World countries has the effect of undermining the production potential of peasant producers. As a result, the United States supplies an increasing percentage of the Third World food supply. This has the effect of weakening the food security of those Third World nations and of making them dependent upon the vagaries of U.S. agricultural production (Freudenberger, Ch. 16).

At the same time, the opening of U.S. agriculture to world markets leaves it unprotected against wide variations in the prices of basic foodstuffs, uncertain as to the prices for next year's production, and vulnerable with respect to foreign protectionism and subsidized competition (Koppel, Ch. 17). In fact, policies with respect to foreign agricultural trade, agricultural aid, domestic farm programs, soil conservation efforts, and public research and extension are largely uncoordinated, and may often work at cross-purposes. In addition, the growing role of what Koppel calls the middle sector--the various groups that operate internationally between producers and consumers--is not well understood.

This brief overview of domestic food security cannot do justice to the extraordinary complexity of the issues. While we may treat soil erosion, marketing, nutrition, and other issues as distinct, they are in fact highly interrelated. The grower who produces oranges under contract for a food processing firm employs the necessary pesticides to ensure that those oranges reach the processor blemish-free. As a result environmental pollution may be increased with no net gain in quality to the consumer. Similarly, tax structures that encourage investors to finance large-scale irrigation systems may contribute significantly to soil erosion. Of these interrelationships, we know very little. The very fragmented character of the scientific community itself makes those interrelationships extraordinarily difficult to unearth.

In closing, it is perhaps worth reiterating that we shall not run out of food next year. However, a serious examination of the problem of food security is necessary now if we are to avoid any of a number of possible catastrophic scenarios in the future. We need to rethink just what kind of food system will best serve the American public. As Orville Bentley, Assistant Secretary, Science and Education, U.S. Department of Agriculture, recently stated, "American agriculture is at a crossroads of significant proportions, and all those involved must reexamine ways of collaborating and marshalling resources for the future" (1983:441). We hope that this book will serve as a beginning for that task.

REFERENCES

Bentley, O. G. 1983. "Perspective on Agricultural Research." Science 219(4584):441.

Busch, L., and Lacy, W. B. 1984. "Sorghum Research and Human Values." Agricultural Administration (forthcoming).

Land

1
America's Agricultural Land: Basis for Food Security?

Neil Sampson

It is scarcely necessary to argue that a healthy, productive, profitable agriculture is essential to the security of America's people, as well as the well-being of many millions of other people around the globe. With agricultural exports of around $40 billion in 1982, despite a world-wide recession, farm commodities were one bright spot in an otherwise gloomy economic picture. But while those national economic figures brought smiles to statisticians, they caused little joy on the farm. USDA economists calculate that it cost nearly $6.00 to produce a bushel of wheat in 1982, but that wheat sold on the international market for $3.20-$3.80, and the domestic price was established at the same level. Many other crops had similar ratios during the period.

The result, predictably, was a deepening financial crisis on America's farms. Farmers were in debt $175 billion at the start of 1981, and slipped another $20 billion during that year. Total farm debt stands at over $200 billion today. The outlook for 1983 is even worse, with net profits for this year predicted in the $15-$20 billion range, with virtually all of that money coming from the largest commodity program in the nation's history--the Payment-in-Kind (PIK) program. Without that program, estimated to cost in the range of $20 billion, we faced the prospect of having the lowest real purchasing power on America's farms since the Great Depression of the 1930s.

The connection between this grim financial picture and the survival of the nation's farm lands is devastatingly simple and direct. Farmers are in business, and when any business is losing money, it has only two places to turn: depreciation of its capital assets or extension of its credit line. Farmers have been doing both, and soil damage rates have been accelerating just as fast, and for the same reasons, as debt levels have skyrocketed.

It is critical for agricultural policymakers to understand the basic causes of these problems, and the magnitude of the current afflictions being observed on

America's farm lands. There is always uncertainty about any projection into the future, but in this situation, the fact that American farmers are operating a system of biological deficit financing is clear. Such a system not only has no future, but is a clear threat to the future strength and security of the nation.

SOIL EROSION

Despite a significant investment in conservation practices by the federal government and by private landowners, erosion from agricultural lands continues at a massive rate. Nearly 4 billion tons of topsoil are estimated to be moved each year by sheet and rill erosion, with half the loss occurring on cropland (USDA, 1980). Adding the losses caused by wind, gully, and streambank erosion brings the national total to near 6 billion tons of topsoil moved each year.

No one can fully conceptualize what that many billions of tons of topsoil actually means, of course. One interpretation, based on the general concept of tolerable soil loss, is that on 12 percent of the nation's croplands, and 17 percent of our rangelands, we face the task of either slowing soil erosion or writing those acres out of the productive inventory within a few short decades (Sampson, 1981).

In the past, the adverse effects of soil erosion on productivity have been masked. New and more productive crop varieties coupled with the heavy use of fertilizers, better control of pests and crop diseases, and improved tillage and planting methods resulted in yield increases despite topsoil loss. While those technological increases helped hide the effects of soil erosion, at least temporarily, they did not cancel out the real losses. If we continue to let topsoil slip away at the rates found in 1977, we are doomed to a future of spending more and more trying to coax less and less from a dying land.

Although it is difficult to predict with any certainty, the data at hand suggest that over the next 50 years the loss of productivity due to erosion on cropland will be equivalent to the loss of between 25 and 62 million acres. To get some idea of the magnitude of that loss, 25 million acres could produce 50-75 million metric tons of grain, or half of the total exported from the U.S. in 1980. If we lose 62 million acres, it could represent the loss of virtually all of 1980s exportable surplus. So these estimates are significant to America's productive future (Sampson, 1981).

FARMLAND CONVERSION

Every hour between 1967 and 1977, about 320 acres of U.S. agricultural land was converted to non-farm uses. Not all of that land was cropland, but about 100 acres was, and another 100-120 acres had the physical capability to be used for cropland. Losing 220 acres of existing and potential cropland an hour means more than 5,000 acres every day--the equivalent of losing 23 average farms. It means an area the size of the average Midwestern county every 2 and 1/2 months, and the equivalent of a whole state by the year 2000. It means paving over a one-mile wide strip from New York to California every year (NALS, 1981).

The loss can be calculated in economic terms, as well. Each day, America's potential to produce corn was reduced by some 5,800 tons, worth more than $620,000 when corn is $3 a bushel. Each year we lost the capacity to produce about 2 million tons of corn, worth some $220 million at the same market price. And each year's losses must be added to the previous year's losses, and the next year's losses will add to both, until the total economic impact from wasteful land use practices adds up quickly to an appalling economic toll.

Urban uses take the best land. Between 1967 and 1975, about 39 percent of all the land converted to urban use was prime farmland. Since prime farmland makes up only about 25 percent of the nation's nonfederal lands, it is clear that urban development tends to concentrate on the best lands (Sampson, 1981).

The rate of land use change affecting agricultural lands speeded up in the 1970s. By 1977, there were 90 million acres of urban and built-up areas larger than 10 acres, up from 61 million acres in 1967 and 51 million acres in 1958.

Nor is there much evidence yet to suggest that the land use trends have slowed, although there has clearly been a slowing of building activity in the past five years. Whether this will rebound as the economy recovers, as has been the case in the past, is a question. There are some signs that Americans, while they may not think small is better, are beginning to believe that it is all they can afford. The current economic situation may be much more than a short-term belt tightening. It could signal the start of a very different society where we build fewer roads and airports, are less wasteful of land, energy, and community values in our settlement patterns, and settle for smaller living spaces using less land per person.

Even then, however, there will be 11 million more Americans reaching the prime home buying age of 30 in the 1980s than did so in the 1970s. The number of new families will increase 25 percent in the next decade. When added to the number of families occupying substandard housing and the number of housing units destroyed each year, the net result could be the need for 23 million new housing units in the 1980s, according to the National Association of Home Builders (Napolitano, 1980).

SOIL QUALITY

In addition to the losses being inflicted on our farm productivity by soil erosion and farmland conversion, there are other resource concerns as well.

Soil organic matter levels dropped rapidly in most of the soils in the humid cropping zones such as the Corn Belt after the initiation of cultivation, but recent evidence on those trends is mixed. Under continuous corn, for example, it has been estimated that it would take about 6,000 pounds of cornstalks plowed into the soil to maintain organic matter content (Larson, et al., 1972). That would yield about 109 bushels per acre, which is about what the U.S. average yield ran in 1979. Many farmers still feel, however, that modern farming methods do not maintain adequate organic matter levels and soil tilth. Results from no-till farming demonstrations indicate that water infiltration rates are dramatically increased and runoff greatly slowed under this technique. Just how much of that can be attributed to increased surface protection against soil sealing, and how much to the increased organic matter that results under no-till is not certain (Welch, 1980).

In the arid lands, desertification is affecting somewhere in the range of 225 million acres, ranging from the southern tip of California east through southern Nevada, Arizona and New Mexico, then almost all the western half of Texas and northward through the Oklahoma panhandle into southwestern Kansas. This is not a problem of "deserts on the march," as many would think, but is a slow deterioration of soil, moisture, and plant communities that results mainly from overgrazing of rangelands (Sheridan, 1981).

Increasing salinity will impair the future productivity of some irrigated cropland in the eleven western states. Salt build-up in the soil is difficult to reverse. It can be done, but it will increase costs and result in that land becoming less economically competitive, particularly for crops of lower value.

In some parts of the nation, acid rain is affecting crop growth. Data assembled in 1974 showed that acid rain covered most of the northeastern U.S., in addition to large areas of the West around Los Angeles, Oregon's Willamette Valley, Tucson, and Grand Forks.

The damage potential, especially in humid areas, is massive, and the problem is spreading (NALS, 1980).

Fragmentary research indicates that changes in soil biota may have been the result of recent trends toward heavy chemical and fertilizer usage. It seems possible that modern intensive production techniques have altered soil micro- and macro-biology to the extent that, in effect, the soils are "addicted" to those chemical inputs for most, if not all, of their productive capability. Natural recycling of nutrients, along with factors that assist with weed, disease and insect control, have been suppressed or destroyed.

Thus, soils that are suddenly denied the artificial supports provided by agricultural chemicals do not just drop back to production levels experienced on soils that have never received chemicals, but fall far below those "benchmark" levels of production. A few years of treatment and rebuilding may be required before normal biological--and yield--conditions return. Studies underway now at the Rodale Research Center near Emmaus, Pennsylvania, are showing that there are ways to hasten this transition by selecting low-nitrogen using crops. In corn, those studies have indicated that yields of 50 percent or less can be expected in the first year of transition away from commercial fertilizers (Harwood, 1981). This situation, to the extent that it is borne out by further research, could mean that America's productivity is vulnerable (at least in the short term) to disruption of the industries that supply chemical inputs. If farmers were denied fertilizers, pesticides, etc., national production of crops such as corn and wheat, that rely heavily on nitrogen fertilizers, or cotton, that requires both heavy fertilizer and pesticide use, could crash for at least a year or two before substitute methods were developed.

THE WATER SITUATION

Agriculture is by far the nation's largest single consumer of water, accounting for 83 percent of total use, some 89 billion gallons of water each day (Frederick, 1982). Irrigated agriculture produced 27 percent of the value of farm crops harvested in 1980, on only 12 percent of the harvested acres.

The Water Resource Council has estimated that water consumption will increase almost 27 percent between 1975 and 2000, mostly from growth in manufacturing and mineral industries, steam electric generation, and agriculture (Walker, 1980). In regions where water supplies are limited, that means increasingly intense competition for the available surface and groundwater.

Our ability to increase water supplies has about run its course. The Water Resources Council has estimated that by the year 2000 there will be severely inadequate surface water supplies in 17 subregions, located mainly in the Midwest and Southwest. During dry weather, more subregions, including some in the East, will also be faced with a water shortage (Walker, 1980).

In recent years, groundwater has been available to fill in the gaps where surface supplies were overtaxed, but this source is dwindling as well. Spurred by improvements in pumps and reductions in energy costs, groundwater use increased to the point where it was providing 39 percent of all irrigation withdrawals in 1975. But much of that water is being pumped from pools where the water supply is replenished more slowly than the current rate of withdrawal. This "mining" of groundwater is now estimated to be depleting the nation's water supplies at the rate of 21 billion gallons a day (WRC, 1978).

In total, it is doubtful if irrigation will make the contribution to the expansion of agricultural yields that it has in the past 30 or 40 years. Rising energy prices are expected to double the costs of irrigating with a standard center pivot system between 1980 and 2000. Efforts to improve system efficiencies may drop that cost somewhat, but there will still be an economic disadvantage facing irrigated agriculture in the future.

THE POTENTIAL FOR NEW CROPLAND

Throughout history, the U.S. has always had a reservoir of fertile land to meet new crop demands. The existence of this land surplus has dulled arguments for improved methods of soil conservation or farmland protection. It has always been too easy to argue that, with more good land just waiting for the plow, there is little need to worry about some topsoil loss or conversion of some land to urban uses. But a commonsense look at the landscape tells us there is a limit to the amount of land that is good for crop production.

In the 1975 Potential Cropland Study, and again in the 1977 National Resource Inventories, the Soil Conservation Service (SCS) identified land with potential for conversion to cropland. On lands that were not being cropped, field samplers estimated how feasible it appeared, under current economic conditions, to convert the land. Comparisons with similar land nearby that may or may not have been converted were part of the judgment, as were the local experiences of SCS technicians, County Extension Agents, and farmers. Each sample was rated on the basis of high, medium, low, or zero potential for conversion to cropland in the near future. The results indicate that there were around 127 million acres with high and medium potential for conversion in 1977.

Those 127 million potential crop acres were not just "lying there," waiting for the plow. The land was being used for the production of meat, milk, and wood products. If that 127 million acres were all plowed up, the result would be the loss of 51 million acres of pasture, 39 million acres of rangeland, and 31 million acres of forest. Indications are that farmers have converted 40-50 million acres to crops since 1977, but we will not know the full story of that situation until the new set of national resource data is released by SCS some time in 1983.

Clearly, however, the nation has a limited ability to add new land to its cropland base, and there will be significant--and rising--costs of doing it. Most of our better lands for crop production are being used today. In a world of virtually unlimited demand, we cannot totally depend on new acres as an unlimited source of future food expansion.

TOTAL DEMANDS ON THE LAND

In 1977, American farmers used 413 million acres of cropland, including all the land that was in summer fallow, rotation hay and pasture, orchards, vineyards and other similar uses. Adding the 127 million acres of potential cropland makes a total of about 540 million acres of land that could be counted as our total cropland resource pool. From that pool, we lose each year between one and three million acres as farmland is converted to non-farm uses, another one to three million acre-equivalents to soil erosion, and an unknown amount to such factors as soil compaction, desertification, salinity, alkalinity, water logging, and the loss of irrigation water supplies (Sampson, 1981).

Based on the rates of loss experienced in the decade between 1967 and 1977, that 540 million acres could shrink to somewhere in the range of 500 million acres or so by the year 2000. If average crop yields continue to rise at the rate of about one percent per year, producing the basic food and fiber that we might need in 2000 is expected to take somewhere around 450 million acres, providing that farm exports do not grow significantly between now and then.

These projections still do not reflect all of the demands for agricultural crops. Grain for alcohol, it was calculated in 1981, could add from 8 to 30 million acres of new demand in the next two decades, although that estimate could shrink considerably in the face of extended petroleum supplies and the development of new technology for converting woody plants directly into ethanol. Both of these developments may mean that the use of good cropland for energy crops will be less than was thought only a few years back. In truth, however, any production of energy crops would add to the current USDA projections of land needs for agriculture.

The production of crops for rubber, pulp, and oil could add up to 55 million more acres, but well over half of that growth could be on lands not used for traditional crops, such as semi-arid or desert soils. Without much data to rely on, it seems reasonable to project new demand for cropland to grow these crops in the range of 10-20 million acres by the year 2000. Adding alcohol gives a range of 18 to 50 million new acres needed. The range is wide because of uncertainty, but some amount of new demand from these crops seems likely.

Comparing the projected needs to the potential supply of cropland based on current trends of soil erosion, farmland conversion, and other land losses, leads to the conclusion that the nation's land supply will be fully utilized prior to the turn of the century. There is no cropland to waste. This nation cannot afford to lose good cropland; we may soon need to find more.

TECHNOLOGY AND THE FUTURE

The power to produce is the most vital and underrated asset of American agriculture. It is the measure of the system's strength. Like the horsepower in the family car, productive capacity isn't all needed all the time, but when the hills get steep and the going tough, nothing else can replace it.

There are only two ways to produce more food--use more acres of cropland or get more from each acre. It appears that we will need to double the amount of food and fiber grown in the U.S. by 2030 if we are to meet the needs of our own people plus maintain a significant export business (USDA, 1981). To do that, we must either double crop yields per acre, double the acres of cropland, or achieve some combination of the two. Farming twice as many crop acres would mean halting current farmland losses completely, then using all the 413 million acres of current cropland, plus the 127 million acres of high and medium potential cropland, plus some 260 million acres that now rate a low potential for growing crops. The enormous difficulties involved in that leaves the other option--raising the yields on each acre so that fewer acres are needed.

If average crop yields rise by 2 percent a year, we could meet USDA's demand projections in 2000 with around 400 million acres of cropland, or we could sell a great deal more product overseas. That would just about eliminate any need for concern about our farmland supplies, at least for this century. If we do not get those steady yield increases, however, we are in trouble. If average crop yields in 2000 turn out to be no higher than those of today, the land demand will be in the range of 570 million acres, far more than could be economically developed unless the price of food more than doubles.

The past has seen dramatic yield increases in U.S. crop yields. Will those trends continue? The question is critically important, but the evidence is not clear, and there are equally logical projections that portray far different visions of the future.

For instance, if grain and soybean yields continue to rise at 2 percent per year, it will take 224 million acres to meet the predicted demand for those crops in 2005. If, however, yields rise at only one percent per year, it would take an additional 69 million acres to meet the same demand. Choosing to predict the high growth rate leads to the conclusion that farmland will be adequate in the future; choosing the lower leads to a prediction of land constraints.

This means, therefore, that technology for increased productivity is critically important to the future of U.S. agriculture. Unlike land and water, technology is man-made, created by research and development. If the nation is concerned about lagging productivity, we can increase the public funds for research and development programs.

The potential of properly directed agricultural research is tremendous. Agricultural scientists in Colorado, for example, have developed a new small-farm approach to growing apples in the western part of that state. Dwarf apple trees are planted in a hedgerow pattern with 20 times as many trees per acre as would normally be planted. A trickle irrigation system waters 40 trees with the same amount of water that it took to flood-irrigate one tree before. The result--4,000 bushels of apples per acre, compared to 1,200 before. A 20-acre farm becomes a feasible commercial family operation, and land and water are used efficiently where they were probably too limited to allow commercial orchard production under conventional management systems.

Part of our current dilemma may be the result of doing the wrong kinds of research in the past, and adopting the wrong kinds of technologies. Not all technological gains help maintain the productivity of soil and water; not all contribute to a sustainable agriculture. Some kinds of technology, such as those that feature high-intensity, clean-cultivated, one-crop monocultures, are but a thinly masked method of mining topsoil. If that is the kind of new technology produced by research scientists, the nation would be wise not to purchase too much.

New agricultural research, if it is to contribute to the long-term strength of U.S. agriculture, must focus on technologies that build a sustainable agriculture without soil or water depletion. We must work at ways of keeping the soil healthy and productive, as well as finding ways to efficiently use the food we can extract at nonexploitive rates. There is no such thing as strength through resource exhaustion, regardless of the compelling arguments that are sometimes made for such a strategy as a necessary short-term expedient.

There is also no cause to think that research and technology can be used as substitutes for an aggressive program to prevent topsoil erosion and the conversion of prime farmland to non-farm uses. Some writers have pointed out that the past payoffs from research and development indicate that they might be more cost effective solutions than conservation or farmland protection in meeting the food needs of the future (Crosson, 1982).

Such conclusions represent a misunderstanding of the fundamental relationships between topsoil thickness, soil quality, and the cost-effectiveness of research. Studies of yield responses to new technological innovations lead to a conclusion that no agriculturalist finds too startling: new innovations increase production more on thick topsoils than on thin, more on prime farmlands than on marginal lands (Walker and Young, 1982).

The implications of that relationship are clear: a research program done in the absence of a conservation thrust will find, in 25 years, the fruits of its efforts providing far less benefits in terms of added average yield than would the same program if the quality of the resource base were maintained during that quarter-century. In fact, as Walker and Young clearly point out, when topsoils get thin, or much of the land is marginal at the outset, letting our croplands sustain much more deterioration can quickly eat up all the benefits of a research program. In that circumstance, our unwillingness to properly conserve resources means that we condemn ourselves to spend more and more on a research program that does not yield net benefits, but becomes vital to our ability to maintain even our productivity instead of suffering serious losses.

We are faced, then, with the need to change the course of resource depletion upon which we have embarked. The current arguments about the accuracy of today's data and different predictions for the future simply camouflage the real issues. Is the real rate of farmland loss one million or 3 million acres per year? Is the real rate of soil erosion 2 billion tons from cropland per year or 4 billion tons? The resolution of those questions, even if we could answer them with pinpoint accuracy, would only help us identify the generation upon which the real costs of our negligence will fall. Would that really make a difference, in terms of our failure to provide responsible stewardship?

The real issue is that we are wasting land and water in ways that are not in keeping with a society as technologically and scientifically brillant as we are, or as moral as we would like to pretend to be. Reversing that wasteful situation will not be easy. It will take a great deal of courage from both scientists and political leaders. First, it seems to me, must come the leadership from the scientists, who have a responsibility to see beyond their own narrow technical scope, understand the situation whole, and tell the people what that vision entails. Only then will the American people have the factual basis needed to meet the emerging crisis in American agriculture, and be able to create a "safety net" that precludes the permanent loss of either productive farmland or skilled farmers.

NEEDED: LONG-RANGE POLITICAL SOLUTIONS

The most immediate problem facing the nation as it tries to create that safety net may not be the financial or environmental crunch affecting the land, but the political stampede to "do something" about the situation. That the soil erosion issue will come to the forefront is not in doubt; that policy responses will be devised seems equally certain. That they will be misdirected is highly likely.

The reaction of public policymakers to long-range natural resource problems is, unfortunately, to devise short-run solutions. To quote Dennis Meadows, in an assessment of the energy situation:

> Two policy options are open to the federal establishment. The first causes energy prices to decline over the short term but leads to rapid escalation of price later on. The second increases prices now but lays a foundation for stability later. Most forecasters overlook the fact that the federal establishment will systematically pick the first policy over the second so long as the most difficult consequences only begin to manifest themselves after the next election (Meadows, 1982).

We see clear indications of that tendency in the agricultural arena. Despite the financial disaster imposed on American farmers by a market that follows the world price, in an era when a combination of conditions has resulted in foreign grain sales at prices well below production costs, the official USDA policy today is to push for increased sales abroad. Increasing sales, without increasing other people's ability to pay, or without changing the relative value of the dollar compared to other foreign currencies, can only be done by forcing prices still lower. Efforts to establish a means of holding out for a fair price for American farmers are rejected for various reasons, most of which have to do with a fear of injecting more federal control in the marketplace or encouraging foreign countries to expand agricultural output. While all these arguments have merit, they overlook one thing: if the current method of doing business bankrupts too many farmers or forces them to ruin the land, it may be the worst choice of all, in the longer view.

Selling more than can be produced within the limits of sound land and water management creates a treadmill where the more we grow, the lower we drive prices, and the more land we damage. As the land's productivity is damaged, it costs more to produce, so we are forced to try to produce still more at still lower prices, to try and pay the rising costs of production. It is a losing game, where we are forced to run faster and faster just to stay even.

The answer to the land damage created by this dilemma does not lie in new or different soil conservation programs, but in developing a coordinated market/conservation system that provides price levels adequate to repay not only the costs of production, but also the necessary re-investments in the land that will conserve, protect and enhance its long-term productivity. Methods and institutions for doing this have been suggested by policy analysts, but so far organized agricultural groups and national farm policymakers have been unwilling to seriously study how such a system could actually be implemented.

Creation of a marketing system that rewards farmers who choose to apply appropriate soil and water conservation management must fall to the federal government. It is only at the federal level that marketing programs have any chance of functioning at all. But in this age of "new federalism," where the conventional political wisdom is to get the central government out of activities rather than into new ones, this idea is definitely out of style. So, in spite of the fact that most international food buyers and sellers are governments rather than individuals, the U.S. is reluctant to move into this arena.

Identifying the farmers who are carrying out appropriate management systems on their land is a job that is too localized for the federal government, but the combination of state and local action created by the soil conservation district system could handle the task. In Iowa, for example, state law sets up a system of allowable soil loss standards, which are then refined and administered at the local level through conservation districts. Several other states have followed this lead, but it is not an easy course (Garner, 1982). Difficult political issues arise when local citizens begin to run programs that have a real impact on their neighbors. In the long run, of course, we must establish a policy framework from which a sustainable agriculture can emerge.

A sustainable agriculture is one that produces the food, fiber, energy, and other crops that the nation needs, including a marketable surplus that can be sold abroad. It produces this in an average year, not just during times of good weather. It weathers a bad year by drawing on stored fertility and moisture in the soil; stored water in reservoirs; stored wealth in financially secure farms; and stored food products in the granaries of farmers, industries, and government. It profits from a good year by setting aside extra commodities or making an extra effort to see that they are sold abroad, without driving prices through the floor and creating financial hardship or ruin among producers. In addition to meeting domestic food needs and a substantial export market, a sustainable, regenerative agriculture could reduce the waste and pollution of water, provide better wildlife habitats, slow the advance of desertification and soil salinization, reduce the loss of prime farmlands and fragile topsoils, and, in general, make rural America a far more healthy, satisfying, and financially rewarding place to live. It is a utopian goal, perhaps, but an essential goal nonetheless.

This will require the nation to establish production levels--along with accompanying agricultural technologies--that can be indefinitely sustained with the land, water, technology, and capital at hand. It will mean a concerted effort not just to develop new lands or technologies, but also to carry out the needed maintenance and upgrading that existing resources require. In economic analysis, it may be inconsequential whether we spend dollars on new development or spend them on maintenance of existing systems. From the political standpoint, it is often far more attractive to promote new development. But in terms of real economic and social progress, we will be limited by our willingness to conserve and maintain today's farm lands and the farmers on them.

Converting today's depleting agriculture to a new type of sustainable agriculture will not be a quick or painless process. But some things can be identified as starting points:

(1) We can take immediate action to overhaul national commodity programs, so that people who practice soil conservation are rewarded, rather than penalized, in the market place.

(2) We can double federal spending for incentive programs to help farmers install soil and water conservation measures. Those programs today attract less than 1/10 of one cent out of each dollar in the federal budget, a share far lower than the critical importance of those lands to our future would suggest.

(3) We can encourage the development of soil and water conservation practices such as no-till or minimum-till farming that result in both soil and cost savings.

and we can help farmers learn to carry out these new technologies without indiscriminate or unsafe pesticide usage, which could negate the cost savings involved and replace soil erosion and sediment pollution with chemical pollution--a trade nobody wants.

(4) We can create ways to help ease farmers off of lands that are too marginal for sustained agriculture, and eliminate financial incentives, such as the federal tax breaks that currently encourage people to attack such lands with a plow.

(5) We can set up an intermediate financing bank to prevent a massive wave of bankruptcies created by the debt spiral in which farmers are currently caught. The bank could offer long-term loans at subsidized interest rates so that farmers could re-finance their debt in a manageable plan. Those farmers who accept low-cost loans from the federal government should also accept a requirement that they manage their land within local conservation standards. If they are unwilling to make that commitment, the public should be unwilling to keep them in business at public expense.

(6) We can accelerate research and extension work in development of on-farm techniques that help farmers wean themselves from technologies that have created the current cost-inflation treadmill. That would include techniques like crop rotations, organic farming techniques, integrated pest management, and conservation tillage. Some aspects of these techniques often have too little commercial potential to be attractive to industrial research and advertising campaigns, so they should be the prime target of public research in USDA and the agricultural universities and experiment stations.

(7) Finally, we can be more aggressive at demanding, as a people, a more sensitive use of the land and water that sustains life. When an industrial user bulldozes up an acre of prime farmland for a strip mine, or covers it with concrete or asphalt, or poisons it with toxic wastes, he has taken an action that is more than private. He has robbed countless people--for countless generations--of an essential part of their opportunity for a constructive life. Such behavior is immoral, and it is time the American people made it plain that it can no longer be tolerated. We know how to do better, and now it is time to show the moral and political courage to do what must be done.

REFERENCES

Crosson, P. R., ed. 1982. The Cropland Crisis: Myth or Reality? Baltimore: Johns Hopkins University Press.

Frederick, K. D. 1982. "Irrigation and the Adequacy of Agricultural Land." In The Cropland Crisis: Myth or Reality?, ed. Pierre R. Crosson, pp. 117-64. Baltimore: Johns Hopkins Univ. Press.

Garner, M. M. 1982. "Innovative Strategies for Conserving Water." The Agricultural Law Journal, 3(4):543-68.

Harwood, R. R. 1981. Organic Farming Research at the Rodale Research Center. RC-81-1. Emmaus: Rodale Press.

Hassebrook, C. 1982. "Tax Incentives for Irrigation Development on Marginal Land." Paper transmitted in private correspondence. Center for Rural Affairs, Walthill, NE.

Larson, W. E.; Clapp, C. E.; Pierre, W. H.; and Morachan, Y. B. 1972. "Effects of Increasing Amounts of Organic Residues on Continuous Corn: II. Organic Carbon, Nitrogen, Phosphorus and Sulfur." Agronomy Journal 64:204-8.

Meadows, D. L. 1982. "Fallacies in Resource Planning." In: Forests in Demand: Conflicts and Solutions, eds. Charles E. Hewett and Thomas E. Hamilton, pp. 99-107. Boston: Auburn House.

Napolitano, F.J. 1980. "Statement of the National Association of Home Builders." Testimony before the Senate Agriculture Committee's Subcommittee on Environment, Soil Conservation, and Forestry, September 16.

National Agricultural Lands Study. 1980. Soil Degradation: Effects on Agricultural Productivity Interim Report No. Four. Washington, D.C.: U.S. Government Printing Office.

National Agricultural Lands Study. 1981. Final Report. Washington, D.C.: U.S. Government Printing Office.

Sampson, R. N. 1981. Farmland or Wasteland: A Time to Choose. Emmaus: Rodale Press.

Sheridan, D. 1981. Desertification of the United

States. Washington, D.C.: U.S. Government Printing Office.

USDA. 1980. 1980 Appraisal Part I, Soil, Water, and Related Resources in the United States: Status, Condition, and Trends. Washington, D.C.: USDA.

______. 1980a. Appraisal 1980: Review Draft, Part II. Washington, D.C.: USDA.

U.S. House of Representatives. 1981. Export Grain Bank. Committee print of Hearing before the Subcommittee on Wheat, Soybeans and Feed Grains of the Committee on Agriculture on H.R. 2901, September 29.

U.S. Water Resources Council. 1978. The Nation's Water Resources 1975-2000: Second National Water Assessment, vol. 1. Washington, D.C.: U.S. Government Printing Office.

Walker, D. J., and Young, D. L. 1982. "Technical Progress in Yields Is Not a Substitute for Conservation. Draft in preparation for publication in Current Information Series. Moscow: Cooperative Extension Service, University of Idaho.

Walker, L.D. 1980. "An Assessment of Water Resources in the United States, 1975-2000." Resource-Constrained Economies: The North American Dilemma. Ankeny, IA: Soil Conservation Society of America.

Welch, L.F. 1980. "Productivity of Soil as Related to Chemical Changes." Paper commissioned for U.S. Office of Technology Assessment study, Impacts of Technology on U.S. Cropland and Rangeland Productivity. Springfield, VA: National Technical Information Service.

2
An Alternative to Till Agriculture as the Dominant Means of Food Production

Wes Jackson and Marty Bender

Long-term food security for the world cannot be connected to non-renewable resources such as fossil fuels and mined minerals. We must look elsewhere for the energy and materials to power civilization, including agriculture. We can only speculate as to the extent that the recent interest in an ecologically-sound agriculture stems from a growing uneasiness about the vulnerability of the industrial world, including our industrial agriculture which requires high quality fuels. Although industrial, or what we will call production, agriculture is now the dominant means of food production in the developed countries, it is increasingly recognized as antithetical to sustainable, regenerative, or ecological agriculture.

The differences in points of view between "production" and "sustainable" agriculturists ultimately have to do with both time and faith. Production agriculturists either discount more of the future than sustainable agriculturists or have the faith that adjustments can be quickly made when the supply of food and materials runs low. In other words, some production agriculturists believe we can quickly shift to solar energy or be saved by fission or fusion nuclear power. Others do not care or are willing to let the future take care of itself. Promoters of a sustainable agriculture, on the other hand, discount less of the future and are more conservative or skeptical that society can adjust quickly and safely to a changing future.

When we explore other major features in a comparison of production to sustainable agriculture, we soon discover that few absolutes exist. For example, proponents of production agriculture have to be interested in some measure of sustainability, even if it is just for one year. Plant breeders must have an eye to some level of sustainable yield as they select for resistance to various insects or pathogens. In this sense, they are interested in more than production for yield; they simply have a shorter time frame than some of us would like. Sustainable agriculturists, in turn, must have some interest in production, for there is little purpose in farming if it does not yield something of human benefit. When we get too specific or insist on absolutes, distinctions fade.

It will be disastrous for both agriculture and culture if we allow these distinctions to fade, because they are both real and crucial. We are interested in a regenerative or sustainable agriculture and, as such, want to talk about a conceptually different way of growing food. Those who share this value hope and expect that it is a way that we can run agriculture on sunlight, cut soil loss below or equal to biological replacement levels, and in addition keep the soil healthy.

THE UNIFYING CONCEPT OF MODERN BIOLOGY

Before On The Origin Of Species was published in 1859, the world of the professional biologist stood before him like a large post office. The preserved and living plants and animals arriving from around the globe in crates and boxes at the docks and depots of such places as London, Munich, Paris, and Cambridge were eventually sorted into what were regarded as the proper slots by the museum biologists of the time. If different enough, taxonomists assigned them a new name and a new slot. Species which differed but little were given slots adjacent to one another, much as a stamp or coin collector would order his collection. Individual representatives of thousands of species looked out on this world like unsigned portraits in a huge art gallery. It was an assembly with incomplete sense, for while there were thousands of clusters, the discontinuities between clusters were more striking than the similarities within. In the schools devoted to medicine, professors and students of anatomy and physiology were no better off. They, too, noted similarities and differences but mostly left it at that. In one sense, the living world stood as vertical strata--pigeon holes--in the human mind, and for the Western mind, at least, the major gap was between the human and the rest of creation, a gap created by the Creator's pause. For on the authority of Genesis, after the Creator had feasted His eyes on the creation before Him, He declared it good and created the human to be His foreman with explicit instructions to protect His interests.

And so, whether these creatures were scurrying crabs in a rocky tide pool or insects pinned inside a fumigated tray, green things gracing an arboretum or moth-balled in the herbarium, or ungulates running wild on an African savanna or mute and still in a glass case in London, they were understood first as representing individual species and secondarily as fitting into some ecological relationship. But that is as far as it went. Mostly they were regarded as autonomous units ordered on shelves or in trays and cabinets by the minds of men, more for the sake of convenience than for any larger understanding. Biologists of the Western world had more or less accepted the idea of instant creation. But as the naturalists went forth and returned, internal inconsistencies in this idea began to emerge. Large questions were asked and went unanswered: how, for instance, did all those marine fossils get buried in the strata high in the mountains if everything began around 4000 B.C.? As internal inconsistencies continued to pop up in the "instant creation" paradigm, the minds of biologists were being prepared for what was to become the era of modern biology.

The dyspeptic and shy Charles Darwin, drawing on his Galapagos experience, the essay of Parson Malthus, the geology of Lyell and Hutton, and his voluminous notes, suddenly united all these life forms with the thread of time in his well-documented explanation of the workings of evolution through natural selection. Darwin tilted the vertical strata in the minds of cataloging biologists exactly ninety degrees. With these strata resting comfortably now in a horizontal position, the time dimension was quickly added and, for the most part, gracefully internalized in the biologists' mind. Gaps which stood as discontinuities representing God's attention to specifics in creation when vertical, became understandable, though often bothersome, breaks in the horizontally-laid fossil record. The new horizontal mind of biologists quickly fell into synchrony with the ancient strata of geology.

Not only did Darwin's unifying concept for biology bring more complete sense to the world of the cataloger, but the implications of this worldview also leaked over into physiology and anatomy and all the other disciplines of biology. No longer would the living world be simply regarded as consisting of individuals or species or even as part of an ecological setting. The creatures of the living world were more than that. They were products of forces which had shaped them over time. The dawn of modern biology arose in a London bookstore in a November morning in 1859.

THE NEED FOR AN INTERNALLY CONSISTENT UNIFYING CONCEPT FOR AGRICULTURE

Darwin may have changed the minds of "pure" scientists in biology, but if he changed the minds of applied biologists working in agriculture, it was not reflected in their work. The approach of agricultural researchers to crop improvement scarcely budged until after 1900, after the rediscovery and expansion of Gregor Mendel's laws of heredity, nearly a half-century after On The Origin Of Species was published. It was Mendel's work which caught their fancy. They were more interested in the manipulation of heredity for crop improvement than they were in the ecological implications of Darwin's ideas. By paying more attention to Mendel's contribution than to the implications of the contributions of both Darwin and Mendel, they continued to operate as though agriculture could be understood in its own terms. The rules of heredity simply gave them the necessary knowledge to better manipulate domestic plant populations--our crops, not ecosystems.

It is easy to appreciate how these researchers believed that agriculture could be understood in its own terms. After 18,000 years and more, since the beginning of agriculture, crops and livestock had become so streamlined for special human purposes through breeding that their roots in a former ecological setting seemed irrelevant. The demands placed on these creatures by human ecology had taken precedence over the demands of natural ecology for so long that crops were regarded more as the property of humans than as relatives of wild creatures.

But the problem runs even deeper. Like many species, we have rather specialized food requirements. Unlike any other species, however, we have the ability to radically alter the environment to meet those requirements. When we began to substitute single species populations (crops) for diverse ecosystems we became a problem for the earth and all of its life forms. An area planted to plant populations met our specialized demands better than the generalized ecosystems of nature, especially in temperate regions.

We seldom appreciate how narrow our food requirements really are. But of the 350,000 plant species world-wide, only 20 or so are of particular importance to us for food. Of the top 18 sources, 14 come from but two flowering plant families, the grasses and legumes (Table 2.1). Nine of the top 14 are grasses, and in all but one case (sugar cane) we are interested exclusively in the seeds. Furthermore, Homo sapiens, who would more appropriately be named Grass-seed eater, also relies heavily on land vertebrates for meat, which in turn depend heavily on grass as forage. It is no mystery why the prairie states, the grassy states, from Ohio to the Rockies, from Canada to Texas, supply most of the food for the U.S. and ship most of the tonnage of a huge export market each year. More than any other family of plants, it is grass which supports us, for when we get serious about food production for meeting basic human needs, most of our field acreage features grasses.

Although polycultures have been important to many indigenous peoples in the tropics, in the Orient, and among American Indians who grew corn-squash-bean

TABLE 2.1
Twenty Plant Species (Out of 350,000 World-wide). Grouped by Family, Ranked by Importance.

Poaceae (grasses)	Fabaceae (legumes)	Solanaceae (nightshades)	
1. rice	8. soybeans	4. potatoes	
2. wheat	15. peanuts		
3. corn	16. field beans	Convolvulaceae	Musaceae
5. barley	17. chick peas	6. sweet potatoes	19. bananas
9. oats	18. pigeon peas		
10. sorghum		Euphorbiaceae	Palmaceae
11. millet		7. cassava	20. coconuts
12. sugar cane			
14. rye		Chenopodiaceae	
		13. sugar beets	

Source: Adopted from Wittwer, 1981.

associations, over most of the landscape we have featured monocultures. This is because, in most cases, we can better understand and manipulate populations than diverse ecosystems.

Nearly all the crops listed in Table 2.1, especially those which are the most important, are grown in monoculture. Some have been grown in highly simplified polycultures such as corn-bean-squash associations. Overall, the most extreme thinking beyond the population level has been to think of these crops as representatives of a particular plant family, not as members of an ecosystem. We want to emphasize our earlier point that farmers and agriculturists alike have regarded crops more as the property of humans than as relatives of wild things. Two principal domestic ecosystem types feed humans--highly-simplified prairies and highly-simplified marshes. The vegetative structure of most of our farmed acreage, even in the wooded areas of our country, more nearly resembles the prairie than any other wild ecosystem. In the corn belt a corn field is a "tall grass" monoculture, a substitute for the grasses of the tallgrass prairie which grew there before the great plowing. A wheat field is a "midgrass" monoculture substituted for mid-grass prairie. Where wheat is grown in short grass prairie country, it requires irrigation or summer fallow.

Marsh ecosystems are places where the biomass turnover is high (Odum, 1971, p. 52). It is no wonder then that rice, a high-yielding grass, grown in a "domestic marsh," is the primary crop of humans in the most crowded parts of the world. The problem is that if nature were fully in charge where rice and wheat are grown, the biotic diversity would be staggering. We homogenize the environment in order to feed from extremely narrow domestic prairies and marshes. We don't live by either wheat or rice alone, but we come uncomfortably close, and we could very well elect to homogenize the earth's habitat even more than it now is to meet the needs and demands of a growing global population.

It may be all right to have food demands that rely most heavily on grasses and legumes, but we can also seek to maximize the variety within those two plant families to better enhance the long-term stability of agroecosystems. There is enough generic and species diversity within those two families alone to have sustainable domestic marshes and prairies over the landscape providing most of the necessary ingredients for our diet. We are fortunate that prairies feature grasses. The native prairie plants of Central Kansas are 95 percent grasses; half of the remaining 5 percent are legumes (Weaver, 1954, pp. 210-11). A Wisconsin prairie may be only 60 percent grass, but grass still predominates (Ahrenhoerster and Wilson, 1981). Although the legumes do not dominate any major ecosystem, they have an abundance of generic and species diversity to draw on. There is no reason, therefore, to assume that we cannot begin to think of developing a sustainable agriculture at the ecosystem level rather than at the population or crop level in order to meet both the needs of the land and the needs of the people. We are fortunate in this, for what if no ecosystem existed in which at least one of our most important plant families were the major component? We would then have to rely on the population level of biological organization to sustain us in the future, rather than the ecosystem level. We can thank nature that grasses, at least, are featured at the ecosystem level.

BRIDGING THE GAP BETWEEN "APPLIED" BIOLOGISTS (AGRICULTURISTS) AND CERTAIN PURE SCIENTISTS (POPULATION BIOLOGISTS AND ECOLOGISTS)

What is most needed now is to discover a starting point for bridging the gap between those who work with the species which have already been domesticated for the human purpose, the agriculturists, and those who study nature and natural systems, the biologists and ecologists. Very few of the latter have had any eye at all to the practical applications of their knowledge. There are some important principles from the two synthetic fields of biology, population biology and ecology, which so far have been little applied to agriculture.

Population biology is the product, primarily, of three traditions of biology, all of which go back to a synthesis of the work of Mendel and Darwin. It has drawn heavily on the traditions of the biosystematicist, the population ecologist, and the physiological ecologist. This knowledge has been accumulated, and it continues to accumulate, mainly for its own sake. Agriculturists cannot take full or immediate advantage of the knowledge accumulated by scientists who have studied the population biology and ecology of wild populations, because our domestic species have been too streamlined to meet the demands of human ecology rather than natural ecology. Too many of the ensembles of genes tuned to work well in a natural environment have been stripped away in the process of selection. We should not abandon these "genetic paupers," these domestic crops, in favor of the wild species, but an ecological agriculture on a large scale seems possible only if the natural integrities of wild ecosystems are put to use. If agriculturists could spend more time learning about wild species whose genetic profiles are sufficiently broad to make it on their own in the environment, we might begin to imagine possibilities for the domestic species we use for food and fiber, possibilities which include resistance to insects and pathogens and favorable competition against weeds. Furthermore, we have to think about a sunlight sponsored fertility.

It would be useful if population biologists and ecologists could work together with agriculturists in developing one or more domestic ecosystems using certain wild species. It is not just an exercise in science fiction to imagine that we will one day take this knowledge and later turn to many of our domestic crops and begin to reconstitute their "ecological genes" by crossing them with their wild relatives. In a biological sense, we are probably closer to developing new crops from wild species than to reconstituting the old, but there is the cultural reality associated with the traditional foods--the reluctance to change, to adopt a new food. In this respect, the cultural barriers may be more powerful than the biological barriers. Because of this, it will be necessary to more completely explore the ecology of agriculture. Such a study would scrutinize the ecological interactions of the various crops and crop rotation patterns in use today. It would be useful to know more of the residual interactions which are a carryover from pre-agricultural times. Such knowledge could give us an idea of how far we have to go in reconstituting the "ecological genes" which were eliminated long ago in our major crops. These genes will be necessary if we are to develop an ecological agriculture around our traditional species. In our future research, it will be helpful if we explain whether our research helps us better understand the ecology of agriculture or contributes to an ecological agriculture.

The interface between "pure" population

biology/ecology and agriculture has scarcely been explored. If some of the researchers in ecology and in the subdisciplines of population biology--the biosystematicists, population ecologists, and physiological ecologists--were to take the information which has been accumulating for its own sake and relate it to the need to develop an ecological or sustainable agriculture, more progress could be made in the long run than if we restrict ourselves to studying the "ecology of agriculture."

And now we come back to the importance of Darwin. Darwin gave biology an internally consistent unifying concept which was rooted in ecology. Regardless of the fine efforts of countless numbers of agriculturists who have resisted such over-emphasis on production, the principle which unites most modern agriculturists has been production, at almost any cost. But an internally consistent unifying concept for the area of applied ecology we call sustainable agriculture does not exist. Once we have said that there is a need to develop a sustainable agriculture, we have made an important leap--we have said that we are no longer willing to discount the future. An agriculture dependent upon high energy inputs, therefore, is no longer consistent with our interests. An agriculture which requires the nearly universal use of chemicals with which our cells have had no evolutionary experience, against which they have evolved no protections, is no longer consistent with our interests. An agriculture which removes people from the land so that the ratio of eyes to acreage is severely reduced is not consistent with our interests either. And most important of all, an agriculture which allows two, three, four, or five billion tons of U.S. top soil to run toward the sea each year is unacceptable. Sustainable agriculture needs a unifying concept which will be internally consistent. The long-term expectations of the land must be considered along with our food needs in our search for an ecologically sound agriculture. The "ecology of agriculture," generally speaking, should be studied in order to diagnose the problems of agriculture rather than to improve the short-run efficiency of food production.

A UNIFYING CONCEPT FOR AGRICULTURE WHICH FEATURES INFORMATION OVER ENERGY, SMALL SCALE OVER LARGE

Perhaps the most easily observable physical feature of modern agriculture is the consequence of our denial of the opportunity for the land to experience species succession. To stall succession, the farmer either manipulates or tricks nature. Ground is worked and weed seeds germinate and then the ground is worked again to destroy the seedlings right before the crop is planted.

Many crops are planted at a certain time to avoid the peak of an annual insect infestation. Whether it is management or trickery, the goal is to give an edge, usually to a single favored plant population. All this imposes certain costs, many of which go beyond the field. Loosening the soil every year causes it to erode. When we do not adequately trick the insects or pathogens, we must use more chemicals. Left alone, life would abound. It might not be life that we would regard as having economic importance, but it would abound. The area would feature several stages of plant and animal interaction as it moved toward a dynamic equilibrium recognized as a climax community--tallgrass prairie, oak-hickory forest, or whatever.

This progression does not happen in the same way that a rock rolls down a mountain slope toward the inevitable rest at the bottom. Biological progression happens under the direction of a molecular information system. What may have begun as bare ground soon becomes a complexity which is scarcely comprehensible. What we see when we walk through prairie is the manifestation of billions of biological bits of interacting information--the molecular DNAs and RNAs of species we will never completely count.

Not only does each individual carry enough instructions to cope in the environment, but those instructions also resonate against the collective background of instructions guiding the other members of the same population. By the time we add in the other populations of a natural community, the total fund of information is awesome. If we were to magnify the molecular "letters" residing in a square mile of tallgrass prairie to the size of print fit for books, we would probably need more space than is present in all the libraries of the world. The biological instructions necessary to produce a corn plant, on the other hand, might fill a large room. After all, one-third of the U.S. corn crop comes from only four inbred lines (Zuber and Darrah, 1980).

Viewed this way, the most obvious difference between crop monoculture and natural polyculture is that the former is information-poor while the latter is information-rich. Furthermore, it is not miscellaneous information, for much of it is keyed to ensuring a graceful transition of one ensemble of species to the next. When species succession is denied and the juvenile stage is insisted upon each year by the agriculturist, the human-nature split is the most profound. We believe, therefore, that as we begin to maximize information in agriculture, we will eventually feature succession. In such an agriculture, a high percentage of the entire genetic program of a "domestic prairie," from birth to death, would be put to use. We expect this emphasis on information will be the starting point of a new paradigm for agriculture which will one day replace the energy-intensive agriculture of today.

Biological information is energy cheap. It has had to depend on sunlight, a dispersed source, and after millions of years of evolution, it has had a chance to evolve toward operating close to thermodynamic limits of efficiency. Compared to the miniaturized and energy efficient information system of biology, our most sophisticated chip-based hardware is large, clumsy, and awkward. In succession agriculture, we would be trying to take advantage of the "combination of circuits" which nature has built in over millions of years in "systems" which are self-producing, something we have so far never done. In depending on natural ecosystem integrities, such an agriculture would therefore rely on the DNA language of various species, which operates at an energy cost 10^{21} less than the energy cost for keypunching a comparable "bit" of information on Hollerith cards (Delin, unpublished manuscript).

Such a saving of energy is not really important. It only illustrates the capability of life forms to become highly efficient in miniaturization. More important are the efficiencies achieved by the products of the code--the parts of the plant which display remarkable efficiencies in structure and function. At the individual level, these life forms cooperate with one another, to the benefit of each.

With such an agriculture we would not call ourselves "designers," but "imitators" of ecosystem designs which are the culmination of millions of years of trial and error in nature. The distinctly human effort in all this would be to direct these ecosystems toward human use, to encourage yields comparable to those of our traditional crops while avoiding the devastating consequences of monocultures. Such an agriculture would have to be run by naturalists, not industrialists.

Production agriculturists continue to argue for more research money on the grounds that "we must feed the world." Sustainable agriculturists would say instead that "the world must be fed." The way to ensure that it will not get fed is to make production so expensive that the poor and starving cannot buy the food and the producers cannot afford to give it away. A more sensible strategy would be to develop an agriculture based on the principles of nature's ecosystems and to take advantage of nature's resilience and her efficient use of sunlight once it has been trapped by the chlorophyll molecule. An expensive fossil-fuel or nuclear infrastructure should not be necessary to feed people around the world. We should be exporting the principles which emerge from working on the interface between the "pure" sciences of biology and the applied science of agriculture.

We are not completely condemning the researchers who

developed production agriculture; sustainable agriculturists will stand on their shoulders and will owe them a debt. But if their successes of the past 50 years so dazzle us that we fail to develop a sustainable agriculture, then future generations will become their victims.

Perhaps no one should be blamed for not starting earlier. The kind of agriculture we are advocating is dependent on interactions between ecology and population genetics, which began around 50 years ago. Modern ecology began around 1910, but population genetics became a distinct discipline only in the 1930s and did not mature until the 1960s, for it was not until then that enough research had been done to give us an understanding of the dynamics of the evolutionary process (Stebbins, 1979). This marriage of these two disciplines has now been widely publicized and somewhat more refined. But most agricultural researchers are either unaware that it exists, or they see it as mostly irrelevant. It seems inevitable, however, that practical minds will eventually ponder the implications of such knowledge for agriculture.

The discrepancy between the two economies, the human economy and nature's economy, is real and seems to be widening by the year. Of course, our socioeconomic system controls much of agriculture. As bleak as it now seems, it is worth remembering that the human economic system is full of wild cards. Eighty dollar a barrel oil or the fall of the Saudi Arabian monarchy can change the economic picture, literally overnight. Even if no wild card is ever played, the externalized costs of contemporary agriculture will eventually be more fully understood. When that time comes, it will be important to have a few sustainable alternatives to offer.

We believe that we already know enough to move toward a partial solution of "the problem of agriculture" (Jackson, 1980). Although some of the basic knowledge has been used to fine tune the traditional crops for increased yield, it has not been used to reduce soil erosion or the use of farm chemicals. It has not decreased the dependency on fossil fuel-based fertilizers or reliance on huge capital investments for farm machinery and equipment. Agriculture will benefit from the inherent virtues of plant mixtures or plant communities--reduced soil loss, resistance to insects and pathogens, appropriate water and nutrient management, etc.

This is where an understanding of ecosystems and the dynamics of evolutionary processes are essential. We will be relying on the same tools developed and used by the pioneers in these disciplines, the mathematical and statistical models, the microscopes and greenhouses. We need only change the emphasis.

HERBACEOUS PERENNIAL POLYCULTURES: A VISION FOR THE FUTURE

From the literature, from our own experience in the field, and from plants alive in our herbary at The Land Institute, we have surveyed hundreds of species, looking for the best perennial candidates to produce high yield. During this phase of our investigation, we cared little whether such plants were directly useful to the human enterprise or not. We simply wanted to know whether sustained high yield in perennial flowering plants is possible.

The literature survey revealed the following: Buffalograss (*Buchloe dactyloides*) (Ahring, 1964), which had been fertilized and irrigated, yielded 1,727 pounds per acre. Although this yield included the burs, of which seeds are a small part, the yield represents fruit/seed material. Fertilized and irrigated Alta fescue (*Festuca arundinacea*) averaged 1,460 pounds per acre (USDA, 1976). A native stand of sand dropseed (*Sporobolus cryptandrus*), under dryland conditions at Hays, Kansas, yielded 900 pounds per acre (Brown, 1943).

Irrigated legume yields were also encouraging. A five-year average for Illinois bundle flower (*Desmanthus illionensis*) amounted to 1,189 pounds per acre (USDA, 1978). Fertilized cicer milkvetch (*Astragulus cicer*) yielded 1,000 pounds per acre (USDA, 1966), as did sanfoin (*Onobrychis viciaefolia*), which received 70 pounds of nitrogen (Thornberg, 1971).

Two species of sunflower produced exceptional yields when irrigated. The grayheaded coneflower (*Ratibida pinnata*) yielded 1,600 pounds per acre (personal communication with Robert Ahring). Maximilian sunflower (*Helianthus maximillianus*) yielded 1,300 pounds per acre.

Although these high yields from relatively unselected plant materials are encouraging, there are cases where yield may be somewhat less important than the quantity of protein per acre. Based on their per acre yield and respective protein content, Illinois bundle flower, cicer milkvetch, and sanfoin all have a protein yield of about 400 pounds per acre (Earle and Jones, 1962; Jones and Earle, 1966). This exceeds the protein from an acre of wheat yielding 31.6 bushels in which the protein, by weight, amounts to 228 pounds. It even compares favorably with 100-bushel-per-acre corn, which yields about 500 pounds of protein per acre.

As more data on perennial seed yields is collected, numerous perennials with high-yield potential may be discovered. For example, almost all grass seed yields are from grasses ordinarily used for forage and hay. Such grasses have been selected to send much of their energy to the leaves rather than to the seed. Perhaps there are perennial grasses that are poor for forage but good for seed yield that have not been studied, such as the dropseed genera _Sporobolus_ and _Elymus_.

The direction of future perennial grain crop research may depend upon the results of research yet to be done, which could determine which perennials should be grown in rows and which ones in solid stands. Because prevention of soil erosion is the prime motivation for perennial grain crop research, erosion studies are needed to determine the relative ability of perennials in rows or solid stands to control erosion. Rows of perennials may not reduce soil erosion to a sustainable level. In that case, seed yields of perennials in solid stands will need to be recorded in order to find candidates for further yield improvement.

Although the selection program to improve seed yields of herbaceous perennials is practically nil compared with the extensive breeding done on annual crops, improvement in perennial yields will almost surely be more dramatic than that from a comparable effort with annuals from here on. These gains might be expected to come from the same sources breeders historically have called upon to increase yield, by reallocating the resources within the plant and by increasing plant performance overall.

Deciding whether a particular species is a good candidate or not depends on whether a species (1) demonstrates high variation so the breeder has a basis for selection; (2) shows promise as a potentially high seed-yielder, unless it serves another role (such as being a strong nitrogen-fixer or natural repellent of insects and pathogens for associated species); (3) suggests some potential for determinate seed set, shatter resistance, and a favorable reproduction-to-vegetation shoot ratio; (4) promises a relatively high yield for a minimum of three years in order to accommodate itself in a three-year or more replant cycle; and (5) exhibits meiotica stability. For example, if the species is of polyploid or hybrid origin, the probability of regular meiosis should be very high or show some promise of high regularity. With perennials, meiosis need not be as regular as with annuals (Grant, 1971).

Benefits of Perennial Polycultures

An agriculture based on a perennial polyculture would produce a number of dramatic benefits for society and the nation's natural resource base. The resource-oriented benefits include the following:

- Soil loss would first be cut to zero, then soil would begin to accumulate. The fossil energy savings for fertilizer to replace this lost soil would be significant.
- The indirect consumption of fossil energy in

agriculture would decline substantially. For example, the application of commercial fertilizer would drop greatly because, like a prairie, a perennial polyculture would use nitrogen more efficiently.

- The efficiency of water use and water conservation by the perennial ecosystem would be near maximum, and springs, long since dry or short-lived during the year, would return. Irrigation would decline, reducing fossil energy use.
- Commercial pesticides, especially those with no close chemical relatives in nature, would be absent or nearly so. In addition to the reduction in chemical contamination, a minor fossil energy savings would result.
- The direct consumption of fossil energy at the field level would be greatly reduced.

Among the social benefits of perennial polyculture would be the following:

- Because of efficiency in water use, long before such an agriculture is completely in place, major policy questions surrounding irrigation projects that promote aquifer mining or water diversions that lead to soil salting and silting problems would become more manageable.
- Policy considerations that include equitable land ownership rather than major corporate control of land would become more manageable due to the reduced need for high capitalizaiton in machinery, energy, farm chemicals, irrigation, and seed.
- More than 100 million acres, currently set aside as marginal land largely because of the serious erosion potential under current cropping systems, could be brought into production, thereby reducing the pressure for ever-increasing land prices. This increases the opportunity for people to have a farm as a place to promote and experience right livelihood rather than having a farm as a food factory.

Soil Savings

The primary purpose for developing such a radically new agriculture is to save soil from being eroded or chemically contaminated. What would be the effect on soil conservation if herbaceous perennials occupied the 316 million acres in the United States currently planted to the top ten crops? Results vary, of course, depending upon the estimate of soil loss chosen and the number of years between replanting. Let us assume all 316 million acres, including the cotton and hay acreage, were planted to mixed perennials. By replanting fields every five years, 253 million acres of the 316 million acres in any given year would remain unplanted. On the basis of a soil loss of five tons per acre, that would save 1.27 billion tons. Given a 12 ton per acre soil loss, the savings would exceed 3 billion tons a year, an amount equal to the annual loss when SCS was established.

These estimates are conservative. They are not adjusted to account for the soil build-up during the unplowed years, nor are they adjusted to reflect that plowed perennial fields, if they behave like plowed perennial pastures, would lose less soil the first year than a field that has experienced continuous cropping (Moldenhauer, 1977).

It is also possible to calculate the energy required to replace the nitrogen, phosphorus, and potassium in the eroded soil on these 316 million acres. If that loss is 5 tons per acre, the energy value is about 96.4 million wellhead barrels; at 12 tons, the energy value is 231 million barrels; and at 15 tons, 289 million barrels (USDA, 1980).

The energy value of the nitrogen, phosphorus, and potassium lost through the harvest of 100 bushels of grain from an acre roughly equals the energy lost in 9 tons of eroded soil on an acre--more than half a barrel of petroleum at the wellhead in each case. Assuming a five-year replant cycle, the savings could be equivalent to 80 million wellhead barrels if the loss is 5 tons per acre or 239 million barrels if the loss is 15 tons per acre (USDA, 1977).

Substantial energy savings are also possible as a result of reduced seedbed preparation and field work, pesticide and fertilizer use, and irrigation (Jackson, 1980; Pimentel, et al., 1973; Pimentel and Terhune, 1977; Roller, et al., 1975; Sloggett, 1977; Steinhart and Steinhart, 1974; USDA, 1974; U.S. Water Resources Council, 1978; Vaughan, et al., 1977; Zaprozec, 1979).

PROSPECTS FOR RESEARCH

Two constraints impinge on the prospect for development of perennial polycultures. The first constraint is that promoters of conventional agriculture believe the traditional methods of soil conservation--terracing, grass waterways, check dams, contour farming, stubble mulching, conservation tillage, crop rotation, etc.--are adequate for the task. Many apologists for conventional agriculture believe that where these methods are not practiced it is either the result of a farmer's ignorance or bad character. At the deepest level of contemplation, soil loss is regarded as a lack of exercise of stewardship. However, any exercise of stewardship that employs these traditional measures may be beyond the ethical stretch for _most_ farmers (Jackson,

1980).

The second constraint is the consequence of the somewhat critical acceptance of conservation tillage, which is chemical-dependent, for growing traditional row crops, particularly corn. This acceptance indicates that most agriculturists believe a technical fix is already at hand to reduce soil loss significantly.

What can be expected of agribusiness and government in the way of research on perennial polycultures? First, little help is likely to come from private seed companies. Who can blame them for not producing a perennial or groups of perennials that could put them out of business? To leave them out of the action, however, may be unfortunate. These companies have important resources, particularly human power for extensive analysis, an absolute necessity if new crops are to be developed. But if these companies are to be involved, what they will likely want in return are patents on their products.

Many people are concerned that plant patenting might lead to an unacceptable level of centralized control over food production. On the other hand, if a seed company were granted a patent on a particular crop or crop mix that was perennial, then there would be a strong incentive to increase the variety of useful plants. This could be managed in several ways. One possibility would be to grant the private seed company the right to accept royalty payments on a particular crop or crop mix. The payments could be adjusted according to yield and, at least in the early stages, subsidized by the government in order to encourage their adoption by the nation's farmers. This raises the question as to how much the government can afford to spend to assure food supply. But that question might better be asked by considering how much it can afford not to spend.

The agribusiness firms that sell fertilizers and pesticides would likewise appear to be poor candidates for promoting cropping systems that reduce the sales of their products. How they would be involved requires a stretch of the imagination. Perhaps instead of paying them for applying "medicine," some of which has harmful side effects, they might be paid for contributing to the maintenance of soil health and balance. Nutrients will have to be added, but why should the nutrient and pesticide bill be based on quantity alone? It is in the spirit of the recent emphasis on integrated pest management to recognize that a species mix is automatically a chemical mix, and if that chemical mix discourages insect and pathogen spread, the organizations and individuals responsible could be rewarded. The entire idea is somewhat like paying a doctor a fee while the individual is healthy with payments ceasing or slowing when an individual is sick. In our increasingly service-oriented economy, "package plans" are sold more and more.

This is not a promotion of agribusiness involvement. The burden of defining their role in the development of a sustainable agriculture should be on their shoulders, not on the shoulders of the rest of the people in a free-market economy. But neither should private seed companies or other agribusiness firms be written off before they are given a chance to show how they can participate in developing a sustainable agriculture.

The U.S. Department of Agriculture has the greatest potential of any institution for defining the mission of a sustainable agriculture. If it were decided internally that such an agriculture would be appropriate, the machinery exists within the department to give it the necessary emphasis and financial support.

Private research agencies perform tasks very well when they are clearly defined. Many of the biochemical studies and much of the cytogenetic screening could be done by them.

Probably the most competent of all would be the national laboratories, although the nature of their work does not lend itself to the type of research considered here. The most ideal situation would be the establishment of a mission-oriented agency similar to the National Aeronautics and Space Administration.

REFERENCES

Ahrenhoerster, R.; and Wilson, T. 1981. Prairie Restoration for the Beginner. Prairie Seed Source, Box 83, North Lake, WI 53064.

Ahring, R. M. 1964. "The Management of Buffalograss for Seed Production in Oklahoma." Technical Bulletin T-109. Stillwater: Oklahoma Agricultural Experiment Station.

Brown, H. R. 1943. "Growth and Seed Yields of Native Prairie Plants in Various Habitats of the Mixed Prairie." Kansas Academy of Science. Transactions 46:87-99.

Delin, S. G. 64, 151 52 Sodertolje, Sweden. "Technology or the Art of Production." Unpublished manuscript.

Earle, F. R.; and Jones, Q. 1962. "Analyses of Seed Samples From 113 Plant Families." Economic Botany 16(4):221-50.

Grant, V. 1971. Plant Speciation. New York: Columbia University Press.

Jackson, W. 1980. New Roots for Agriculture. San Francisco: Friends of the Earth.

Jones, Q.; and Earle, F. R. 1966. "Chemical Analyses of Seed II: Oil and Protein Content of 759 Species." Economic Botany 20(2):127-55.

Moldenhauer, W. C. 1977. "Erosion Control Obtainable Under Conservation Practices." In Universal Soil Loss Education: Past, Present, and Future. Madison: Soil Sci. Soc. Am.

Odum, E. P. 1971. Fundamentals of Ecology. 3rd ed. Philadelphia: W. B. Saunders Co.

Pimentel, D.; Hurd, L. E.; Bellotti, A. C.; Forster, M. J.; Oka, I. N.; Shoels, O. D.; and Whitman, R. J. 1973. "Food Production and the Energy Crisis." Science 1982:443-49.

Pimentel, D.; and Terhune, E. 1977. "Energy and Food." Annual Review of Energy 2:171-95.

Roller, W. L.; Keener, H. M.; Kline, R. D.; Mederski, H.

J.; and Curry, R. B. 1975. "Grown Organic Matter As a Fuel Raw Material Resource." Wooster: NASA Contractor Rep. 2608, Ohio Agricultural Research and Development Center.

Sloggett, G. 1977. "Energy Used For Pumping Irrigation Water in the United States, 1974." In Agriculture and Energy, ed. William Lockeretz, pp. 113-30. New York: Academic Press.

Stebbins, L. 1979. "Fifty Years of Plant Evolution." In Topics in Plant Population Biology, eds. Otto T. Solbrig, et al., pp. 18-41. New York: Columbia University Press.

Steinhart, J. S.; and Steinhart, C. E. 1974. "Energy Use in the U.S. Food System." Science 184:307-16.

Stern, K.; and Roche, R. 1974. Genetics of Forest Ecosystems. New York: Springer-Verlag.

Thornberg, A. 1971. "Grass and Legume Seed Production in Montana." Bulletin No. 333. Bozeman: Montana Agricultural Experiment Station.

USDA. 1966. "Technical Note 12/1/66." Bridger, Montana: Plant Materials Center, Soil Conservation Service.

______. 1974. "Energy and Agriculture: 1974 Data Base," Vols. 1 and 2. Washington, D.C.

______. 1976. "Technical Report 1975-6." Coffeeville, Mississippi: Plant Materials Center, Soil Conservation Service.

______. 1977. "Cropland Erosion." Washington, D.C.

______. 1978. "Annual Report." Knox City, Texas: Plant Materials Center.

______. 1980. "Soil and Water Resources Conservation Act, 1980 Appraisal." Rev. draft, part 1. Washington, D.C.

U.S. Water Resources Council. 1978. "The Nation's Water Resources: 1975-2000." Second National Water Assessment, vol. 1, summary. Washington, D.C.

Vaughan, D.H.; Smith, E. S.; and Hughes, N. A. 1977. "Energy Requirements of Reduced Tillage Practices for Corn and Soybean Production in Virginia." In Agriculture and Energy, ed. William Lockeretz, pp. 245-59. New York: Academic Press.

Weaver, J. E. 1954. North American Prairie. Lincoln: Johnson Publishing Co.

Wittwer, S. 1981. "The Twenty Crops that Protect the World from Starvation." Farm Chemicals 144 (September):17-28. (International Plant Protection Issue).

Zaprozec, A. 1979. "Changing Patterns of Ground-water Use in the United States." Ground Water 17(2):200.

Zuber, M. S.; and Darrah, L. L. 1980. Annual Corn Sorghum Research Conference Proceedings 35:234-49.

Water

3
Water: Its Changing Role in U.S. Agriculture

Burton C. English

The interrelationship between the food security of the United States and a diminishing water resource balances on the condition of that water resource, the quantity of other substitute resources, and the level of future technology. Whether food supplies in the United States can be threatened by a given resource such as water cannot be determined solely by examining the resource in question. Rather, the interactions of all resources used in the production of agricultural commodities must be analyzed.

In the recent past, the United States had an abundance of land with yields increasing faster than demands. With real commodity prices decreasing, for the most part, over the past century, supply control programs have been initiated so that price supports are gained.

On the other side of the ledger, embargos have been placed on agricultural commodities when domestic food prices spiralled upwards (1973-1974 period for instance). Weather conditions have at times dampened yields. The quality of presently unused inputs, such as land, does not seem as good as what is already in production.

A glance over the past two decades of agricultural history yields some interesting observations. The 1960s were characterized by chronic surpluses in most nonperishable agricultural commodities. Various land retirement programs were in evidence. As much as 63 million cropland acres were withheld from production during the '60s.

The turn of the decade, however, brought about reduced supplies, increased input costs, and higher commodity prices. During the early to mid-'70s, world grain production greatly fluctuated. A world food crisis began. The United States called upon the farmers to plant fence row to fence row. Experts were predicting famine for much of the less developed world (Cochrane, 1979). In the space of five years (1969-1974), the United States nearly doubled its wheat exports. Farmers responded to the increased worldwide demands by dramatically increasing production.

The water resource played an important role in this change. Wells were sunk in the Great Plains. Previously unproductive lands became extremely productive as water became available.

Since then, much concern has been expressed as to resource availability, technological development, and resource quality. Attempts are being made to analyze whether the United States has the resources available to meet future agricultural demands (both domestic and export). Water availability and quality are both being studied.

WATER SUPPLY AND DEMAND

In the long run, the consumption of water must be less than or equal to the supply of water. In 1975, the United States withdrew 338 billion gallons per day of fresh water (both surface and ground) for such offstream uses as agriculture, manufacturing, etc. By the year 2000, this amount is projected to decrease to 306 billion gallons per day or a 9 percent reduction. This decline in water demand will be caused primarily by a more efficient use of water (U.S. Water Resources Council, 1978). On the supply side, an average of 40,000 billion gallons per day of water passes over the United States in the form of water vapor. Approximately 10 percent or (4,200 billion gallons per day) falls as precipitation. Nearly two-thirds of this precipitation evaporates immediately or is transpired by vegetation. Thus, 1,450 billion gallons per day accumulates in ground and surface storage, flows to the oceans or into Mexico or Canada, is consumptively used, or evaporates from reservoirs. Only 675 billion gallons per day can be considered as a reliable water supply (U.S. Water Resources Council, 1978).

From these nationwide figures, one would conclude that the reliable supply exceeds projected demand by a good margin. However, regionally this is not the case. An estimated 392.8 billion gallons per day were withdrawn to meet 1975 water demands with 81.2 and 254.2 billion gallons per day coming from ground and surface supply sources, respectively (Table 3.1). More than 20 percent of the groundwater used in the Missouri (24.6 percent), Arkansas-White-Red (61.7 percent), Texas-Gulf (77.2 percent), Rio Grande (28.1 percent), Lower Colorado (48.2 percent), and Great Basin (41.5 percent) water resource regions result from groundwater mining. Thus, in these regions, supply is less than demand at the present time.

As of yet, little has been attributed to agricultural reliance on water. It is estimated that 68 percent of the groundwater withdrawn and 35 percent of all water is used for irrigation. Table 3.2 shows that withdrawals of water are assumed to decline between 1985 and 2000 by nearly 12 million gallons per day, with many of the river basins in the western United States reflecting a decline in water withdrawn.

TABLE 3.1
Water Withdrawn by Source and Percent of Groundwater Overdraft by Water Resource Region, 1975

Water Resource Region	Ground Total Withdrawn (mgd)(c)	Overdraft (a) (percent)	Surface Total Withdrawn (mgd)	Total Withdrawn (b) (mgd)
New England	635	0	4,463	10,314
Mid-Atlantic	2,661	1.2	15,639	37,925
South Atlantic	5,449	6.2	19,061	31,970
Great Lakes	1,215	2.2	41,598	42,813
Ohio	1,843	0	35,091	34,934
Tennessee	271	0	7,141	7,416
Upper Mississippi	2,366	0	10,035	12,401
Lower Mississippi	4,838	8.5	9,729	15,820
Souris-Red-Rainy	86	0	250	336
Missouri	10,407	24.6	27,609	38,016
Arkansas-White-Red	8,846	61.7	4,022	12,868
Texas-Gulf	7,222	77.2	9,703	26,088
Rio Grande	2,335	28.1	3,986	6,321
Upper Colorado	126	0	6,743	6,869
Lower Colorado	5,008	48.2	3,909	8,917
Great Basin	1,424	41.5	6,567	7,991
Pacific Northwest	7,348	8.5	30,147	37,626
California	19,160	11.5	20,476	54,705
Total	81,240	25.0	254,169	392,826

Source: U.S. Water Resources Council, 1978

(a) Overdraft/total withdrawn
(b) Includes water from saline sources
(c) Million gallons per day

Although the volume of groundwater greatly exceeds surface runoff, the increasing demands on this resource are straining the supply. Groundwater mining is occurring in the Ogalala Aquifer from Nebraska to

TABLE 3.2
Projected Irrigation Withdrawal and Consumption by Water Resource Region, 1985 and 2000

Water Resource Region	Withdrawal 1985	Withdrawal 2000	Consumption 1985	Consumption 2000
	(million gallons per day)			
New England	41	46	29	33
Mid-Atlantic	366	481	269	354
South Atlantic	4,008	4,509	3,184	3,597
Great Lakes	211	282	169	232
Ohio	68	91	53	74
Tennessee	18	21	14	17
Upper Mississippi	283	387	230	323
Lower Mississippi	4,559	4,444	3,204	3,272
Souris-Red-Rainy	144	434	116	350
Missouri	39,376	36,236	17,597	17,607
Arkansas-White Red	10,483	9,776	7,468	7,125
Texas-Gulf	9,333	7,427	7,597	6,100
Rio Grande	5,498	4,873	3,920	3,570
Upper Colorado	7,223	6,672	2,657	2,741
Lower Colorado	7,299	6,343	3,962	3,720
Great Basin	6,120	5,825	3,082	3,196
Pacific Northwest	34,639	29,961	13,362	13,213
California	34,863	34,764	25,134	26,311
Total	164,532	152,572	92,047	91,835

Source: U.S. Water Resources Council, 1978.

Texas, in south-central Arizona, and parts of California.

As water tables decline, energy costs for pumping increase. Real energy prices have increased in the past decade and will continue to increase in the foreseeable future. These real energy price increases, along with declining water tables, will influence the quantity of water that can be economically pumped. Before this occurs, alternative sources of water must be found, artificial recharge methods must be developed, water-using activities must be relocated, and/or water use must be reduced through conservation and improvement of managerial techniques. The U.S. Water Resources Council's assessment (1978) indicates that the potential savings in irrigation withdrawals range between 30 and 45 billion gallons per day. This can be achieved by lining and/or covering canals, monitoring and scheduling of water release using the computer, and other technological advances. Applying these technologies is estimated to increase efficiency 10 percent in moving waters from the source to the field. On-farm efficiencies are estimated to range between 10 and 40 percent. These technologies include closer scheduling of water application in meeting crop needs, irrigating at night, improving the irrigation system, and preparing the land better.

However, these increased efficiencies will not significantly decrease the amount of mining that currently occurs. The Interagency Task Force (1979) stated that only about 15 percent of the current groundwater mining would disappear if the Soil Conservation Service's accelerated water conservation programs were implemented. Thus, other methods of conservation, such as capturing water before it becomes runoff, increasing the levels of snow management, and controlling undesired vegetation along waterways will have to be undertaken if water use levels are to be maintained.

EXAMINATION OF SOME FUTURE ASSUMPTIONS

Even though we are certain that the supply of water for agriculture in the United States will decline over the next three decades due to increased pumping costs (because of lowered water levels and higher energy costs) and increased competition with other water users, we can best dampen the impact of declining water supplies through a vigorous R&D program which meshes the corresponding resource scarcities and prices and technologies which can substitute for them.

In examining how critical water is in meeting national food demands, it would be useful if we could select specific future dates, set all of the exogenous and some of the endogenous variables, vary water supplies, and predict the resulting expected increases in commodity prices. Statistical, econometric, and other methods or models for these predictions do not exist, nor will they in the near future. Changes in the variables and institutions that affect water supplies and demand may change gradually or dramatically (i.e., the 1974 energy shortage). Meanwhile, these types of changes do not exist in historical data, so we cannot use the past to statistically predict future impacts.

For some years, economists at the Center for Agricultural and Rural Development (CARD) have incorporated both ground and surface water sectors into large-scale interregional programming models of U.S. agriculture. This modeling technique does not attempt to predict; rather, it examines what should occur. Generally, these models have examined alternative demand levels and increasing real water prices, considered trend and other levels of yield improvement over time, allowed land not currently in crops to be transferred to cropland, and studied the impacts of alternative water price levels. These variables provide the means by which projections can be made as to the potential changes in regional and national production and resource use in agriculture. An analysis can be conducted as to the national as well as regional importance of the water resource required to meet national demands. Finally, a means of studying resource abundance and resource scarcity can be conducted.

Recently, a set of models was analyzed for the Soil and Water Resources Conservation Act (RCA). The level of exports, the level of technological change assumed, and the amount of land incorporated in the model are the primary factors in determining the ability of the nation to meet production goals. The models determine the quantity of water and its marginal price as the above factors are allowed to vary.

In the RCA, the two major resources of concern to agriculture are land and water. The natural resource, land, is covered elsewhere in this volume. Suffice it to say that the Soil Conservation Service (SCS) has estimated that there are 127 million acres of high or moderate range and forest land that could be converted to cropland. Thus, an additional increase of 30 percent in the cropland base could occur in the future. This land is not presently available for conservation due to economic, physical, and sociological factors. However, over a long-run time frame of 20 to 50 years, these factors tend to change.

Based on current existing technology, by the year 2000 wheat yields could increase by 50 percent, soybean yields by 60 percent, and corn by 40 percent. Over the same time frame, the amount of production gains per breeding female is predicted to increase 25 percent for beef, pork, and dairy products (English, Maetzold, Holding, and Heady, 1983). These estimates are much higher than the "moderate" technology scenario used in the 1980 Soil and Water Resources Conservation Act analysis(1).

Given these data, land does not appear to be a scarce resource and the notion of production plateaus cannot be supported. Thus, it would seem that demand levels would determine whether water will be a scarce enough resource over the long run to threaten the nation's food supply. However, several land estimates and technology levels are used in the RCA analysis.

An agricultural programming model assuming projected demands and yields for 2030 in conjunction with the estimated quantity of resources available is used in the RCA analysis. In the 2030 CARD-RCA model, a set of shadow prices is generated(2). These shadow prices suggest that a general interaction among these variables exists. An index for required water use and the related shadow prices is presented in Table 3.3 for (1) a base-1 solution (BASE-I) which assumes a 380 million acre cropland base, (2) a base-2 solution (BASE-II) which allows an additional 127 million acres identified by the SCS as having the potential of conversion, (3) a high technology scenario with the increased land base (III), (4) a low technology scenario with the increased land base (IV), and (5) a maximum production scenario assuming a high technology level and increased land base.

TABLE 3.3
Indices of Selected Variables for Five Different Scenarios, 2030

Variable	BASE-I	BASE-II	III	IV	V
Water use	100	91 (100)(a)	65 (71) [100](b)	109 (120)	210 (231) [323]
Water shadow price	100	63 (100)	44 (70) [100]	63 (100)	280 (444) [636]
Corn price	100	50 (100)	37 (74) [100]	64 (128)	167 (334) [451]
Wheat price	100	64 (100)	54 (84) [100]	74 (116)	176 (275) [326]
Soybean price	100	71 (100)	58 (82) [100]	59 (83)	179 (252) [309]
Cotton price	100	70 (100)	71 (101) [100]	77 (110)	139 (199) [196]

(a) ( ) compares the technology solution (III, IV, V) to BASE-II.
(b) [ ] compares solution V to III, with demand levels as the only item that is changing.

The figures indicate that if technology continues to increase, and 127 million acres of land is available for commodity production, water use would decline by 10 percent with the water value(3) declining by 37 percent and commodity shadow prices(4) decreasing 50, 36, 29, and 30 percent for corn, wheat, soybean, and cotton, respectively. Assuming high technology and an additional 127 million acres of cropland (III), a 29 percent decrease in the quantity of water used and a 30 percent decline in the value of water is projected to take place when compared to BASE-II. If the technology levels are lower than the BASE-II levels, an increase in water use of 20 percent is projected but the value of water does not change at all when compared to BASE-II. The production capacity in IV is still great enough so that the marginal value of water is similar to the BASE-II solution. However, if the water supplies were reduced below this level, one should expect the commodity shadow prices to move upward considerably and eventually exceed the shadow prices reflected in the BASE-I solution.

The demand level also influences the levels of change that these variables undertake. The maximum production alternative examines the production potential of this nation assuming a high level of technology along with the 507 million acre cropland base. To fully use this land and to maximize production, water use would increase 223 percent from III and 110 percent from BASE-I. An increase of 280 percent in the water shadow price is projected when this solution is compared to BASE-I, with commodity shadow prices increasing an average of 70 percent. These higher prices result from the increased production level, complete use of the available land, and greater demand on the water resource. If this amount of water available in 2030 were reduced, commodity prices would increase, or the nation would not be able to meet the prescribed production levels.

Another assumption that one must examine when determining whether water will be a critical resource in meeting the nation's food and fiber demand is the impact of prices of that input on the nation's ability to produce. The "market" price of surface water, historically, has not been subject to normal market forces because of publicly subsidized rates. Recently, however, some of the water previously used for agriculture has been purchased by users and market transfers have taken place. However, the observations and data base of these transfers is so incomplete that statistical estimation of a water demand function that incorporates commodity supplies and prices as well as resource or input prices is currently impossible. Thus, it seems that a normative analysis of the demand for ground and surface water is the only means for evaluating the national impacts incurred due to increased water prices. A normative study undertaken by Christensen, Morton, and Heady (1981) set prices for ground and surface water at farm levels for each with the initial price of surface water equaling the 1975 Bureau of Reclamation costs and the price of groundwater being estimated by the 1975 pumping costs plus 15 percent of the fuel costs for maintenance (Dvoskin, Heady, and English, 1978). These costs or prices were then doubled (SW2, GW2), tripled (SW3, GW3), and quadrupled (SW4, GW4), so that 16 price combinations were derived (Figure 3.1). The normative water demand responses indicate the level of water demanded for each set of prices.

Relative to the base solution (GW1, SW1), water use decreases from 50.5 to 24.7 million acre-feet for endogenous crop and livestock production when both ground and surface water prices are quadrupled. While water use declines by 50 percent, very little impact on commodity prices is projected (Table 3.4).

TABLE 3.4
Indices of Selected Variables Under Four Price Combinations for Water

Item	Water Price Levels			
	GW1SW1	GW2SW2	GW3SW3	GW4SW4
Groundwater use	100	59	46	44
Surface water use	100	85	65	51
Corn shadow price	100	100	100	105
Wheat shadow price	100	104	106	110
Soybean shadow price	100	102	103	105
Pork shadow price	100	102	103	105
Beef shadow price	100	103	104	106

CONCLUSIONS

It is inadequate to simply study the supply and demand for water in a localized area when examining food security of the United States. Numerous resources are used in the production of agricultural commodities. The resources serve, at least at a national level, as substitutes. As one becomes limiting, more of another is used.

As agriculture approaches production capacity, changes in the prices of a resource, or a reduction in the quantity of the resource, will have a larger impact on the nation's ability to produce the necessary commodities if some resources are left idle. It would appear that given 127 million acres of potentially available cropland, changes in levels of technology not less than those experienced in the past, and modest exports, water is not an input whose scarcity will impact on the food security of the United States.

It is entirely conceivable that the long-run aggregate agricultural commodity supply function is of the following nature:

Figure 3.1 The 4 by 4 Alternative Price Matrix

Groundwater Prices →

Surface Water Prices ↓

GW1SW1	GW2SW1	GW3SW1	GW4SW1
GW1SW2	GW2SW2	GW3SW2	GW4SW2
GW1SW3	GW2SW3	GW3SW3	GW4SW3
GW1SW4	GW2SW4	GW3SW4	GW4SW4

> Over some range, it may remain highly elastic as opportunities remain to convert more land to crops and to further adjust the allocation and technology of water use. But eventually, with complete use of all potential cropland which can be converted at reasonable costs, higher prices and smaller supplies of water and exhaustion of reallocation possibilities for water, the supply elasticity may decline greatly with a sharp upturn in the commodity supply function (Heady, 1982:22).

Many scientists would argue that this point has already been reached and that the corner has been turned. Under this scenario, the major agricultural problem of the future will be how we can produce more at reasonable prices. However, given our recent experiences, characterized by large crops and low commodity prices, others will argue that the corner has not been turned and we are still in the elastic portion of the supply function. Supplemental irrigation techniques, double cropping, land conversion, etc., all act as substitutes for scarce resources, such as water. Even within the scarce resources, substitutes or increases in efficient use of the resource can be found. As the price of water increases due to diminished supply and greater demands, more incentive will exist to make more efficient use of this resource. Almost no evidence can indicate that this corner has been turned. The substitutes available in agriculture still are present in great quantities.

However, the impact of declining water tables can have severe regional consequences. Regions forced to dryland farming as the water situation changes will be looking at a reduced output but little price increase. Hence, revenues will decline. As revenues decline communities serving those agricultural areas will be affected and hardships placed upon them.

NOTES

1. This scenario assumed 1.1 percent per year productivity gain by the year 2000 as the most likely alternative.
2. Shadow price refers to the cost of producing the last unit. Hence, it is a marginal cost concept. It is a set of crop prices which indicate what the producer must receive if the assumed demand is to be met.
3. Water value is determined from a shadow price resulting from a programming model that is specified so that the given alternatives can be examined. The prices of water indicate the value of water, at the margin, to produce the nation's output under the combination of conditions outlined.

4. The supply prices for the commodities show the levels necessary to attain the prescribed level of production under the resource and technology conditions analyzed.

REFERENCES

Bekure, S. 1971. "An Economic Analysis of the Intertemporal Allocation of Groundwater in the Central Ogalala Formation." Unpublished Ph.D. dissertation, at Oklahoma State University, Stillwater, Oklahoma.

Christensen, D.; Heady, E. O.; and Morton, A. 1981. "Potential Water Use and Agricultural Production Patterns Under Alternative Water Prices." American Water Bulletin 17(2):844-50.

Christensen, D.; Morton, A.; and Heady, E. O. 1981. Potential Effects of Increased Water Prices on U.S. Agriculture. CARD Report No. 101. Center for Agricultural and Rural Development, Iowa State University, Ames.

Cochrane, W. W. 1979. The Development of American Agriculture. A Historical Analysis. Minneapolis: Univ. of Minnesota Press.

Coomer, J. J. 1978. "Projected Irrigation Adjustments to Increasing Natural Gas Prices in the Texas High Plains." Unpublished M.Sc. thesis, at Texas Tech University, Lubbock, Texas.

Dvoskin, D.; Heady, E. O.; and English, B. C. 1978. Energy Use in U.S. Agriculture: An Evaluation of National and Regional Impacts from Alternative Energy Policies. CARD Report No. 78. Center for Agricultural and Rural Development, Iowa State University, Ames.

English, B. C.; Alt, K. F.; and Heady, E. O. 1982. A Documentation of the Resources Conservation Act's Assessment Model of Regional Agricultural Production, Land and Water Use, and Soil Loss. CARD Report No. 107T. Center for Agricultural and Rural Development, Iowa State University, Ames.

English, B. C.; Maetzold, J. A.; Holding, B. R.; and Heady, E. O. 1983. "Future Agricultural Technology and Resource Conservation Executive Summary." RCA Symposium, Center for Agricultural and Rural Development, Ames, Iowa.

Heady, E. O. 1982. "National and International Commodity

Price Impacts of Declining Irrigated Agriculture." Presented at Impacts of Limited Water in the Arid West: A Major Interdisciplinary Conference. September 8-30. Asilimar, California.

Interagency Task Force. 1979. Irrigation Water Use and Management. U.S. Department of the Interior, USDA, and EPA.

Lacewell, R. D.; Condra, G. D.; Hardin, D. C.; Zazaleta, L.; and Petty, J. A. 1978. The Impact of Energy Shortage and Cost on Irrigation for the High Plains and Trans Pecos Regions of Texas. Technical Report No. 98. Texas A&M University, Texas Water Resources Institute, College Station, Texas.

U.S. Department of the Interior. 1973. West Texas and Eastern New Mexico Import Project, Reconnaissance Report. Washington, D.C.

U.S. Water Resources Council. 1978. The Nation's Water Resources 1975-2000. Summary, vol. 1. Washington, D.C.: U.S. Government Printing Office.

4
Domestic Food Security and Increasing Competition for Water

E. Phillip LeVeen

The decade of the 1970s witnessed some dramatic changes in the world food economy. Between 1973 and 1975 a series of events sent U.S. and world farm commodity prices soaring to spectacular heights, forcing a sharp rise in the real costs of food. Suddenly the world seemed to have run out of food as demand for U.S. farm exports jumped dramatically. With the coming of the "world food crisis," modern-day Malthusians were quick to claim that population growth had finally outstripped food production. Earl Butz, Secretary of Agriculture in the Nixon Administration, saw the crisis as a golden opportunity for the U.S. and advised farmers to plant "fence row to fence row" to make the most of the expected increases in exports to a hungry world. Even usually cautious economists, such as Edward Schuh, one-time president of the American Agricultural Economics Association, predicted a "new era" of increasing pressure on agricultural resources in which all of the problems of excess capacity that prevailed in the 1950s and 1960s would be replaced by shortages and rising prices (Schuh, 1976).

Subsequent events have shown these Malthusian predictions to be, at least, premature; the old problems of overproduction and low farm prices returned after 1976, with the slowing of international demand for U.S. farm commodities and a substantial expansion of domestic production. Once again, government price support and acreage diversion programs were instituted to support farmers who had followed Butz' advice by investing heavily in new equipment and land, only to find it impossible to meet debt repayments from their declining incomes. Yet by the early 1980s, even the increased public support was insufficient to prevent a deep depression in U.S. agriculture rivaling that of the 1930s.

In recent years, this crisis of overproduction has shifted public attention away from an important debate over whether the United States can continue to expand its agricultural production to accommodate the growing international demand for food without straining its resource base and threatening long-term productivity. (USDA, 1981; Wessel and Hantman, 1983). Now the severe drought of 1983, which has depleted a substantial portion of the nation's grain stocks and in turn will soon trigger higher food prices, reminds us that agricultural production is indeed subject to limits and it should serve to renew the debate.

This chapter examines one aspect of this debate; namely, the possible relationships between food production capacity, the limits on additional irrigation development, and overall food security. It will be argued that there is little potential remaining for expanding irrigation and limited irrigation development will, in combination with other factors, reduce the potential growth of the nation's food production capacity. If important reforms are not undertaken to reduce foreign demand for U.S. grown food, food prices will likely rise substantially, threatening domestic food security.

This is not intended to be a Malthusian argument, however, for our limited production capacity could be used much more efficiently to maintain domestic food security than is now the case. Indeed, limited productive capacity will critically test those institutions that determine how and what kinds of foods are produced and, even more important, how available food supplies are distributed. If limits on production capacity should lead to Malthusian-like food shortages, the fault will lie in our inability to change these political and economic institutions, and not with inadequate food and nutrition production capacity.

INCREASING WATER SCARCITY FOR AGRICULTURE: GENERAL CONSIDERATIONS

In the past two years, many popular magazines have carried articles with the theme that the energy crisis of the 1970s will be replaced by an equivalent water crisis of the 1980s. The scientific literature on water supplies also confirms the dawning of a new era of "scarcity," especially in the arid regions of the West(1). Current water use exceeds average stream flows in most of the West's major watersheds, while groundwater is being depleted in many important basins. Table 4.1 provides a perspective on the over-appropriation of most significant Western rivers. Only five of these streams, located along the northwestern Pacific coast, can provide enough water to meet current water demand during dry years. A critical concern of most water users is reliability of supply, and dry-year access to water is a key characteristic of water supply. While 50 percent of these rivers have more than enough water in average rainfall years to meet current demand, further expansion of water consumption would reduce reliability of supply during dry years and therefore such expansion would be detrimental(2).

TABLE 4.1
Total Water Use As a Percent of Stream Flow in Average and Dry Years

Name	Average	Dry
SOURIS-RED-RAINY	62%	110%
MISSOURI	87	120
Missouri-Milk-Saskatchawan	82	105
Missouri-Marias	82	104
Missouri-Musselshell	81	102
Yellowstone	96	117
Western Dakotas	84	108
Eastern Dakotas	82	102
North and South Platte	140	160
Niobrara-Platte-Loup	103	122
Middle Missouri	91	107
Kansas	123	191
Lower Missouri	87	120
ARKANSAS-WHITE-RED	83	138
Upper White	84	126
Upper Arkansas	134	175
Arkansas-Cimarron	114	243
Lower Arkansas	83	152
Canadian	122	261
Red-Washita	129	180
Red-Sulphur	83	133
TEXAS-GULF	101	197
Sabina-Neches	85	163
Trinity-Galveston Bay	89	176
Brazos	142	327
Colorado (Texas)	119	188
Nueces-Texas Coastal	96	183
CALIFORNIA	82	113
Klamath-North Coastal	65	95
Sacramento-Lahontan	76	106
San Joaquin-Tulare	109	131
San Francisco Bay	91	152
Central California Coast	83	169
Southern California	107	116
Lahontan-South	243	290
RIO GRANDE	136	180
Rio Grande Headwaters	110	159
Middle Rio Grande	140	165

TABLE 4.1 (continued)
Total Water Use As a Percent of Stream Flow in Average and Dry Years

Name	Average	Dry
RIO GRANDE (continued)		
Rio Grande-Pecos	148%	176%
Upper Pecos	144	177
Lower Rio Grande	136	180
UPPER COLORADO	84	112
Green-White-Yampa	87	114
Colorado-Gunnison	80	106
Colorado-San Juan	84	112
LOWER COLORADO	225	239
Little Colorado	80	103
Lower Colorado Main Stream	225	239
Gila	304	315
GREAT BASIN	125	158
Bear-Great Salt Lake	102	125
Sevier Lake	186	204
Humboldt-Tonopah Desert	177	222
Central Lahontan	116	165
PACIFIC NORTHWEST	84	102
Clark Fork-Kootenai	62	73
Upper/Middle Columbia	79	94
Upper/Central Snake	91	119
Lower Snake	78	96
Coast-Lower Columbia	85	102
Puget Sound	81	96
Oregon Closed Basin	101	161

Source: U.S. Water Resources Council, 1978.

Table 4.2 describes the "mining" of groundwater throughout the West. "Mining" means pumping groundwater faster than it is naturally recharged, so that the resource is depleted. These extraction rates cannot be sustained, either because the water resource is eventually depleted or because the additional pumping costs incurred as the water table drops make it uneconomic to continue pumping. As indicated by Table 4.2, most aquifers are being depleted, so eventually the groundwater use in most areas will decline from current levels. Since more than 50 percent of the water supply in the West comes from underground, the reduction in pumping rates will have substantial effects on overall water availability.

TABLE 4.2
Groundwater Mining As a Percent of Annual Consumption

Name	Mining Rate	Name	Mining Rate
MISSOURI	17	RIO GRANDE	16
Missouri-Musselshell	1	Middle Rio Grande	21
Western Dakotas	2	Rio Grande-Pecos	46
Eastern Dakotas	7	Upper Pecos	16
North and South Platte	13	Lower Rio Grande	1
Niobrara-Platte-Loup	13		
Lower Missouri	5	LOWER COLORADO	53
Middle Missouri	16	Little Colorado	7
Kansas	41	Lower Colorado Main Stream	27
ARKANSAS-WHITE-RED	68	Gila	61
Upper White	2		
Upper Arkansas	3	GREAT BASIN	16
Arkansas-Cimarron	103	Bear-Great Salt Lake	3
Lower Arkansas	2	Sevier Lake	40
Canadian	85	Humboldt-Tonopah Desert	27
Red-Washita	55	Central Lahontan	3
Red-Sulphur	1		
CALIFORNIA	8	PACIFIC NORTHWEST	5
Sacramento-Lahontan	4	Clark Fork-Kootenai	2
San Joaquin-Tulare	10	Upper/Middle Columbia	8
Central California Coast	10	Upper/Central Snake	4
Southern California	8	Lower Snake	7
Lahontan-South	43	Coast-Lower Columbia	2
		Oregon Closed Basin	2

Source: U.S. Water Resources Council, 1978.

Agriculture now uses more than 85 percent of the West's total water supply, and therefore any reduction in the supply will have its greatest impact on this sector. This conclusion is reinforced by the fact that agriculture faces intense competition for water supplies from the growing urban centers, from new businesses anxious to develop western coal and oil resources, and from environmentalists wishing to preserve remaining undeveloped streams for recreational and other purposes. Many of these competitors can afford to pay considerably higher prices for water. In short, the increase in demand for a declining water supply will mean that agriculture will have a difficult time maintaining its current supply, let along procuring additional water supplies.

In the past, competition for water would have led to pressures on the public sector, both federal and state, to develop new water supplies and thus to satisfy all interests; indeed, pressures for development are still strong. However, the potential for new development is limited. Additional water development is technically feasible, especially in the Northwest, but the amount of water that can be captured by new dams and canals is small relative to the overdraft of groundwater basins and the overcommitment of streams. Furthermore, the location of the remaining water supplies is generally far from those regions that have the greatest water deficits. Moving the water long distances is technically difficult, requires large amounts of energy, and is as politically sensitive as an issue can be in the West.

Perhaps the most important impediment to new development is the costs. Private capital has little incentive to develop this potential because water cannot be sold at the prices required to make such projects profitable. For example, in water-rich California, where 30 percent of the total water supply still runs into the ocean, the cost of developing additional water supplies starts at $100 per acre-foot (an acre-foot amounts to about 326,000 gallons) and goes to more than $400 per acre-foot. The costs of delivering this water could well add another $50 to $100 per acre-foot. For comparison, the average price currently charged agricultural water users is less than $10 per acre-foot, and the maximum price that growers can pay is less than $50 per acre-foot for all but a few crops, such as strawberries, lettuce, or avocados.

In order for farmers to afford the higher costs of new water developments, real commodity prices of important crops would have to rise substantially more than they have at any time in this century and remain at this high level. This prospect seems very unlikely (3), so unless the price of new water can be reduced, farmers throughout the West cannot afford new projects.

The federal Bureau of Reclamation continues to press for new water projects. Even though all of its projects are supposed to have benefits greater than costs, most recent federal projects do not meet this economic test(4). The lack of profitability does not reduce the political appeal of water projects because the federal government does not require agricultural users to repay most of the development costs. This means that even though projects are unprofitable from the overall perspective of society, they are very profitable for project beneficiaries who do not pay a large portion of their costs (which, instead, are paid by taxpayers or by users of federally-produced hydroelectricity). The water subsidy creates strong political demands for uneconomic projects. Indeed, the demands have escalated to include such massive projects as diverting the Yukon and Mississippi Rivers through thousands of miles of aquifers, over mountain ranges, throughout the entire arid West.

For all the political pressure, public support for new water projects is declining. The Reclamation program has become a major fiscal drag on an already tight federal budget and the government cannot afford the luxury of large pork barrel expenditures with very limited payoff. While the Bureau of Reclamation continues to press for new projects, other federal agencies, concerned with the overall budget, have been increasingly successful in their efforts to reduce the water subsidies. For example, when the Reclamation Act was rewritten in 1982 to increase the maximum holding that could receive federal water from 320 to 960 acres, new pricing provisions were also added that will significantly reduce the water subsidy to many agricultural users. In addition, the federal government is no longer willing to underwrite the entire costs of new projects and is insisting that individual states pay a share. Given the budget crises in most Western states, this requirement further weakens the position of agricultural interests in obtaining new projects. In sum, it is unlikely that the Reclamation program will reduce the growing pressures for water in the West.

Energy and Water Supplies

In addition to increased competition for water supplies, water users will also be forced to contend with sharply rising water delivery costs. Because water is often used far from its source, getting water from where it originates to the farm involves a substantial amount of pumping. Even when groundwater is located directly beneath the farm, it must be pumped to the surface. Water is heavy, and thus pumping requires large inputs of energy. For instance, irrigation pumping uses more energy than any other use in California.

Rising energy costs mean that irrigation water costs must also rise, although the effects of higher energy costs have yet to be felt in many federal and state irrigation projects because publicly subsidized energy has kept electricity rates far below true cost. But these subsidies are slowly being eliminated, and so surface water costs are going to rise dramatically over the next decade. For instance, farmers in the southern part of the San Joaquin Valley of California will experience a doubling of their current water prices, from $20 to more than $40 per acre-foot, when new contracts for the energy used in the state's major irrigation project will increase the cost of pumping energy roughly five times.

Increased Irrigation Efficiency and Food Production

The combined effects of rising water prices from increased competition and higher pumping costs will serve to prevent the continued expansion of the irrigated land base, causing an overall decline in agricultural productivity that may also imply less food production and higher food prices. Rising water prices will effect important changes in the productivity of already irrigated land. Some farmers will experience sharply lower incomes as a result of higher water prices and will quit. The number giving up will increase if markets arise allowing farmers to make substantial profits by selling their water rights to the nonagricultural users who can afford to pay much higher prices.

There will be important changes on the land in irrigated agriculture as well. Heavy water-using crops, such as alfalfa, that produce low value per acre-foot, will become unprofitable under the new water cost structure, and there will be strong incentives for growers to replace these crops with ones producing greater value per acre-foot. Markets for the highest value crops, especially fruits and vegetables, are limited so it is possible that growers will shift to crops such as wheat that produce relatively low returns per acre but do not require much water and will have expanding foreign markets. Major shifts in irrigated cropping patterns may also affect, both positively and negatively, the fortunes of farmers in the humid regions through their impacts on prices.

Similarly, higher prices will create conservation incentives, encouraging growers to invest in new technologies that reduce water consumption. Competition and rising prices will give rise to strong pressures for reforming water rights laws, making it possible to set up water markets so as to facilitate transfers of water from low to high valued uses, and to allow current right holders to capture the true value of their water. Finally, declining water tables will encourage efforts to rationalize the use of these underground sources, forcing users to recognize the long-term effects of their pumping on all other users.

The effects of conservation will be more efficient use of declining water supplies in agriculture. Greater efficiency will mitigate the effects of scarcity; higher efficiencies will permit farmers to produce more food with less water, and therefore food production on irrigated land can be expanded or at least maintained at current levels even while water supplies remain at current levels or even decline.

Clearly, the magnitude of the potential savings from irrigation efficiency is critical in any assessment of the effects of water scarcity on food production.

Unfortunately, there are no reliable or widely accepted estimates of this potential for the arid West. Moreover, because water has been so cheap there has been little incentive to develop or implement new technologies; the era of water scarcity is only now dawning and researchers have just begun gearing up to deal with the problem.

The U.S. is at the same stage in its development of irrigation efficiency as it was in energy efficiency in the early 1970s. A 1972 projection of the potential savings from energy conservation would certainly have understated the actual conservation that has taken place; indeed, during the 1970s most of the utilities (not to mention American automakers) did make such miscalculations and suffered accordingly. The experience with energy conservation strongly suggests that when economic incentives arise, entrepreneurs will find new ways of reducing water requirements that are not now understood.

Even at this early stage of development, there exists a wide array of alternatives that can, in many cases, substantially lower water use in agriculture. These alternatives include such mundane changes as careful monitoring of water use to eliminate excessive application, reusing tail water (water that runs off before sinking into the ground), or lining canals and aquifers to eliminate seepage and evaporation. There are also "high-tech" alternatives, such as laser leveling that permits a field to be flood irrigated without waste, or new computer systems that will allow surface water to be delivered "on demand" when it is needed for optimal application rather than according to a fixed schedule. Perhaps the greatest water savings will come from new plant varieties having lower transpiration rates. Finally, there are capital-intensive technologies, such as drip and sprinkler systems that reduce evaporation and save water.

Some of these alternatives, such as more careful monitoring of use, can be accomplished without large expense, but most will involve substantial additional costs, both in fixed capital and in annual operation and maintenance. The more efficient application devices have the further disadvantage of increasing energy consumption; e.g., sprinklers and drip systems under pressure generated from pumps. Water savings must be balanced against higher energy costs.

While it is generally much cheaper to generate an acre-foot of water through conservation than to develop the remaining water resources, conservation will increase production costs over current levels, and unlike new water development, water conservation is not subsidized by government programs. Thus, growers have been, and still are, more willing to invest their resources in the political process that they hope will provide them with additional cheap water rather than undertake more costly (from their perspective) conservation measures. The gradual elimination of water subsidies and the inability of the Bureau of Reclamation to generate the financial support needed to underwrite the massive projects that would divert the Mississippi or Canadian rivers will encourage much greater interest in conservation.

In sum, water conservation in agriculture will not be a general panacea for declining water supplies; it will not permit the continued expansion of the irrigated land base (and there may be some contraction of this base), nor will more expensive conservation save the low-value, marginal crops such as irrigated pasture and hay. Therefore, the contribution of irrigated agriculture to overall U.S. food production capacity will be diminished. However, conservation will probably allow Western agriculture to accommodate declining groundwater supplies and increased competition from nonagricultural water users without giving up the higher valued crops. Indeed, the remaining irrigated land will be more intensively cultivated and the productivity of this land will continue to rise.

Implications of Water Scarcity for Commodity Production

The reduction in the contribution of irrigated agriculture *may* imply an overall reduction in U.S. crop production, *but* this possibility depends on how the agricultural economy reacts to increasing water prices and declining water use. Land and energy are substitutes for water; if crop prices should begin rising because of water scarcity, farmers will experience greater incentives to use more land (in humid regions) or to intensify production by increasing their use of fertilizer or other chemicals (i.e., energy) to increase yields. If there is a ready supply of land that can be cultivated at reasonable prices, or if additional fertilizer applications produce sufficiently large increments in yields to warrant their cost, then as output falls in irrigated agriculture, the resulting pressure on prices will call forth additional crop supplies and these, in turn, will reduce or eliminate any upward pressure on commodity prices.

In other words, increasing water scarcity may simply reallocate food production from irrigated to nonirrigated land, with potentially great regional economic consequences for owners of agricultural resources, but with no important consequences for consumers, or, presumably, for domestic food security. The basic issue, then, is whether there is sufficient surplus capacity in the overall agricultural production system to absorb any loss of output in irrigated agriculture while meeting the demand for commodities at prices that do not rise substantially.

The chapter by Burton English, found elsewhere in this volume, specifically addresses this critical question by surveying the results of three different models that simulate the possible relationship between water and agricultural production capacity. His findings are not very useful, for by assuming that adequate resources are available to substitute for increasingly scarce water, they beg the critical empirical issue. English provides no explanation for his apparent acceptance of such an assumption. In fact, he side-steps the problem by saying that while Earl Heady thinks the U.S. will run up against the limits of its productive capacity in the next fifteen years, he does not. The reasons for his optimism and his differences with Heady are not disclosed. English does hedge this conclusion, however, by acknowledging that if his assumptions are not met and if surplus capacity is not available, then commodity and food prices could rise significantly as a result of declining water supplies.

The discussion in following sections of this chapter challenges English's perspective on surplus capacity. The position developed is that the productive potential of U.S. agriculture is sharply limited and that while the U.S. is now experiencing a temporary period of surplus because of worldwide economic depression, the condition will not last. And, in the resource-scarce world of the next twenty years, declining water supplies will have a significant impact on output and food prices.

IRRIGATION AGRICULTURE AND U.S. AGRICULTURAL PRODUCTIVITY

This section develops the proposition that over the next two decades U.S. agricultural productive capacity will not grow as fast as demand; therefore, the limits on further development of irrigated agriculture in the West will help to increase already growing pressure on agricultural commodity and retail food prices. The expansion of irrigated land played an important role in allowing the U.S. to accommodate increased demand during the past decade; without this source of productivity, the U.S. will be required to increase the amount of nonirrigated land under cultivation to meet future demand. While there is a considerable stock of uncultivated cropland, it is less productive and more expensive to farm. The lack of new irrigation development will also reinforce the adverse effects on productive capacity of soil erosion and the loss of agricultural land to nonagricultural uses.

The Importance of Irrigated Land in U.S. Agriculture

On the surface, irrigated land would appear to play a minor role in overall U.S. agricultural production. The 1978 Census of Agriculture determined that only 51 million of the 380 million acres in crop production (14 percent) were irrigated. Of the irrigated cropland, 44 million acres (or 86 percent) were located in the arid West.

Even though irrigated land is a small fraction of all cropland, it is much more productive than the average nonirrigated land. In general, one irrigated acre produces as much as two nonirrigated acres. We conservatively estimate that irrigated agriculture contributed 25 to 30 percent of total agricultural production. As can be seen from Table 4.3, irrigated agriculture accounts for more than half of total production in several crops.

The Contribution of Irrigated Land to Productive Capacity

The concern here is not the absolute magnitude of the past contribution of irrigated agriculture, but rather how its past growth has influenced the rate of growth of overall U.S. production. As argued above, irrigated agriculture will probably be able to maintain its current level of output, even with declining water supplies, but it will not be able to expand. If past expansion of irrigated agriculture contributed materially to the expansion of overall productive capacity, then the restrictions on additional water development could have an impact on the ability of the U.S. to meet future demands for its commodities.

The data in Table 4.3 compare some important changes related to U.S. agricultural production between the 1960s and 1970s. The most important difference in the two decades is the growth rate of overall crop production, which more than doubled from 1.2 percent per year in the 1960s to 2.7 percent in the 1970s. This increase in growth was a response to the dramatic surge in U.S. agricultural exports, the physical volume of which increased more than two and a half times between 1970 and 1980.

The increased output of the 1970s was the product of two factors: higher land productivity (higher yields per acre) and increasing the cropland base. Here, too, the 1970s were unlike the 1960s. In the 1960s, increased demand was more than met by rising land productivity; in fact, government programs restricted the cropland base to prevent output from rising faster to maintain the viability of commercial farmers. In the

TABLE 4.3
Productivity of Irrigated and Nonirrigated Land
(Selected crops, 1979)

Crop	Percent of Crop Irrigated: Acreage (a)	Percent of Crop Irrigated: Output (b)	Yield Per Acre: Irrigated (Units/Acre)	Yield Per Acre: Nonirrigated (Units/Acre)	Yield Per Acre: Irrigated/Nonirrigated (Percent)
Wheat	3.9	6.6	60 bu	34 bu	+176.5
Rice	96.4	99.6	4950 lb	4500 lb	110.0
Corn	8.5	10.4	133 bu	105 bu	126.7
Sorghum	40.8	56.1	95 bu	51 bu	186.3
Barley	15.5	22.7	71 bu	41 bu	173.2
Hay	17.5	25.8	3.54 tons	2.16 tons	163.9
Cotton	29.3	46.5	778 lb	408 lb	190.6
Sugar Beets	73.2	78.4	23 tons	17 tons	135.3
Dry Beans	35.0	45.9	1730 lb	1096 lb	157.8
Vegetables	53.9	65.0	210 tons	97 tons	216.5
Citrus Fruit	100.0	100.0	13.4 tons	not reported	
Non-citrus Fruit	63.6	77.7	5.4 tons	2.0 tons	270.0

Sources: U.S. Dept. of Agriculture, 1980b,c,d.

(a) Percent of total crop acreage that is irrigated.
(b) Percent of total crop output produced from irrigated land.

1970s, the growth of land productivity slowed by more than 25 percent, from 1.57 percent per year to 1.18 percent. However, slowing productivity growth was more than offset by the rapid expansion of the land base, made possible by government policies that relaxed 1960s restrictions on cropland diversions.

Nevertheless, in spite of increasing the land base and higher yields, the long-term trend in commodity prices was reversed, and, in the 1970s, food prices became a major source of inflation, in sharp contrast to the 1960s, when they held down inflation. These price trends suggest that the productive capacity of U.S. agriculture grew more slowly than demand(5).

During both periods, irrigated agriculture expanded significantly; 6 million acres were added to the irrigated land base in the 1960s and more than 11 million were added during the 1970s. The contribution of this land to overall output can be estimated by making the following assumptions. During the 1960s,

TABLE 4.4
Crop Production, Land Use, Yields, and Prices

Category of Change	Regression Analysis (a) of Trend 1960-1970	1970-1980
	Annual Percent Change	
Crop output	1.2	2.7
Physical volume of exports	2.5	7.6
Cropland harvested	-0.5	1.5
Crop production per acre	1.57	1.18
Increase of irrigated acres as a percent of all cropland	0.18 (b)	0.38 (b)
Crop prices received by farmers	-0.3	7.3
Prices paid by farmers for production inputs	2.4	9.1
Consumer food prices	2.4	8.0
Consumer price index, less food	2.6	7.3
Average farm real estate value, per acre	5.8	12.7
Real Farm Income	-0.1	-3.2

Source: U.S. Dept. of Agriculture, Agricultural Statistics, various years.

(a) Ordinary least squares regression of annual values against time. Percentages are calculated by dividing the time coefficient by the mean annual value.
(b) These figures are not derived from regression analysis; they are obtained by dividing the average annual increment to the irrigated base by the average cropland acreage for the ten-year period. The incremental increase in irrigated acreage averaged about 545,000 acres for 1959 to 1969 and 1,160,000 acres for 1969 to 1978.

newly irrigated land did not encourage any expansion of the land base, given government policies; indeed, the increased output of irrigated land probably caused greater pressures for increased land diversions. Assuming that irrigated land was twice as productive as nonirrigated land(6) and that each new irrigated acre substituted for a nonirrigated acre, increased irrigation contributed about 15 percent of the observed annual growth in total output during the 1960s.

With the relaxation of government land diversion programs in the 1970s, the expansion of irrigated acreage both promoted increased productivity of the land base and helped to bring new land into the overall cropland base. Most of the 11 million acres added to the irrigated land base were already part of the cropland base; irrigation allowed the land to be farmed more intensively and with much greater productivity. However, some newly irrigated land, probably less than 2 million acres of the 11 million acres, had not been farmed and was a net addition to the cropland base. These figures lead to the conclusion that expanded Western irrigation contributed a minimum of 16 to 18 percent to the overall annual growth in output during the 1970s. Significantly, irrigation development accounted for 35 percent of the observed increase in land yields.

These estimated contributions are conservative because they do not include land already included in the irrigated land base that received additional water, allowing it to become more productive. For example, Bureau of Reclamation projects provided supplemental irrigation to an additional million acres during the 1960s and another half million acres during the 1970s. The State of California provided supplemental water to an additional 300,000 acres in the 1970s.

Other Sources of Productive Capacity

If all other factors remain the same, one implication of this analysis is that the limits on new irrigation development will significantly reduce the growth of land productivity, perhaps by as much as a third. Furthermore, if water conservation is not effective enough to allow agriculture to accommodate increased competition, then the productivity of the existing irrigated land base could also decline, putting an additional damper on overall productivity growth(7).

The loss of this source of productivity may have no consequence for food prices, however, if yields can be made to increase from other sources, or if the land base can be expanded to compensate for slower growing yields. Unfortunately, other evidence suggests that yields are unlikely to increase fast enough to offset slower irrigation growth, and that the land base is also limited in its potential growth as well.

Looking first at yields, the evidence from Table 4.4 indicates that the long-term trend in yields is down more than 25 percent, in spite of the substantial boost provided by expanding irrigated acreage. In fact, if the effects of irrigation on productivity are removed, the decline in the growth rate of yields from the 1960s and 1970s is more than 30 percent.

The reasons for the observed slowing in land productivity are complex; partly it reflects increased cultivation of an additional 50 million acres of much lower productivity, which depressed the overall growth rate. However, the downward trend in land productivity also reflects the fact that the U.S. is approaching the limits of existing biological and chemical technologies that are the basis of land productivity. New seed varieties offer relatively modest yield increases in comparison to the increases of the 1950s and '60s; additional applications of fertilizer do not produce the large increases in yields that we experienced in those early years. Pesticides lose their effectiveness over time. Perhaps the manipulation of DNA will produce a new "green revolution," but until this or some other major breakthrough comes, the growth in land productivity will continue to slow down. This downward trend will be reinforced by the lack of new irrigation development that will encourage the cultivation of more marginal land. Similarly, increasing rates of soil erosion, described in detail elsewhere in this volume, will depress yield growth rates.

These considerations point to a growth rate of land productivity well below one percent per year. If demand grows anywhere near the 2.7 percent rate of the 1970s, the slow productivity growth will bring about shortages and sharply higher crop prices unless the land base can be expanded at least as fast as during the past decade. In fact, relatively more land will be needed because the productivity of the new land will be lower than the land brought into production in the 1970s.

Therefore, if the additional cultivation of 50 million acres, including 11 million newly irrigated, during the 1970s was not sufficient to prevent sharp increases in commodity prices, it would appear that maintaining adequate productive capacity over the next two decades would require at least 100 million acres of new cropland, if not more, assuming no loss from the existing cropland base to nonagricultural uses. But, the existing cropland base is also being depleted.

In recent years, the conversion of agricultural land to nonagricultural uses has received increasing attention. The USDA estimated that in 1975 about 675,000 acres of actual cropland and 225,000 acres of potential cropland were lost through such conversions (U.S. Dept. of Agriculture, 1981:Ch. 5). If such conversion continues, it will imply that still greater amounts of less productive land must be brought into agriculture if output is to increase.

The amount of land needed for conversion into agriculture is probably not available. The actual cropland base was estimated to be 413 million acres in 1977 by the National Agricultural Lands Study. This land includes all available nonfederal land with suitable soil, moisture, growing season and accessibility to support crops (U.S. Dept. of Agriculture, 1980a). In 1980, a total of 387 million acres were devoted to crop production, of which 346 million were harvested, 29 million placed in summer fallow to promote soil moisture, and 12 million had failed crops (Frey, 1978). If the NALS estimate is correct, then there remain only about 26 million acres not now used for crops that could be readily converted to crop production. This is far less than the 100+ million that will be required.

Previous land inventories have identified another 120+ million acres now in pasture, range, or forest production with a high or moderate potential for conversion to crop production, so it might be assumed that when the existing cropland base is exhausted, additional land will be converted from this potential base(8). However, the accuracy of these inventories is called into question by two regional studies that carefully identified the potential land for conversion in the Delta states and in Iowa, and found the potential was limited to 10 and 11 percent of the existing cropland base, respectively, in these two regions, which was about one third the land projected by the national surveys (Shulstad and May, 1980). If projections for other regions are as overstated, there may be much less than 120 million acres for additional conversion.

Conversion of land requires more than simply a large stock of available acreage; there must be sufficient economic incentives to warrant the initial investment that such conversion requires. During the 1970s, two thirds of the 50 million acres brought into cultivation came from the existing cropland base, including much of the land held out of production by federal commodity programs; only about 16 million acres were converted from timber or pasture uses (U.S. Dept. of Agriculture, 1981, p. 89). It should be noted that agricultural land values quadrupled between 1972 and 1980, rising much faster than inflation. Yet, these high land values apparently did not provide enough incentive to owners of potential agricultural land to encourage them to make the necessary investments to clear the land and sell it for profit. The logical conclusion that can be drawn from this fact is that the costs of conversion must be considerable, and in order for there to be much greater rates of conversion agricultural prices and incomes must be much higher than they are now.

In other words, while there may be debate over the amount of land potentially available for agriculture, the relevant issue is not the absolute amount of this land, but rather, whether it can be converted fast enough to allow output to grow with demand to keep prices stable. The evidence suggests that while there may be some additional land that can be brought into production, it will be insufficient to allow output to keep pace with demand. There may be a substantial stock of land that could be converted if real commodity prices were to double or triple, creating large economic incentives for additional conversion, but if real commodity prices must rise to these levels, then food prices, too, will be forced to rise dramatically. Price increases of this magnitude would produce substantial pressure on domestic consumers. In short, the availability of excess marginal land is probably irrelevant to the discussion of food security.

The Recent Crisis in Agriculture, Productive Capacity, and the Future

In 1982, the U.S. had accumulated the largest grain and dairy reserves in its history, and real commodity prices fell to historically low levels. In light of this evident surplus productivity, it may appear foolish to portray a future of failing agricultural production capacity and rising crop prices, especially since the government induced farmers to set aside more than 80 million acres of grains and cotton in order to reduce stocks and raise commodity prices. Perhaps one should accept English's judgment (ch. 3) that there remains ample surplus capacity for the future.

But now we have experienced the most severe drought in 50 years that, along with the government land diversion, drastically lowered output and used up a major portion of the food reserves. Another year of drought, even with all land in production, will mean a return to the food scarcity days of the early 1970s. Furthermore, it is important to recognize that much of the farm crisis of the early 1980s was brought on by falling international demand, not uncontrolled increases in productive capacity (as occurred during the 1950s and '60s).

Demand for agricultural commodities has historically been extremely sensitive to the business cycle. Given the fact that U.S. agriculture is critically dependent on foreign demand, and that virtually all nations, including the centrally-planned nations, simultaneously experienced the worst economic depression in 50 years while the value of the U.S. dollar rose, making our agricultural products more expensive, one would expect severe repercussions in terms of sharply declining demand for U.S. commodities such as occurred; unfortunately, the demand declined at the same time farmers were experiencing two of their best crop years ever, thanks to ideal weather conditions. This abundance depressed prices even more.

If the world remains in this economic crisis and there is no recovery of economic growth, then demand for U.S. commodities will not grow and excess capacity in agriculture will become a more chronic problem. Should this happen, productive capacity will not be a constraint on food supply. If, as seems more likely, the present international economic crisis is resolved, with a return of economic growth and rising incomes, then the demand for U.S. commodities will resume its growth and the limits on productive capacity will come into play. The degree to which productive capacity is a future problem thus depends in large measure on overseas economic growth and the willingness of the U.S. government to allow foreign consumers unrestricted access to U.S. food stocks.

The data of Table 4.4 suggest that the growth in demand for agricultural output in the 1980s and '90s will not have to be nearly as great as it was during the 1970s to create substantial pressure on the resource base; even the modest growth rates of the 1960s could produce stress, given that there will not be the additional productivity from newly irrigated lands or the ready availability of surplus cropland as there was during the 1970s. Any additional decline in the growth of land productivity, either from the effects of erosion or from the lack of new biological innovation, will increase the possibility of stress.

Assuming a return to more normal economic conditions, the facts indicate that productive capacity will become a significant problem. This growing pressure cannot be attributed to any one cause; the limits on irrigation development will add to the pressures, perhaps significantly, but pressures will exist even if irrigated agriculture keeps growing as it has in the past. Because the demand for food is price-inelastic, relatively small reductions in supply can have substantial effects on price (just as surpluses drive prices sharply lower). If food production capacity does in fact become a major problem over the next two decades, even relatively small losses in output that can be attributed to restricted irrigation development will have substantial impacts on food prices.

FOOD PRODUCTION CAPACITY AND FOOD SECURITY

This section explores the possible connections between limited irrigation development, slow-growing food production capacity, and food security. In this context, "food security" shall be defined according to the wishes of the workshop organizers; thus, the three characteristics of food security include "availability, access, and adequacy." Two propositions are developed: first, even though limiting water resource development will add to the pressures on productive capacity and will contribute to higher food prices, there is little doubt that the U.S. has more than sufficient resources to feed itself. Food availability should not be an issue of domestic food security.

The second proposition is that rising food prices may eventually threaten domestic food security by restricting low income people's access to food supplies. Diminishing economic influence in food markets may prevent the food needs of these people from influencing the deployment of food production resources. Instead, the resources will more likely be used to produce "nutritionally inefficient" and resource-intensive foods desired by wealthy people all over the world.

Availability and Food Security

Food "availability" is the first dimension of the proposed definition of food security; a domestically secure food system must produce "enough" food for the entire population, both now and in the long run, in such a way that does not impose undue risks on the producer. Evidently, any effort to analyze domestic food security in terms of availability must begin with some notion of what is meant by "enough" food. An absolute measure of "enough" is that minimum quantity of food necessary to sustain healthy life, specified in terms of the food energy, protein, and vitamin requirements of the overall domestic population. The problem here is obvious: the basic nutritional requirements of the population can be satisfied by millions of possible combinations of foods, each involving a different use of agricultural resources and level of cost.

Theoretically it would be possible to restructure the food production system to produce much more nutrition per acre of land or acre-foot of water. The nutritional "waste" in our present system derives mainly from the expenditure of many resources to produce meat and dairy products, which are much more resource-intensive foods than fruits, vegetables, legumes, and grains. Thus, eliminating grain-feeding of cattle in the U.S. would release more than 200 million acres for crops that could be consumed directly by humans. It is estimated that one acre of fertile, rain-fed midwestern cropland now produces enough food energy and protein to meet the yearly nutrition requirements of more than ten individuals. Even assuming that the productivity of all cropland is only half that of the Midwest, the U.S. could now feed up to 2 billion people (Revelle, 1976:174).

California agriculture offers another perspective on the potential productivity of irrigated land. California produces more than half of the nation's fruit, vegetables, and nuts. It might be presumed that most of the state's 8.5 million irrigated acres are devoted to this production, but only 600,000 are used to produce the vegetables and about 1.8 million the fruit and nuts. Vegetables use less than 5 percent of the state's agricultural water supplies, while fruit and nut crops use about 15 percent. If the demand were strong enough, output of fruit, nuts, and vegetables could be multiplied as much as several times the current levels without new water development or significant new pressure on the land base by simply displacing some of the land and water now used to produce alfalfa and irrigated hay.

In summary, there should be no problem of "availability" under any conceivable set of circumstances, save nuclear war or a major shift in climate. Unfortunately, this potential cannot be realized without a major reorganization of priorities in the food system.

It hardly needs to be pointed out that the food system is not organized to produce nutrition efficiently; high incomes, modern food technology, and strong preferences for fat and sugar have combined to create a very resource-intensive and nutritionally-inefficient use of food resources. Moreover, as real incomes have risen around the world, there has been a pronounced trend toward the production of even more resource-intensive foods, such as grain-fed beef. Less than 10 percent of U.S. cropland is devoted to crops, such as fruits, vegetables, and cereals, that produce nutrition efficiently for direct human consumption.

The food system also reflects the preferences of foreign as well as domestic consumers; a major portion of U.S. agriculture's productive resources (i.e., more than 40 percent of the cropland and more than 30 percent of water supplies) are required by these exports. More than 65 percent of the exports are purchased by developed nations, the USSR, and Eastern Europe and are used primarily to support livestock. Even a considerable portion of the exports to Third World nations also go to livestock projects that do not provide significant food supplies to the malnourished.

It may be argued that as long as individuals are free to choose, they will not voluntarily give up their preferences for resource-intensive foods, and therefore the potential for increased food availability will not be exploited. Yet, while it may be true that food preferences are powerful and slow to change, they do change, particularly in response to price and income incentives. For example, as a result of declining real incomes and higher prices over the past decade, annual consumption of meat has fallen by more than 12 pounds per capita, a 10 percent reduction. During the same period, consumption of poultry has increased about 10 pounds per capita, mainly because poultry is a much more efficient converter of grain than is beef, and has become relatively less expensive. Similarly, coffee consumption declined by more than 30 percent when frosts in Brazil caused a tripling of retail prices, and neither consumption nor prices have returned to pre-frost levels.

There have been other significant changes in diet, some of which are the result of economic pressures, some because of concern over adverse health effects. The most noteworthy changes in the past decade are significant reductions in consumption of dairy products (down 25 pounds per capita) and animal fats (mainly because annual consumption of butter is down 6 pounds), and increases in consumption of fresh fruit (up more than 10 pounds) and fresh vegetables (also up more than 10 pounds). Taking all these shifts into account, there has been an overall increase in annual consumption of "crop products" equivalent to 35 pounds per capita and a 28 pound reduction in animal product foods.

Although these changes in diet have been modest, they have still had a significant impact on resource use. For example, a reduction of 12 pounds per capita in beef consumption reduces the use of corn and other feeds by about 500 million bushels, which is the equivalent of the production from about 7.5 million acres. In contrast, the additional 20 pounds of increased fruit and vegetable consumption can be produced on less than 300,000 acres.

If future domestic food production capacity is strained by increasing scarcity of land and water resources, then this pressure will be manifest in all food prices, but the most affected will be the prices of resource-intensive foods. Rising prices of these foods relative to less-resource intensive foods will further encourage the shifts in food consumption patterns that are already underway. Like the V-8 engine, the New York steak may become a relic of the past.

The critical question is, will these market forces work quickly and powerfully enough to ensure sufficient food availability for all members of society? This is the problem of "food access," the second element of the food security issue.

Food Access and Adequacy: Some General Considerations

The working definition of food security used in this volume states that there must not only be "enough" food to go around, but also it must be sufficiently accessible to all members of society so that everyone has adequate nutritional intake to promote healthy life. These are demanding requirements, and are not now met by the existing food system. Presumably the issue here is not whether these requirements are now being met, but rather, whether the pressures on the food system identified above will reduce accessibility in the future.

Since most people cannot grow their own food, food access is through food markets and is determined on the basis of ability and willingness to pay. Consequently, the level and distribution of income and the real price of food are important determinants of access. Individuals and families with low incomes have only limited access to food markets; this access is further constrained by other subsistence needs, including shelter, clothing, medical care, etc. Individuals with no income have no access unless they receive free food stamps or other welfare benefits.

The distribution of purchasing power obviously is a critical element in any analysis of food access; furthermore it will have an important impact on how the society reacts to rising food costs brought on by increased pressures on the food production system. To illustrate, first imagine a society with only one income class. Rising food prices would encourage all consumers to substitute cheaper, less-resource intensive foods, causing shifts in the production process that would increase the availability of these foods, as described above. In such a society market forces would produce the necessary adaptive responses to prevent a loss in food access.

Contrast this society with one in which there are two income classes consisting of a high income and a low income sector. Let us further suppose that the income level of the high income sector is growing as fast as real food prices, while the low income sector's income grows slower than real food prices. In this society, members of the low-income sector will attempt to substitute cheaper, less-resource intensive foods in response to higher food prices, but members of the high-income sector may continue to purchase the resource-intensive foods which they prefer and continue to afford. In this situation, the preferences of the high-income sector will reduce or even prevent any adaptive shifts in food production; hence high-income consumers will effectively capture more of the food production resources at the expense of low-income consumers, whose food access will be diminished. In this case, market forces will not produce the desired shifts in production needed to maintain food access.

Another dimension of the access issue is the issue of food adequacy; that is, individuals may have sufficient income to purchase nutritionally complete diets but they may actually purchase the wrong foods and have inadequate diets. This condition could be true of low- and high-income groups alike.

Income is a necessary but not sufficient condition for food adequacy. For instance, if rising prices of meat and dairy products were to cause increased sugar consumption (produced from the corn no longer used for fattening beef), the consumer might prevent food costs from rising by substituting a less-resource intensive food, but the resulting diet might be nutritionally inadequate. Obviously, food adequacy requires a general understanding of the basics of nutrition. Furthermore, the better this information, the more the food dollar can be stretched without impairing food adequacy. In other words, the better educated the population, especially those with low incomes, the better its ability to maintain adequate food intake in the face of strains on the food production system.

This general discussion leads us to examine in more detail the nature of nutrition information and the formation of food preferences, on the one hand, and the characteristics of income distribution and the demand for food on the other. These issues are now addressed.

Food Preferences, Nutrition Information, and Food Security. The forces shaping food preferences are complex and not well understood; certainly one "learns" them in the course of being socialized. But socialization does not explain how food preferences came into being originally, nor does it account for the fact that preferences evolve over time. Probably in traditional societies, food preferences were shaped by the availability of various foods, but in an industrialized food system, where choice is not limited by the consumer's geographical location, it is clear that other factors must be present in food choice decisions.

Cost is a proxy for availability in an industrial economy, but there are other factors shaping preferences as well, perhaps the most important of which is "taste." There are strong preferences for foods that are sweet, salty, or taste of fat. One might speculate that these preferences are inherited from a simpler world with only limited amounts of salt and fat, and where refined sugar was not available at all. Such preferences may have reflected the desirability of having some high-calorie density foods in the diet. However, in an industrialized food system, they may be dysfunctional; without limits on the availability of fat, salt, and sugar, individuals may be led to consume too many calories from foods with these tastes and consequently they may choose nutritionally inadequate diets even when they have more than sufficient purchasing power to afford better nutrition. In such a food economy, nutrition information and education are essential to food adequacy.

For many, food advertising is the main source of information about food, but such information is strongly biased toward foods with a high degree of processing where food manufacturers can potentially earn the highest profits. Food manufacturers produce foods that appeal to powerful taste preferences, for that is how they ensure a large market. Insofar as good nutrition is not compatible with such foods, the marketing strategies of these advertising firms probably reduce overall food adequacy and impair the food system's adaptability to increasing strains on production resources.

While food advertising may play a role in fostering poor food choice, food corporations play an even more negative role through their efforts to prevent the public sector from disseminating nutrition information that might reduce their markets. Because there is no comparable private interest that profits from the distribution of unbiased information, nutrition education must be a public sector activity if information is to be widely available. Unfortunately, public sector activities are shaped by private economic interests that are translated into political pressures; there has been little "countervailing" political pressure on the public sector from those who want to see better nutrition information since they lack significant economic power. Therefore, the food industry is able to dominate nutrition policy. Even modest efforts to improve nutritional information, such as nutrition labelling, have been defeated. Any public figure who tries to argue for policies favoring reduced meat consumption will be confronted by powerful opposition from the various trade associations of growers, packers, and retailers.

Modifying basic food preferences is a formidable task that will require a major sustained public effort. The benefits of such an effort would be to improve overall nutrition _and_ accommodate the increasing strains on food production resources. Market forces alone probably will not achieve this fortuitous outcome, and unfortunately, the needed public effort will probably not be forthcoming because it runs counter to the interests of those with the greatest economic and political power.

Income Distribution and Food Security. If income is not a sufficient condition for food adequacy and access, it is a necessary condition, as the following evidence suggests. Table 4.5 contains calculations of the percentage of after-tax income of different income groups needed to purchase three basic USDA food plans. While the USDA is quick to point out that the low-cost plan is not indicative of a "subsistence" diet, this plan does provide a benchmark for approximating how much income is needed to ensure adequate nutrition.

These calculations indicate that the lowest income families would be required to spend two and a half times their entire income just to purchase the low-cost food plan. Even families in the next two higher income brackets must spend a substantial portion of their income on the low-cost plan, leaving little for other necessities. At the other end of the income spectrum, families and individuals would need only a small fraction of their incomes to purchase even the more liberal diets. If one takes an average over all families, it may appear that the typical consumer can easily afford food, but this averaging process obviously neglects the important distributional facts. Note that the income measure used here _includes_ all forms of money income, including all welfare benefits and other transfers, as well as wages, salaries, rents, dividends, etc.

The USDA food plan analysis does not show how families in different income groups _actually_ allocate

TABLE 4.5
Percentage of After-tax Income(a) Required by USDA Food Plans by Families and Individuals in Different Income Classes, 1978

Income Class	Percent of All Families or Individuals(b)	Percentage of After-tax Income Required(c) by USDA Food Plan: Low Cost Food Plan	Moderate Cost Food Plan	Liberal Cost Food Plan
Families				
Under $3,000	3.5	252.9	308.6	364.2
$3,000-4,999	5.8	64.6	78.8	93.0
$5,000-6,999	7.3	45.5	55.4	65.3
$7,000-9,999	10.8	33.8	41.3	48.7
$10,000-11,999	7.2	30.2	41.0	43.8
$12,000-14,999	11.4	24.7	29.7	35.6
$15,000-24,999	31.8	19.7	23.2	28.6
$25,000-over	22.4	13.1	14.9	20.1
Unrelated Individuals				
Under $3,000	20.3	49.1	63.6	75.1
$3,000-4,999	21.3	22.1	28.6	33.8
$5,000-6,999	13.7	15.4	19.9	23.5
$7,000-9,999	15.2	11.7	15.2	17.9
$10,000-14,999	16.2	8.5	11.0	12.9
$15,000 over	13.3	5.2	6.8	7.9

Sources: U.S. Dept. of Commerce, 1981, Tables 442, 443, 444, 745, and 749. U.S. Dept. of Agriculture, 1980b, Table 766.

(a) Includes all forms of money income, including welfare.
(b) In 1978 there were approximately 57 million families and 23 million unrelated individuals.
(c) Food expenditures have been adjusted for differences in mean family size between income classes.

their food budgets, nor does it describe the degree to which low-income groups fail to have adequate nutritional intake. Perhaps these groups find cheaper sources of nutrition than is available in the USDA low-cost food plan. Of course, finding cheaper sources of nutrition requires some degree of nutrition education. It also may require access to discount stores or the ability to store bulk foods in a refrigerator or freezer. Unfortunately, low-income is highly correlated with poor nutritional education, lack of access to the cheaper food markets, and inadequate means of storing or preparing foods. In sum, it would be difficult to argue that the low-income groups have sufficient resources to ensure adequate nutritional intake.

The picture of food access presented in Table 4.5 is static and does not provide much insight into the issue of whether food access is deteriorating or improving for people in low income groups. There is, however, indirect evidence that the access problem has been growing over the past decade.

Throughout the 1950s and '60s real food prices fell as a consequence of rapid technical changes in agriculture that produced abundant supplies of farm commodities at declining prices, as can be seen in Figure 4.1a. Lower real food prices are complemented by rising real wages and incomes during the same period as rapid economic growth trickled down throughout all levels of the economy. These two beneficial effects made food relatively cheaper for all income groups (see Figure 4.2) and increased overall food access.

During the 1970s, real food prices began to rise. As can be seen in Figure 4.1a, the critical turning point for real food prices was 1972, the first year agricultural exports dramatically increased. The dramatic shift in food prices came mainly from rising farm commodity prices as world demand pressed against limited U.S. production capacity. The more popular explanation of rising food prices, frequently espoused by farmers and agricultural economists, that food prices rose because of the higher profits of "middlemen" food processors and distributors, does not hold up under scrutiny. During the 1950s and '60s, the farm-retail price spread generally grew faster than overall inflation, even while food prices fell. This trend did not materially change in the 1970s, but what did change was the dramatic increase in commodity prices, as can be seen in Figure 4.1b.

At the same time as real food prices were rising, real wages began to decline with the result that after 1972 an hour's work bought less and less food (see Figure 4.2). Rising food prices relative to wages put new pressure on access to food for those whose wages did not rise with food and other costs.

Paradoxically, even while real wages have declined, average family income has remained relatively stable in terms of purchasing power (although this "no-growth" trend also is in sharp contrast to the average 3 percent growth of the previous two decades). Therefore, the percentage of the average family's budget spent on food does not appear to have risen very significantly after 1972 (see Figure 4.2). This fact is often used to prove that families are at least as well off today as ever before in terms of access to food.

Figure 4.1A

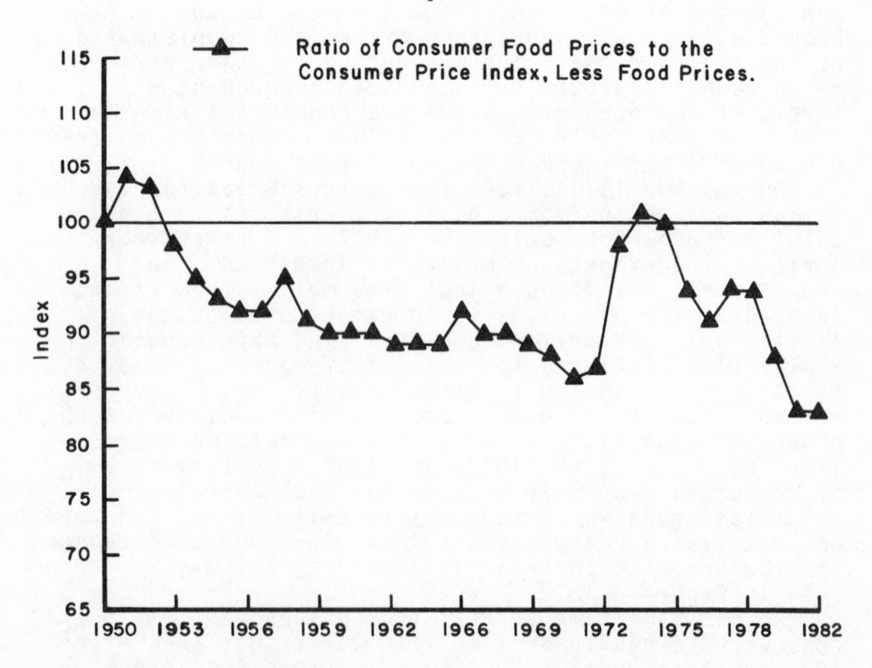
Ratio of Consumer Food Prices to the
Consumer Price Index, Less Food Prices.
115
110
105
100
95
90
85
80
75
70
65
Index
1950
1953
1956
1959
1962
1966
1969
1972
1975
1978
1982

Figure 4.1B

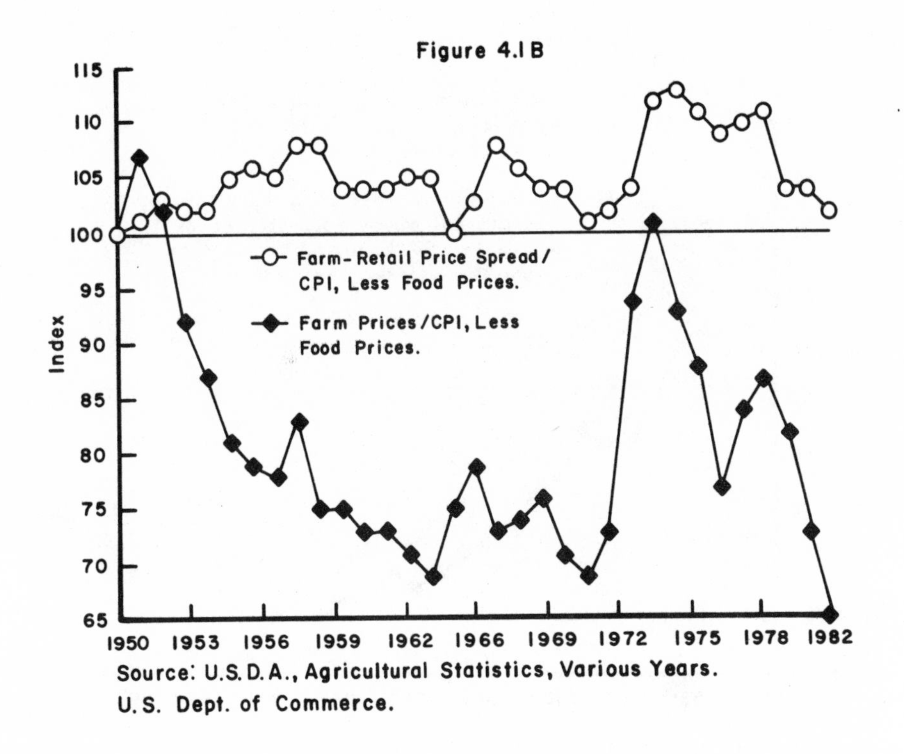

Source: U.S.D.A., Agricultural Statistics, Various Years.
U.S. Dept. of Commerce.

Figure 4.2 Comparison of Two Measures of The Importance of Food to The Consumer.

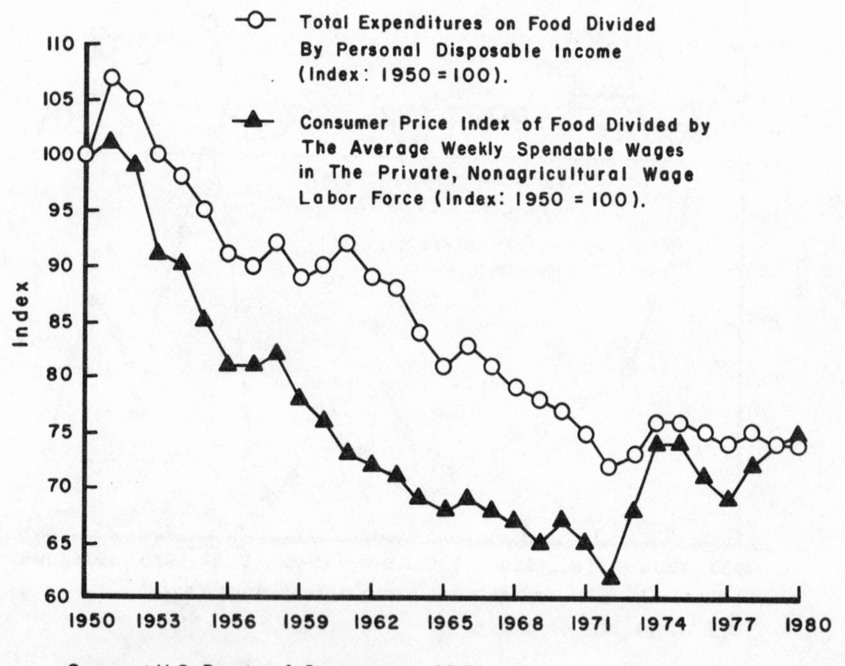

Source: U.S. Dept. of Commerce, 1981.

We have already seen that "average" family expenditures is not an appropriate indicator of the problems of food access for the poor. Moreover, the fact that the proportion of the average family income spent on food has remained constant in the face of declining real wages masks other important distributional changes taking place between families. Such shifts may well imply a deteriorating of food access for many families. To understand why, note that the reason average family income has remained constant in terms of purchasing power while average wages have declined in purchasing power is that many families have found ways of increasing the amount of time they spend in the labor force, either by working more than one job or by increased participation of women family members. This means that, on the average, while families continued to spend the same percentage of their incomes on food, they actually had to work more hours to buy their food. Thus, even the average family was actually worse off, since the additional time spent working for food meant less time for other activities.

The impact of rising real food costs was not spread evenly over all families. At one extreme, families depending on workers with relatively greater economic bargaining power, whose real wages did not decline relative to food prices, may actually have improved their ability to purchase food and other goods. At the other extreme, families with only one wage earner whose real wages did not keep pace with food prices and who could not increase their work effort to compensate for lower wages, were impaired in their food access. The latter characteristics are more likely associated with low-income families, the former with high-income families.

While there are no estimates of the actual number of families in the group with increasing problems of food access, we do know that the number of families under poverty has grown steadily at a time when the benefits of food stamps, which offset rising food prices, have also been reduced. It is not surprising, then, that the issue of hunger in America has re-emerged after more than a decade, and is sufficiently salient to warrant a special investigation from a Presidential task force.

It is apparent that if the trends of the past decade in food prices, wages, and incomes continue in the future, food access will become an increasingly difficult problem for low-income families. It is unlikely that market forces will shift production resources significantly away from their present inefficient uses because there will remain many consumers who can afford the higher priced foods. Many of these consumers will not be Americans, but rather, they will be increasingly wealthy Japanese, Europeans, Russians, and the elite of many poor nations who want to increase their consumption of meat and will be able to pay for such resource-intensive foods.

If food production cannot grow sufficiently fast to meet this increased demand from overseas, food prices will indeed continue to rise and low income consumers will gradually lose access in the emerging world food markets. Ironically, the poor in the U.S. will find themselves in a position parallel to that of the citizens of most Third World nations. They will be forced to compete with the wealthy all over the world for access to food markets. They will become increasingly dependent on government food aid.

CONCLUSION: U.S. AGRICULTURAL POLICY AND FOOD SECURITY

Changes in the world economy have had profound effects on U.S. agriculture and account for the dramatic rise in commodity prices during the 1970s. The "internationalization" of U.S. agriculture was a deliberate policy, initiated by the Nixon Administration, intended to reduce both the cost of government agricultural programs and U.S. trade deficits (that have grown with the export of manufacturing jobs) by promoting agricultural exports. The strategy did increase exports, but it did not eliminate substantial government intervention in agriculture; it marginally reduced trade deficits, and perhaps most significant, it served to reverse the long-term downward trend in food prices and put much more pressure on agricultural resources. In combination with the other structural changes in the U.S. economy, the development of export-oriented agriculture has created the basis for our increasing food access problems. As argued above, it is only a matter of time before agricultural commodity prices again begin rising.

Given the trends in the world economy, it would appear more likely that income will become less rather than more evenly distributed, within both the U.S. and world economies. The world's wealthy will constitute a profitable market for food processors and exporters. In such an environment of scarcity and increasingly concentrated economic power, the "hidden hand" of market forces will not produce desirable adaptive responses within the productive system that could reduce the negative effects of these forces on domestic food access.

What are the policy options for the U.S.? One set of policies might attempt to overcome the obstacles to increasing food production. Technology was the basis of past agricultural success, but it is difficult to believe that future land yields will grow at rates remotely approaching those of the 1950s and '60s.

To return to the discussion of water resources--some might use the excuse of increasing food scarcity to justify the huge water projects that would tap the Yukon or Mississippi Rivers for additional irrigation development in the West. Conceivably food prices might rise so high that farmers could afford dramatically higher water prices and still earn a profit, making such projects economically feasible. One reason why this is unlikely to occur, however, is that rising food prices generally stimulate overall inflation of all goods and services as workers attempt to maintain the purchasing power of their wages, forcing labor costs to rise. General inflation means higher farm input costs, which will absorb much of the additional revenue created by higher commodity prices. Thus, even if food prices rise, the income of farmers will not grow proportionately and therefore the farmers' ability to pay for water will likewise not rise enough to justify such large investment in new water supplies.

Whether or not potential water projects are economically efficient in terms of benefits exceeding costs, some may still argue for continued public subsidies to increase food production on the grounds that the additional food production will improve overall food access, especially to the poor. While there may be some merit to such arguments, it would appear that improving food access could be obtained more cheaply through other kinds of public policies, including increased income supports. Ultimately it may be cheaper to redistribute income than to push the productive system to its outer limits.

Of course, there is much more that can be done to mitigate growing demands on agricultural resources. We have already mentioned the likely gains from new investments in water conservation and the reasons why such investments have not been made. Much more can be done to reform water allocation within agriculture to ensure that water is applied to the "highest and best" uses. Significant administrative, legal, and political barriers impede such rational use of water today.

However, no matter how much we may mitigate the impacts on limited resources, there will be no way to meet growing pressures for food production without incurring much higher costs. These costs will strain our domestic social fabric and increase demands on the public sector. In the long run, the only way to avoid threats to domestic food security will be to limit the pressure on the resource base. Such limits may require agricultural trade restrictions. A more effective alternative would be for the U.S. to underwrite large investments in the food production systems of other nations that have a relatively large undeveloped food production capacity.

One would be naive to expect that politically powerful agricultural interests or the large grain traders will accept any restrictions on agricultural trade which would directly threaten their profits. Similarly, there is no powerful constituency that would support U.S. investments in foreign agricultural development. Thus, it appears that a major and prolonged food crisis must develop before we can expect any significant political response.

Malthusians might argue that the only alternative to a food crisis is drastic reduction of the world's population growth rate, especially among the poor. While there is an inevitable logic to such an argument, this analysis suggests that food production potential is sufficient to meet the needs of the world's population. This potential is left unexploited because of political and economic forces that serve to concentrate economic power and inhibit new patterns of production and consumption that would otherwise permit an equitable and efficient sharing of a more limited resource base. A more effective and just, if equally unlikely, resolution to the impending food crisis would be a more equitable distribution of the world's income so as to make the earth's resources more available to all.

NOTES

1. Economists are quick to argue that "scarcity" is not a useful concept; the availability of a resource is related to its price and if prices rise, availability will increase. Furthermore, from the perspective of an individual farmer, all inputs are equally "scarce." In this chapter "scarcity" is defined according to its more common usage; i.e. water supplies are becoming more "limited." "Limited" means that the costs of developing new water supplies are high relative to the ability of farmers to pay for the water. As a consequence, scarcity means future irrigation water prices will rise in real terms and therefore agriculture will have less access to the resource than it has had in the past.

2. Because rainfall is highly variable, annual variation in stream flow also tends to be high. Dams and reservoirs reduce this annual variation by capturing water during high water years and holding it for use in dry years. Diverting surface flows into underground aquifers can accomplish the same task, in many cases more efficiently because there is no evaporation, which significantly reduces water supplies held in reservoirs. With enough storage capacity, most of the variation in river flows could be reduced, and dry year flows would approach the average flows. Thus, there would appear to be potential for additional development along at least half the Western watersheds. However, maintaining the extra storage capacity necessary to eliminate annual variations may be very expensive.

3. The statement that "real prices must rise" means that commodity prices must rise faster than the prices of other goods, especially those of other production

inputs, so that farm income will rise faster than the nation's overall income. This is another way of saying that the parity ratio must reverse its historic downward trend. There is a strong theoretical basis for arguing that such a reversal probably cannot occur over a sustained period because farmers lack sufficient economic power to increase their prices relative to others. Hence their share of the nation's income will probably not rise. This lack of economic power derives from fundamental structural differences between the production of agricultural commodities and industrial goods, and it is unlikely that the power imbalance will be altered. (For a full explanation, see Kaldor, 1976.) Therefore, it is unlikely that agriculture will be able to compete for water more successfully in the future, even if commodity prices should rise.

4. Of 18 Reclamation projects studied by the Department of the Interior, only six were found to have benefits greater than total costs, and none of these six were built in the past 20 years (U.S. Department of the Interior, 1980). A similar conclusion was reached by the General Accounting Office which concluded, after examining the projects chosen at random, that in none of the projects could agricultural water users afford to pay the "full-cost" of their projects, including interest on capital (U.S. General Accounting Office, 1981).

5. Rising commodity prices did not necessarily make farmers better off; costs of farm inputs rose even faster than commodity prices (see Table 4.4), so after an initial sharp rise in 1973 and '74, farm income declined throughout most of the decade. The failure of farm income to rise even at a time of increasing pressure on food production resources underscores the point made above regarding the farmers' probable inability to pay for new irrigation projects, even under rising prices.

6. The relevant comparison is not between the average productivity of irrigated and nonirrigated land; the correct comparison is between the newly irrigated land added to the base and the marginal land that it displaced. Newly irrigated land was at least as productive as the average of all irrigated land but the marginal land displaced was considerably less productive than the average of all nonirrigated land. Thus the above equation of one irrigated acre to two nonirrigated acres is conservative.

7. Rising soil salinity (and declining water quality) poses an additional, if unquantifiable, threat to irrigated agriculture's productivity. However, in California, over a million acres in the rich San Joaquin Valley will eventually be forced out of production by salinity unless the land is more effectively drained. Conservative drainage costs indicate that the costs of draining the region will be more expensive than the

original irrigation projects that provide its water, and thus farmers cannot afford them. The Imperial Valley of California also is experiencing increased levels of salinity that probably cannot be reduced. Plant biologists are attempting to breed salt-resistant strains of cotton and other crops but little progress has yet been made. If California's agriculture is typical of that of other regions of the arid West, salinity may well further reduce the output of irrigated regions.

8. An analysis of the various national land inventories and their significance is provided in Crosson, 1982.

REFERENCES

Crosson, P. R., ed. 1982. The Cropland Crisis: Myth or Reality? Baltimore: Johns Hopkins Univ. Press for Resources for the Future.

Frey, H. T. 1978. Major Uses of Land in the United States, 1978. U.S. Dept. of Agriculture, Economics Research Service. Agricultural Economics Report No. 487.

Kaldor, N. 1976. "Inflation and Recession in the World Economy." The Economic Journal 86(344):703-14.

Revelle, R. 1976. "The Resources Available for Agriculture." Scientific American 235(3):164-78.

Schuh, E. 1976. "The New Macroeconomics of Agriculture." American Journal of Agricultural Economics 58(5): 802-11.

Shulstad, R., and May, R. 1980. "Conversion of Noncropland to Cropland: The Prospects, Alternatives and Implications." American Journal of Agricultural Economics 62(5):1077-83.

USDA. 1980a. SCS. Agricultural Land Data Sheet. Interim Report No. 2.

______. 1980b. Agricultural Statistics. Washington, D.C.

______. 1980c. ESCS. Crop Production, 1979 Annual Summary. CrPr 2-1(80). Washington, D.C.: January 15.

______. 1980d. ESCS. Vegetables, 1979 Annual Summary. Vg 1-2(80). Washington, D.C.: June 6.

______. 1981. A Time to Choose, Summary Report on the Structure of Agriculture. Washington, D.C.

U.S. Dept. of Commerce. 1979. Statistical Abstract of the United States, 1979. Washington, D.C.: GPO.

______. 1981. Statistical Abstract of the United States, 1981. Washington, D.C.: GPO.

U.S. Dept. of Interior. 1980. Interim Report, Acreage Limitation. Water and Power Resources Service. Washington, D.C.: GPO. March.

U.S. General Accounting Office. 1981. Federal Charges for Irrigation Projects Reviewed Do Not Cover Costs. PAD-81-07. Washington, D.C.: GPO.

U.S. Water Resources Council. 1978. The Nation's Water Resources: The Second National Assessment, vol. 3, app. II, Table II.6. Washington, D.C.: U.S. Government Printing Office.

Wessel, J., with Hantman, M. 1983. Trading the Future: Farm Exports and the Concentration of Economic Power in our Food System. San Francisco: Institute for Food and Development Policy.

Energy

5
Energy Inputs and U.S. Food Security

David Pimentel

INTRODUCTION

The essential resources of energy, land, and water for food production are in short supply throughout the world, including the United States. Fossil energy sources in particular are being rapidly depleted, and it is projected that fossil energy use per capita in the world will reach a peak shortly after the year 2000 (Figure 5.1). After this fossil energy sources such as oil and gas will be mostly depleted, and coal and nuclear fossil energy sources will then be utilized for another 100 to 200 years.

Overpopulation is stressing the food resources of the world. Of about 180 nations in the world (Paxton, 1981), fewer than 10 are net food exporters (Bastin and Ellis, 1980). The U.S. is the largest food exporter and is currently blessed with some food surpluses. The U.S. situation is in sharp contrast to the rest of the world, where about one half billion humans are malnourished (NAS, 1977). Most of these people do not have the resources of energy, land, and water to produce food, and many do not have other resources that would allow them to trade for the food resources they need for their nation.

The lack of economic resources by the poor people of the world is the prime reason that the United States has a surplus of food and last year implemented acreage controls--that is, paying farmers not to produce certain food resources. Because the dollar is strong on the world money market, U.S. food and other products are expensive for other nations to purchase. This hurts the poor nations in particular as well as the United States and its exports. U.S. food exports were $45 billion in 1981, and it is expected there will be a marked reduction in U.S. food exports as well as other exports.

Although a high-value dollar situation hurts U.S. food exports, a high-value dollar aids this nation in purchasing oil on the world market. There is a double advantage with world oil prices because an oversupply of

Figure 5.1 Estimated world population numbers (____) from 1600 to 1975 and projected numbers (....) (????) to the year 2250. Estimated fossil fuel consumption (____) from 1650 to 1975 and projected (....) to the year 2250.

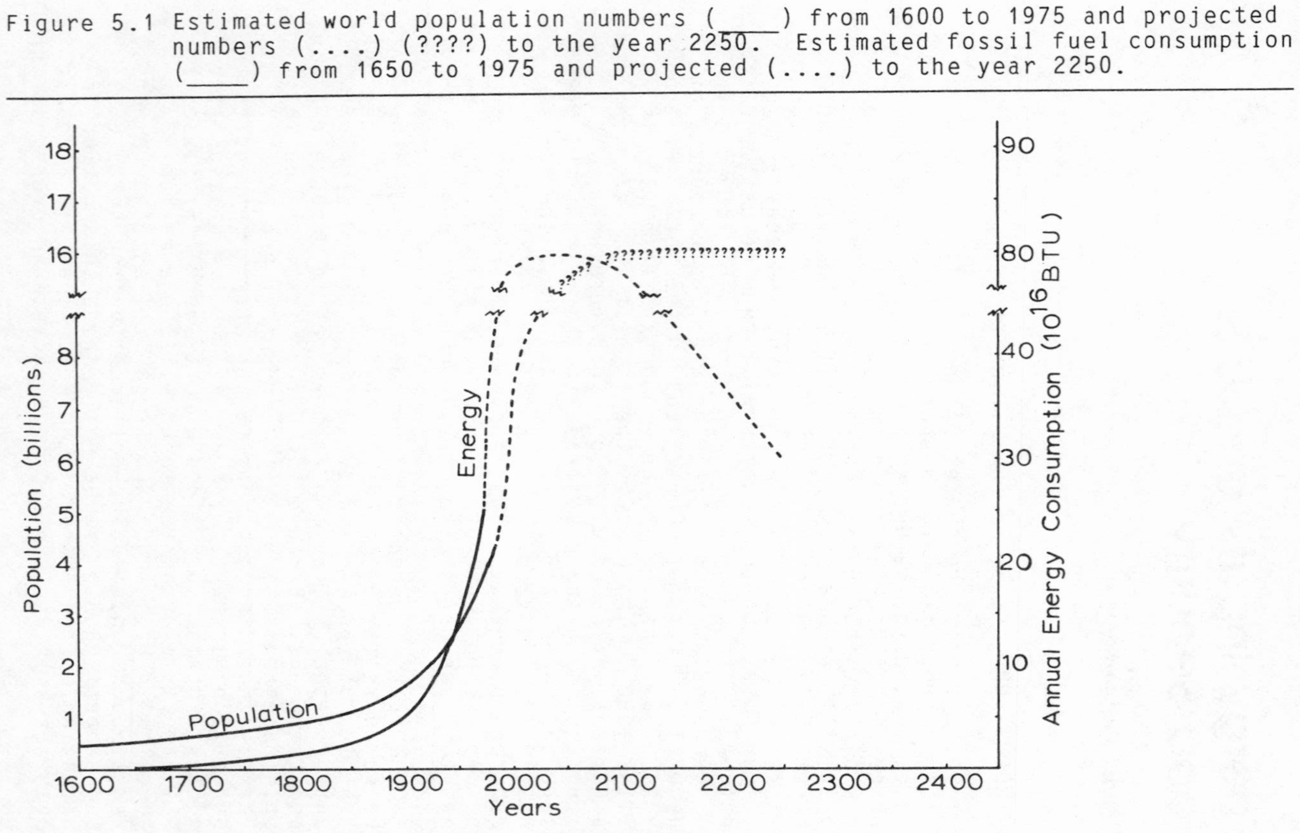

Source: Adopted from Pimentel, et al., 1979

oil currently exists on the world market. As oil resources decline, supply prices will increase. With minimal U.S. food exports, the U.S. will probably have to resort to deficit spending (inflation) to pay for its needed oil. This situation will further stress U.S. food production and security.

In this chapter, I will examine energy inputs in the food system and assess food vulnerability due to the heavy dependence of agriculture on fossil fuels--a fast disappearing resource.

ENERGY INPUTS IN AGRICULTURE

Before the development of agriculture and the manipulation of the natural ecosystems to produce certain food plants and animals, wild plants and animals were the only source of food for humans. Originally humans, like other animals, were an integral part of the natural ecosystem and depended upon the energy flow in these systems for their survival. For most of the time that humans have been on earth, they have "lived off the land" as hunter-gatherers. The density of humans to land was low. Humans were able to remove a small portion of the energy produced by the ecosystem for their food without substantially reducing its productivity. Under these conditions, the ecological system was adequate when the human population in the world numbered fewer than 10 million people.

Slowly, as the human population increased to above the carrying capacity of natural ecosystems, a more structured or planned method of ensuring an adequate food supply evolved. At first, early agriculture was simple and probably entailed sowing surplus seeds on the disturbed soil of the campsite. These seeds were left over from food collecting by a group of gatherers. The eventual harvest was the grains and other food products that survived weed competition and pest insect, plant pathogen, bird, and mammal attacks. Although such a harvest yield was small by present standards, the quantity of desirable food was greater than that obtained by a similar amount of effort and energy invested in searching extensive areas of the natural ecosystem for food. Thus, a slight alteration of the ecosystem and encouragement of desirable vegetation types provided more food for the same effort than had been expended in hunting and gathering.

Gradually, more intense alteration of the ecosystem, employing "slash and burn" agricultural technology, developed and increased food yields. With greater effort and investment in human energy, the seedbeds were improved and losses to pests, especially weeds, were reduced. The greater investment of human energy to manage the ecosystem was rewarded with larger food yields and energy value in the form of food than previously possible.

Basically, the principle of all agriculture is manipulation of the ecosystem to produce plants and animals needed or desired by humans for food and fiber. Encouraging certain species of plants and animals, while concurrently employing methods to discourage the growth of unwanted species, requires significant inputs of human, animal, wood, and fossil energy. Thus it is not surprising that as the alteration of the ecosystem increased and control of the environment intensified, greater quantities of energy inputs, especially fossil energy, also increased.

Energy Inputs in Crop Production

The inputs in crop production can be assessed by examining the inputs in an important crop, corn. In this analysis, three corn production systems will be compared: hand-powered, tractor-powered, and horse-powered.

Corn produced by hand in Mexico employing swidden or cut/burn agricultural technology requires only a man with an axe, hoe, and some corn seed (Table 5.1). The energy input for the human power is calculated at 3,500 kcal per man day and is assumed to come directly from eating the corn grain produced. Since a total of 1,144 hours of labor input goes into the system, a total of 143 man days of fuel is required or 500,500 kcal. This is the single largest energy input for this production system. The labor fuel input would be increased if other inputs needed to sustain the worker were added, such as clothing, housing, transport, schooling, police, and fire protection. Then the energy requirements needed by the worker's family could also be included as a necessary input.

Under certain circumstances for swidden agricultural production, one might also include the energy lost by burning the trees and other vegetation that were present on the site before burning. However, for the present the system will be kept relatively simple and include only the kilocalories removed from the corn grain required to fuel the manpower input.

When the energy for making the axe and hoe and producing the seed is added to human power, the total energy input needed to produce corn by hand is only about 553,678 kcal/ha (Table 5.1). With a corn yield per hectare of 1,944 kg or 6.9 million kcal the output/input ratio is about 12.5:1 (Table 5.1).

In this system, the fossil energy input for production is only for the axe and hoe. Based only on the fossil energy input of 16,570 kcal, the output/input ratio is about 422 kcal of corn produced for each kcal of fossil fuel expended.

The energy flow in tractor-powered agriculture is distinctly different from that of man-powered agriculture. Corn production in the United States typically relies on heavy machinery for power. The total manpower input is dramatically reduced to only 12 hours compared with nearly 1,200 hours in the hand powered system in Mexico (Tables 5.1 and 5.2). Thus, the mechanized system utilized about 1/100 of the human-power of the hand-powered system. The input for labor in the U.S. system is only 7,000 kcal of food energy for the growing season, or substantially less than that of any other input in production.

TABLE 5.1
Energy Inputs in Corn (Maize) Production in Mexico Using Only Manpower

Item	Quantity/ha	kcal/ha
Labor	1,144 h(a)	500,500
Axe and hoe	16,570 kcal(b)	16,570
Seeds	10.4 kg(b)	36,608
Total		553,678
	output	
Total yield	1,944 kg(a)	6,901,200
Kcal output/kcal input		12.5

(a) Lewis, 1951
(b) Estimated

Balanced against this small input of manpower is the significant increase in fossil energy input needed to run the machines that replace man. In this system inputs of several other items, especially fertilizers and pesticides, have also increased (Table 5.2). Nitrogen fertilizer is now the single largest input in corn production and has increased about twenty-fold since 1945. Less than 0.001 kg/ha of pesticides were applied to corn in 1945, but today more than 8 kg/ha are used and this represents more than a one thousand-fold increase in pesticide use. In 1980 the fossil fuel energy inputs required to produce a hectare of corn averaged about 8 million kcal/ha or the equivalent of about 800 liters of oil (Table 5.2). Based on a corn yield of about 7,000 kg/ha, or 24.5 million kcal of food energy, the output/input ratio is 2.9:1.

What is the solar energy input in the corn production system? The solar energy reaching a hectare of land during the year in the temperate region averages about 14 billion kcal. During an average 4-month growing summer season in the corn belt region nearly 7 billion kcal reach an agricultural hectare. Under favorable conditions of moisture and soil nutrients corn will produce, as mentioned, 7,000 kg of grain plus another 7,000 kg/ha of biomass as stover. This total biomass (dry) converted into potential heat energy represents about 0.5 percent of the solar energy reaching a hectare during the entire year, whereas during the 4-month growing season the percentage would be about one percent. Compared with the solar energy input in the system, the fossil energy input represents only 10 percent of the inputs (Table 5.2).

TABLE 5.2
Energy Inputs Per Hectare For Corn Production in the United States

Item	Quantity/ha	kcal/ha
Labor	12 hr	7,000
Machinery	55 kg	990,000
Gasoline	26 l	264,000
Diesel	77 l	881,500
LP Gas	80 l	616,400
Electricity	33.4 kwh	95,500
Nitrogen	151 kg	2,220,000
Phosphorus	72 kg	216,000
Potassium	84 kg	134,000
Lime	426 kg	134,400
Seeds	18 kg	445,500
Insecticides	1.4 kg	119,950
Herbicides	7 kg	777,500
Drying	7,000 kg	1,437,800
Transportation	200 kg	51,200
Total		8,390,750
	Output	
Total yield	7,000 kg	24,500,000
Kcal output/kcal input: 2.9		

Source: Pimentel and Burgess, 1980.

The energy flow in corn production can be examined another way by substituting horse power for fossil fuel and the tractor. To do this, the tractor power in Table 5.2 was removed and 120 hours of horsepower plus 120 hours of needed manpower were included. An estimated 136 kg of corn and 136 kg of hay would be required to feed a 682 kg (1,500 lb) horse for the portion of the year that the horse was used (Morrison, 1956). The corn grain would come directly out of the corn produced. The hay required to support the horsepower for one hectare of corn would have to be produced on hayland but this would represent only 0.2 of a hectare. Thus, based on the current high yields of corn and hay per hectare the additional land area required to support a horse is about 20 percent.

The major impact of this horse-powered system is the ten-fold increase in manpower per hectare that is required over that of the tractor-powered system (Tables 5.2 and 5.3). The total energy input into the horse-powered system is 7 million kcal compared with 8 million kcal in the tractor system (Tables 5.2 and 5.3). The resulting reduced energy input with horsepower makes it slightly more energy efficient than the tractor-powered system. Thus, the energy ratio for a horse-powered system is 3.4:1 compared with the lower 2.9:1 for the tractor-powered system (Tables 5.2 and 5.3). At present it is not economically feasible to use ten times more manpower to produce corn than the mechanized system.

TABLE 5.3
Energy Inputs Per Hectare For Corn Production in the United States Employing Horse Power

Item	Quantity/ha	kcal/ha
Labor	120 hr	70,000
Machinery	15 kg	27,000
Horse		
Corn	136 kg	477,300
Hay	136 kg	409,000
LP Gas	80 l	616,400
Electricity	33.4 kwh	95,500
Nitrogen	151 kg	2,220,000
Phosphorus	72 kg	216,000
Potassium	84 kg	134,000
Lime	426 kg	134,400
Seeds	18 kg	445,800
Insecticides	1.4 kg	119,950
Herbicides	7.8 kg	777,500
Drying	7,000 kg	1,437,800
Transportation	150 kg	38,550
Total		7,219,200
	Output	
Total yield	7,000	24,500,000
kcal output/kcal input		3.4

Source: Based on data from Pimentel and Burgess, 1980; and Morrison, 1956.

Corn grain provides one of the best output/input

ratios (Table 5.4). Fruits and vegetables in general yield less favorable ratios than grains. Of course, fruits and vegetables are mainly produced for vitamins and other nutrients and not for food energy.

Livestock Production

The forages and grains that were considered earlier are utilized for feed livestock in the United States. An estimated 90 percent of all grain utilized in the United States (135 million metric tons) is fed to livestock to produce meat, milk, and eggs (Pimentel et al., 1980). From 10 to 88 kcal of fossil energy are required to produce one kcal of animal protein (Table 5.5).

The major reason that animal protein products are significantly more energy expensive to produce than plant protein food (Tables 5.4 and 5.5) is that first forage and grain feeds have to be grown and then consumed by the livestock. The forage and grain seed that maintain the breeding herd add additional energy costs in production. For example, to produce one feeder calf a total of 2.3 animals have to be fed (1.3 are breeding cattle).

DEPENDENCY AND INTERDEPENDENCY OF INPUTS

Land, water, and energy resources are all essential to crop production and each can be substituted for one another within limits. For example, the yield of corn from two hectares of land, with energy inputs of about 2.2 million kcal per hectare, is about 2,500 kg per hectare (Pimentel and Pimentel, 1979). If the aim were to produce the same amount of corn (5,000 kg) on one hectare of land, then the energy inputs would have to be increased to about 6.5 million kcal (Pimentel, 1980). Thus, to reduce the land area by half and maintain total corn yield, about three times more fossil energy is required.

This problem becomes acute when one looks at what is happening to the supply of arable land throughout the world. As the world population grows, much valuable cropland is being removed from production for urbanization and other human activities. In the United States, for example, from 1945 to 1975 about 18 million hectares of agricultural land were removed from production as highways were built and urban areas expanded. This "black-topping" of agricultural land is the equivalent in area to the entire state of Nebraska (Pimentel et al., 1976).

To compensate for this loss of agricultural land, greater energy inputs are required to maintain high levels of food and fiber production from the remaining agricultural land. Unfortunately, this problem is not confined just to industrialized nations like the United States but is occurring throughout the world. Urgently needed are the formulation and implementation of land use policies that will protect and preserve agricultural land for food production.

Water is another essential resource in crop production that may be substituted to a limited extent for energy and/or land area. For example, with limited rainfall in the wheat growing region of Washington state, the land area must remain fallow every other year to accumulate sufficient moisture to grow a crop of wheat. As a result, two hectares of land are necessary to obtain one hectare of wheat per year. In contrast, under normal rainfall conditions in Kansas, a crop of wheat can be harvested each year from one hectare.

Irrigation of land is an example of substituting energy for water to maintain cropland productivity. Pumping water from the ground-water store is energy intensive. For instance, to irrigate a hectare of corn in an arid region for one growing season may require as much as 2,100 liters of fuel or nearly three times the total energy inputs if the crop were grown under normal rainfall conditions (Pimentel and Pimentel, 1979).

In addition to fossil energy substituted for either water or land, energy is often used to replace labor with machinery. For example, when corn is produced by hand, nearly 1,200 hours of manpower effort are needed per hectare (Table 5.1). With mechanized systems, corn can be produced with an input of only 12 hours of manpower (Table 5.2). Hence, machinery can significantly reduce the labor input.

Indeed, a gallon (3.79 liters) of gasoline, which contains 31,000 kcal energy, has tremendous power and the potential to do work. This can be illustrated as follows. When one gallon of gasoline is used to operate a mechanical engine, which is about 20 percent efficient in converting heat energy into mechanical energy, an equivalent of 6,200 kcal of work can be achieved. This is equal to about 9.7 hp-h of work or the work equivalent of one horse working at capacity for nearly a 10-hour day. Since man produces only 1/10 of a hp-h working at capacity, the gallon of gasoline produces 97 manpower hours of work. This is equivalent to one man working 8 hours a day, for 12 days.

Clearly, the use of fossil fuel has drastically cut the input of manpower or horsepower needed in agricultural production. As long as fuel supplies are abundant and cheap, this will continue to be an effective trade-off. The use of fossil fuels in mechanized agriculture has had little effect on crop yield per hectare, though it does facilitate timing and speed of planting and harvesting. In areas where growing seasons are short this may be an advantage.

TABLE 5.4
Energy Inputs and Returns For Various Food and Feed Crops Produced Per Hectare in the United States

Crop	Crop Yield (kg)	Yield in Protein (kg)	Crop Yield in Food Energy (million kcal)	Fossil Energy Input for Production (million kcal)	Kcal Food/Feed Output/Kcal Fossil Energy Input	Labor Input Manhours
Corn (U.S.)	7.000	630	24.5	6.9	3.5	12
Wheat (North Dakota)	2.022	283	6.7	2.5	2.7	6
Oats (Minnesota)	2,869	423	10.9	2.1	5.1	3
Rice (Arkansas)	4.742	272	14.0	12.5	1.1	30
Sorghum (Kansas)	1,840	202	6.0	1.5	4.0	5
Soybeans (Illinois)	2.600	885	10.5	2.3	4.5	8
Beans, dry (Michigan)	1,176	285	4.1	3.1	1.3	19
Peanuts (Georgia)	3.720	320	15.3	10.9	1.4	19
Apples (East)	41,546	83	23.3	26.2	0.9	176
Oranges (Florida)	40,370	404	19.8	11.8	1.7	210
Potatoes (New York)	34,468	539	21.1	15.5	1.4	35
Lettuce (California)	31.595	284	4.1	19.7	0.2	171
Tomatoes (California)	49,620	496	9.9	16.6	0.6	165
Cabbages (New York)	53.000	1,060	12.7	16.8	0.8	289
Alfalfa (Minnesota)	11,800 (dry)	1,845	47.2	3.6	13.1	12
Tame hay (New York)	2,539 (dry)	160	5.5	0.6	8.6	7
Corn silage (N.E. U.S.)	9,400 (dry)	753	29.1	5.2	5.6	15

Sources: Pimentel and Pimentel, 1979; and Pimentel, 1980.

TABLE 5.5
Energy Inputs and Returns Per Hectare For Various Livestock Production Systems in the United States

Livestock	Animal Product Yield (kg)	Yield in Protein (kg)	Protein as kcal (thousand)	Fossil Energy Input for Production (million kcal)	Kcal Fossil Energy Input kcal Protein Output	Labor Input Manhours
Broilers	2,000	186	744	7.3	9.8	7
Eggs	910	104	416	7.4	17.8	19
Pork	490	35	140	6.0	42.9	11
Sheep (grass-fed)	7	0.2	0.8	0.07	87.5	0.2
Dairy	3,270	114	457	5.4	11.8	51
Beef	60	6	24	0.6	25.0	2
Dairy (grass-fed)	3,260	114	457	3.3	7.2	50
Beef (grass-fed)	54	5	20	0.5	25.0	2
Catfish	2.783	384	1,536	52.5	34.2	55

Source: Pimentel et al., 1980.

INTERDEPENDENCY OF CROPS AND LIVESTOCK IN FOOD SYSTEMS

It is possible over time for the United States to switch to a total grass-fed livestock production system and produce 2.8 million kg of animal protein annually (Pimentel et al., 1980). This is slightly more than half of current production, calculated to be 5.4 million metric tons. The 2.8 million metric tons of animal protein of meat and milk is sufficient protein to provide each person in the United States with 40 g of protein daily. Added to the 32 g of plant protein, this amounts to 72 g of protein per person per day on average. This 72 g is 44 percent greater than the recommended daily allowance of 50 g (NAS, 1980).

If current livestock systems in which both grain and grass are fed were changed to a grass-only system, considerable changes would occur in dietary patterns of the U.S. population. Currently, animal products supply about 69 percent of the daily protein available for consumption in the United States, 33 percent of the energy, and substantial amounts of other nutrients, notably calcium, available iron, riboflavin, niacin, vitamin B6, and vitamin B12 (USDA, 1977). Reducing animal protein intake from 70 to about 40 g, as would be required under the grass-only systems described here, would substantially alter sources of various nutrients in the U.S. diets.

The decline in energy available from animal products would probably necessitate an increase in consumption of grains, nuts, vegetables, and high-protein plant foods such as legumes or processed vegetable protein. Under a grass-only system, such a direct increase in consumption of plant energy sources by the human population of the United States would increase protein intake to more than the calculated 72 g available from animal and plant sources. The protein available would be sufficient on the average for the U.S. population, although distribution among all socioeconomic levels might be a serious problem if the economic forces at play make the limited available meat and milk products expensive protein sources. This could result in some protein malnutrition in some groups.

Meat and milk products not only contain nutrients, they are considered highly desired foods by most of the population. The relative inelasticity of two products, however, will result in significant rise in prices. A rise in prices of meat and milk would be greater than two-fold for each one percent decrease in supply (Cochrane, 1979). Thus, price increases for meat and milk in a grass-fed livestock system initially might rise more than 100 percent. This could place most meat and milk products beyond the economic means of some groups.

The initial severe rise in prices probably would not continue but would be reduced somewhat with time as farmers, with the incentives of higher prices, increased meat and milk production. The market response to higher production would lead to somewhat lower prices.

Currently, calcium and iron represent two nutrients whose consumption frequently falls below the recommended daily allowance (RDA) when food consumption data are analyzed. Dairy products represent the major source of calcium in the U.S. diet (75 percent) (USDA, 1977). Reduced consumption of dairy products could have deleterious effects on the calcium status of the population. Similarly, animal products account for 37 percent of the iron available for consumption (USDA, 1977). Animal sources of iron are generally at least twice as available as plant sources, and in reality, the animal products probably provide more than 70 percent of the available iron in the U.S. diet. Since iron deficiency anemia among women of childbearing age is the most frequently encountered nutritional deficiency disease in most recent nutrition surveys (Cook et al., 1976), this shift in pattern of nutrient availability would need to be examined carefully.

BENEFITS OF LIVESTOCK IN A SECURE U.S. FOOD SYSTEM

If the 135 million metric tons of grain currently fed to livestock in the United States were consumed directly by humans, the number of people that could be fed would be about 400 million or about twice the U.S. population. This has direct implications to the security of the U.S. food system. If the U.S. agricultural production, grain production in particular, were reduced by one half due to climatic or other disaster, the U.S. population would not starve. All that would happen is that fewer animal products would be eaten and more grain and vegetable products would be consumed.

According to the U.S. Senate Committee on Nutrition and Human Needs, U.S. nutrition would improve by consuming fewer animal products, while consuming more grains and vegetables (U.S. Senate, 1977). In particular, however, this confirms that the U.S. food system has a great deal of redundancy in it because of the livestock component of the system.

Comparing soybean production per hectare to pork production suggests that nearly two times more protein may be produced per unit area with soybeans than with pork production (Tables 5.4 and 5.5). Overall, animal protein products require significantly larger inputs of energy than plant protein production.

Livestock production does have a major advantage in the food production system in that livestock can be utilized to make use of marginal land for food production. Thus, forage crops produced on marginal land can be utilized to produce a valuable human food (Pimentel et al., 1980).

CONCLUSION

The U.S. food system is relatively secure from the perspective that significant quantities of surplus food are produced in the United States for export. In addition, about 90 percent of U.S. grain is fed to livestock. If U.S. food production were significantly reduced due to some crisis such as a climatic problem, less grain could be fed cattle and other livestock and more consumed directly by the population.

Three major threats exist for U.S. food production. These include energy shortages, land degradation, and water shortages. About 6 percent of total U.S. energy is used in agricultural production and about half of this is oil. Thus, any interruption in U.S. oil supplies would have a major impact on the food production system. There are few or no alternatives to oil in the short term for U.S. agricultural production.

Soil erosion is seriously reducing the productivity of U.S. soils. Heavy soil erosion is projected to continue and may, with changes in crop cultural practices such as reduced rotations, intensify. In states such as Iowa one-half of the top soil has been lost already due to erosion. Some land with continued erosion loses its productivity while other lands require heavy inputs of fossil energy in the form of fertilizer and other inputs to keep them minimally productive.

Water is also an essential ingredient to crop production. Reduced irrigation water availability, due to either shortages of water or such high energy prices that farmers cannot afford to pump water, may in some regions limit crop production.

Shortages of water and energy and degradation of soil will not prevent food production but would result in a major reduction in U.S. food production and security.

REFERENCES

Bastin, G, and Ellis, J. 1980. International Trade in Grain and the World Food Economy. EIU Special Report No. 83. London: The Economist Intelligence Unit.

Cochrane, W. W. 1979. The Development of American Agriculture: A Historical Analysis. Minneapolis: Univ. of Minnesota Press.

Cook, J. D.; Finch, C. A.; and Smith, W. J. 1976. "Evaluation of the Iron Status of a Population." Blood 48:449.

FAO. 1973. Energy and Protein Requirements. Report of a Joint FAO/WHO Ad Hoc Expert Committee. FAO Nutrition Meeting Report Series No. 52. Rome: FAO, UN.

Lewis, O. 1951. Life in a Mexican Village: Tepoztlan Restudied. Urbana: Univ. of Illinois Press.

Morrison, F. B. 1956. Feeds and Feeding. Ithaca: Morrison Publishing Co.

NAS. 1977. Supporting Papers: World Food and Nutrition Study. Washington: National Academy of Sciences.

______. 1980. Recommended Dietary Allowances. 9th ed. Washington: National Academy of Sciences.

Paxton, J., ed. 1981. The Statesman's Year-Book: Statistical and Historical Annual of the States of the World for the Year 1981-82. 118th ed. New York: St. Martin's Press.

Pimentel, D., and Burgess, M. 1980. "Energy Inputs in Corn Production." In Handbook of Energy Utilization in Agriculture, ed. D. Pimentel, pp. 67-84. Boca Raton: CRC Press.

Pimentel, D.; Oltenacu, P. A.; Nesheim, M. C.; Krummel, J.; Allen, M. S.; and Chick, S. 1980. "Grass-fed Livestock Potential: Energy and Land Constraints." Science 207:843-48.

Pimentel, D., and Pimentel, M. 1979. Food, Energy and Society. London: Edward Arnold.

Pimentel, D.; Terhune, E. C.; Dyson-Hudson, R.; Rochereau, S.; Samis, R.; Smith, E.; Denman, D.; Reifschneider, D.; and Shepard, M. 1976. "Land Degradation: Effects on Food and Energy Resources." Science 194:149-55.

Pimentel, D., ed. 1980. Handbook of Energy Utilization in Agriculture. Boca Raton: CRC Press.

USDA. 1977. "Agricultural Charts." Food and Home Notes, No. 7, February 14.

U.S. Senate. 1977. Dietary Goals for the United States. Select Committee on Nutrition and Human Needs. Washington: U.S. Government Printing Office.

Plant Breeding

6 The Influence of Plant Breeding and the Farm Seeds Supply Industry on Food Security

Donald N. Duvick

Basic to the nation's food supply is the seed planted by its farmers. Annually, a ritual of faith and hope is performed: our farmers plant their fields with faith in the productivity of the seed and hope for good weather. For 10,000 years humankind has mistrusted the weather--the growing season--and has attempted to influence it with charms, magic, and religious rites (Large, 1940). The seed usually was left to itself. But in recent decades, since the maturation of scientific plant breeding, we have begun to worry about the seed as well. Is it productive enough? Can it become still more productive? Is it dangerously uniform? Can we be assured of sufficient supply? Will there be monopoly control of our seed production? Can we trust our nation's seed supply to the decisions of private, profit-oriented businesses? What can today's magic--genetic engineering--do for, or do to, plant breeding and the nation's food supply? In sum, can the nation trust its plant breeders and its seed producers?

GENETIC DIVERSITY

Genetic diversity and its opposite, genetic vulnerability, have been a concern for 40 years or more, dating back at least to when scientists first were able to demonstrate that certain epidemic disease and insect disasters were the consequence of uniform genetic susceptibility of the affected crops. Thus, Victoria blight of oats in the 1940s was recognized as a disease specific to a certain widely distributed oat variety and its descendants. From such diagnoses came the theory that epidemics are caused by the practice of growing only a few crop varieties, and a corollary theory, that epidemics can be prevented by growing numerous genetically diverse varieties. When race T of southern corn leaf blight caused large and in some cases disastrous yield losses to cytoplasmically uniform corn in the eastern and southern U.S. in 1970 (Tatum, 1971), the theory and its corollary were regarded as proven. Pressure has been placed on plant breeders and on the seed supply industry ever since to provide many varieties of genetically diverse crop seeds; in other words, to broaden the genetic base of our farm crop seeds (NAS, 1972, 1978), thereby preventing future epidemics.

The present nationwide, in fact worldwide, trust in genetic diversity and mistrust of plant breeders' ability to provide it is based on certain inadequate perceptions of reality. For instance, it often is assumed that U.S. crop varieties used to be more numerous and genetically diverse and that therefore they were free of epidemic disease and attack. This assumption is not necessarily correct. For example, in the late 1800s Turkey wheat--a single variety--completely dominated the hard red winter wheat belt of the western Great Plains. In another example, wheat production was driven out of Massachusetts in the early days of the republic, due primarily to repeated epidemics of "blast," or black stem rust, to which the local wheats of that era were uniformly susceptible (Bidwell and Falconer, 1941).

A second common misconception is that epidemics cannot arise in genetically diverse populations. This is not so; Dutch elm disease in the 1960s and chestnut blight in the 1930s devastated the wild and surely genetically diverse populations of the American elm and the American chestnut.

A third misconception is that present-day crop varieties are dangerously few in number and that therefore U.S. farm crops are in a precarious state of genetic vulnerability. This supposition is true only in part; half a dozen cultivars or inbreds of each of our major field crops--corn, wheat, soybeans, sorghum, and cotton--do cover about half the surface area of each crop in the U.S. (Duvick, 1984). But this is only a part of the picture. Our proper concern should be: what is the total available genetic diversity on the farm and in reserve, and how rapidly is it turned over? The answer to this question is that for our major field crops (corn, wheat, soybeans, cotton, and sorghum) a total of several hundreds of commercial experimental cultivars are grown on U.S. farms each year; several thousands of cultivars are in advanced trials each year, ready to serve as replacements for the commercial cultivars if needed; and hundreds of thousands of genetically diverse new selections are in the plant breeders' nurseries and yield trial fields, every year, available for future selection as needed. Furthermore, the most widely grown cultivars of each crop are replaced with new, different cultivars every five to seven years on the average, thereby providing useful genetic diversity in time. (If cultivars are replaced before pathogens specific to them have been able to increase to epidemic levels, one has provided useful genetic diversity in time.) Thus, the total available genetic diversity in U.S. crop plants is large, much greater than is commonly perceived (Duvick, 1981, 1984).

Plant breeders are very much aware of the need for genetic diversity; they are attempting to provide it, but they cannot and should not control what farmers plant. Farmers choose crop varieties based on their experience. Experience with epidemics drives them away from vulnerable varieties and vulnerable species; experience with reliable top performers leads them to concentrate on the best available cultivars. Plant breeders and seed companies simply make available the diversity of cultivars from which farmers make their choices. I suspect many critics of U.S. plant breeding greatly underestimate the ability of the U.S. farmer to critically compare cultivars for needed traits, and then to make shrewd decisions about which ones to plant. American farmers are not led by the plant breeders or the seed companies; the farmers usually are the leaders, and often they are teachers as well.

FUTURE INCREASES IN GENETIC DIVERSITY

The above statements should not be construed as saying that we need no further improvement in genetic diversity of the nation's farm crops. Improvements can be and are being made every year. For example, many present-day crop cultivars are very closely related. To help farmers avoid planting only closely related cultivars, breeders are beginning to point out to farmers the best choices among cultivars to achieve desired levels of diversity.

Another item: although breeders continually introduce exotic germ plasm from diverse sources into their breeding pools, the ensuing necessarily rigid selection for productive, stress tolerant, and pest resistant materials usually pulls out a narrow segment of related, superior genotypes. Breeders now are doing research intended to tell them how to introduce and to keep more exotic germ plasm in elite materials, and how to do so with more efficiency.

Breeders also know that they need to learn more about the best ways for defining and recognizing useful sources of diverse germ plasm, and they now are developing techniques to do this. Unfortunately, they hardly know where to look for specified kinds of diverse germ plasm in our numerous germ plasm banks today because, with only a few exceptions, materials in the germ plasm repositories are poorly catalogued, poorly described, and often are not even maintained in viable or uncontaminated condition (Duvick, 1981, 1984). The blame for these deficiencies, in my opinion, lies with those responsible for funding of the repositories, and not with the curators, who do as well as they can with the funds at their disposal.

In sum, although plant breeders continually are broadening the genetic base of their breeding populations and can cite many examples of successful introduction of useful diverse germ plasm into widely grown cultivars, they can, should, and will do more in the future.

EXCHANGE OF GERM PLASM: PUBLIC, PRIVATE

The germ plasm repositories--the seed banks--are not the only ones to maintain germ plasm. Plant breeders in both public and private institutions have extensive collections of elite adapted materials in their active breeding populations. As was noted above, these pools of adapted, elite materials are immediately useful in case of urgent need. Complaints have been made in recent years that interchangeability (nationwide availability) of these pools has decreased dangerously or has ceased entirely due to the growth of the private seed industry, especially since the advent of the U.S. Plant Variety Protection Act (Mooney, 1979). But, as with genetic diversity, this fear is based on certain inaccurate perceptions about plant breeding and plant breeders.

It is well known in plant breeding circles that the development of hybrid corn and hybrid sorghum have progressed well in the commercial plant breeding organizations, even though these organizations do not exchange breeding materials directly. Actually, there is a great deal of germ plasm movement: publicly employed corn breeders make both basic and finished materials available to all; privately employed corn breeders turn over to public agencies broad-based collections and composites; and through these means, efficient but highly competitive corn breeding programs have been able to make consistent progress through the years. Several independent measurements indicate a one percent per year (or more) improvement in genetic yield potential of hybrid corn; this rate has been maintained consistently over the past 40 to 50 years (Russell, 1974; Duvick, 1977). (The genetic portion amounts to 50-70 percent of total achieved yield gains.) There is no sign of any let-up in this genetic rate of gain; it is constant or even increasing in velocity. Furthermore, the new hybrids are less dependent than the old ones on favorable weather or high soil fertility to make satisfactory yields; the new hybrids are tougher. Nevertheless, the new hybrids also can make better use of good growing conditions; they will add more yield in response to extra fertility or highly favorable weather, as compared to the old hybrids.

It also is well known in plant breeding circles

that, before the growth of the private breeding industry, publicly employed breeders of the self-pollinated crops such as wheat and soybeans did not release or give away breeding materials that were in the final stages of cultivar selection. When superior selections had been identified and were in the process of retesting, purification, and increase, they were closely held by their developers. Public plant breeding was, and is, highly competitive. However, an efficient system did exist in public plant breeding circles for the exchange of advanced lines during the year or two before their release. Its purpose was to allow comparative yield testing of the advanced lines. It was understood that competitive breeders would not appropriate such lines, entrusted to them for observation and testing only. Further, in some--but not all--circles, it was understood that these superior new selections also could be used for hybridization, to start new breeding populations. Interestingly enough, both of these practices usually have been extended to include breeders with private organizations as well. In effect, this means that barriers to germ plasm exchange are put up only during the few final generations in which highly promising selections are being reevaluated to be sure they merit final testing and release. Before and after this stage, exchanges can and do occur. As long as mutual trust is maintained it looks as if this system will continue to work well.

Mutual trust also is needed to maintain genetic diversity in the hybrid crops, corn and sorghum. Private seed companies, breeding and selling their own hybrids of corn and sorghum, need to be able to trust each other not to pirate (that is, to steal) inbred lines from each other. If it should become common practice for seed firms, large or small, to appropriate for their own use and sale the best inbred lines and thus the best hybrids of the leading two or three seed companies, the genetic diveristy of hybrids sold in the U.S. could become unnecessarily and dangerously narrow. In the past, pirating apparently was uncommon. At present, according to word-of-mouth commentary, it is becoming more common. Perhaps, in the future, maintenance of genetic diversity among hybrids will require plant variety protection on proprietary inbred lines plus vigilant enforcement of the law.

SHARING OF RESPONSIBILITIES: PRIVATE AND PUBLIC

Note that due to the nature of their missions, public breeders tend to develop and distribute more broad-based materials, and the private breeders more finished materials (Duvick, 1982). This sharing of responsibilities is most clearly seen in cotton breeding; cotton breeders, public and private, proudly state that they are informally organized into a single efficient unit, in which public breeders are especially charged with introgression of useful exotic germ plasm into adapted breeding pools, and private breeders have as their chief duty the development of finished cultivars.

Despite the present tendency to depend on public breeders to develop broadly based germ plasm pools there is some fear that in the future the best broad-based pools will be privately held and thus the nation, in effect, no longer will be in control of its plant breeding destiny. This seems to me to be a reasonable concern and one which private breeding organizations should recognize and forestall by deposition of significant numbers and kinds of basic germ plasm in the appropriate germ plasm repositories. This actually has been done in several instances already, but I think it should be done more often in the future.

AGRICULTURAL CHEMICAL COMPANIES AND THE SEED BUSINESS

Another recent concern is that a relatively small number of international corporations, with their main business base in agricultural chemicals, have purchased several of the independent seed companies. The consequence would seem to be that the seed supply business will be less competitive than in the past. It also is feared that the agricultural chemical companies will develop and sell only cultivars with unique, genetically specific tolerances for their proprietary herbicides, thus forcing the farmer to buy both herbicide and cultivar as a package, and further restricting his freedom of choice.

This last point has in the past been considered as a useful goal by some of the agricultural chemical companies. But I think that most of them, by now, realize that this was an impractical, simplistic goal. Farmers choose cultivars based on total performance. They consider many traits, yield foremost, and herbicide tolerance is far down the list. Plant breeding progress would be needlessly slowed if cultivars were selected first for herbicide tolerance and then for yield, disease resistance, and other traits of primary importance.

The consequence of continuing incorporation of independent seed companies into large diversified corporations with no past experience in the seed business or in plant breeding is a more serious concern, however. Well-managed, diversified corporations can and have maintained and improved newly-purchased seed companies. Often the large corporations have provided much-needed skills in marketing, or have been able to provide funds needed by the seed company as it regroups and rebuilds its breeding department over a several-year period. But sometimes the parent corporation misunderstands the nature of its newly-acquired seed business; short-term profits are emphasized at the expense of needed long-range breeding programs; efficient plant breeding teams are disbanded or misdirected. Worst of all, I have seen instances in which the seed company apparently was bought only for short-term cash gains; it was quickly sold off in pieces and effectively destroyed.

In general, however, large corporations are in business for long-term profits; if they genuinely intend to run their seed companies at a profit they will learn how to operate them efficiently, which in the seed business means they will breed and sell superior cultivars, recognized as such by the farmers. I, of course, am highly prejudiced in favor of independent seed companies, believing they can make the best decisions when left to themselves. But in some cases the stability of a large corporation's funding may allow more long-term research for seed companies of moderate or small size with fluctuating income. As plant breeding research becomes more expensive and technical, this argument may carry increasing weight.

FOREIGN OWNERSHIP OF AMERICAN SEED FIRMS

Purchase of American seed firms by foreign international corporations presents possible problems. Foreign ownership of American businesses is, of course, not unique to the seed industry, but there is a general feeling that one should worry about foreign control of our seed supply more than about foreign control of our petroleum supplies, our automobile production, or our electronics industry. To date, American seed firms which have come under foreign control have moved ahead with their breeding and seed production operations much as they did when they were independent and American owned, although they have experienced some of the problems inherent to being part of any large organization with distant headquarters. One should remember that American seed firms also have overseas branches, subsidiaries, or wholly-owned companies. The tendency worldwide for such business relationships to develop cannot be ignored. I would guess that as time goes on, governments all around the world will take certain steps to assure the stability of seed supplies and of breeding stocks, from a nationalistic point of view. It is hard to predict what these measures will be; I hope they will be reasonable. But it does seem that it would be worthwhile for all countries, including the U.S., to provide the laws, discipline, and rewards that would encourage stable, long-term breeding programs--private or public. Public words and public deeds are needed for such encouragement. In the end, I think that responsible foreign or domestic ownership of American seed firms will make little difference in the goals of the seed firms: to provide American farmers with superior cultivars, at a reasonable profit to the seed companies. The important question will be: can seed companies operate efficiently as part of large, diversified corporations?

THE SMALL SEED COMPANY: PAST, PRESENT, AND FUTURE

But in the seed business there also is, and has been for the past 50 years, a place for a large number of successful small, and very small, seed companies. In the hybrid seed corn business for example, about one-third of sales have been made by the small companies for at least the past 10-15 years, and their proportion of total seed corn sales has grown larger over the past decade(1). Also, since passage of the U.S. Variety Protection Act, a large number of very small soybean seed companies have appeared, and several small wheat companies have recently come into existence as well(2). The small seed companies exhibit an interesting kind of economic evolution; they are continually changing. Some buy out others, some go out of business, new ones appear. But the total mass persists and the basic cadre of seedsmen stays the same as well, giving experience and continuity to the small company sector.

In most cases these very small companies have no breeding staff, or their staff is so small and is assigned so many non-breeding chores that it cannot provide all the cultivars sold by the company. Small hybrid seed corn companies, in the past, have depended on inbreds and hybrids developed by public breeders. Over the years, several foundation seed companies have developed efficient staffs to test hybrids and counsel the small seed companies regarding the best hybrids to sell. More recently, as public breeders have done less inbred and hybrid development, the foundation seed companies have added well-qualified breeding staff and are now, in effect, acting as breeding departments for the small hybrid seed corn companies. Similar developments are taking place in the soybean and wheat seed business. There seems to be some factor in the market place that favors existence of small, agile, low-margin seed companies, side-by-side with the large, reliable, but comparatively ponderous major companies. To some extent, seed sales of the small companies are made on the basis of price and kinship, but I suspect that the small companies also provide a certain amount of imagination, initiative, and high-stakes risk that farmers understand and are willing to try, on at least a portion of their acres. I think the seed industry needs the small companies; I think we always will have them.

NEW BIOTECHNOLOGIES

The new biotechnologies--usually called genetic engineering--sometimes have been predicted as eventual replacements for conventional plant breeding. Some predictions even have it that we no longer will have conventional crop plants, or at any rate we are told that some very unconventional ones will take over a high proportion of our farm crops area. Molecular biology and cell and tissue culture are the tools of genetic engineering; the premise of genetic engineering is that DNA is the prime mover of all biological organisms, that mastery of DNA manipulation means mastery of biology, and that molecular genetics is the best way to manipulate DNA.

Genetic engineers already have learned how to clone genes at will--to make unlimited numbers of copies of selected bits of DNA. They now are learning how to insert the cloned genes into higher plants, and in a few years it is probable that scientists will be able to put selected genes from almost any organism--plant, animal, or microbe--into our major crop plants such as corn, wheat, or soybeans. This will be a truly outstanding scientific achievement, but it will not overturn conventional plant breeding.

Breeders already transfer single genes with relative ease, using ordinary hybridization and backcrossing techniques. By and large, geneticists have had very little success in trying to transfer genes from too great a distance, taxonomically, because the genes will not function well in strange surroundings. Thus there is only a little interest in being able to transfer genes that conventional plant breeding cannot already transfer.

More important, plant breeders have only marginal interest in manipulating genes one at a time, or even in groups of three or four. The really important traits for selection--yield, durable pest resistance, and stress tolerance--are governed by large numbers of genes, with unknown locations and unknown primary products that interact in numerous, intricate, and unknown ways. Molecular genetics, today, is too specific to be of much use in solving problems such as these, which involve polygenic inheritance.

The science of population genetics, and plant breeders' rule of thumb methods, provide useful and economical techniques for efficient selection for traits governed by many genes with complicated interactions. The techniques depend on approximations and are based on the science of probabilities. It is likely that even with computer assistance, molecular genetics always will be too precise and ponderous to be used to plan entirely new cultivars, or even new desired traits.

On the other hand, as genetic engineering skills

improve in power and speed, plant breeders will welcome opportunities to move certain genes or gene blocks with greater speed, and with freedom from undesirable linkages. New, efficient gene transfer tools some day will be as indispensable to plant breeding as are today's mechanical plot planters, grain combines, and computers. Also, in the fairly near future, I expect that molecular genetics will let us learn about some of the highly important single genes in such detail that we can specify improvements in their structure as, for example, in a gene for resistance to wheat stem rust. We then will be able to manufacture the gene according to the new design, clone it up to a useful copy number, and place it in desired cultivars. I don't look for these techniques to be available tomorrow, but they will be welcomed and will be used by plant breeders when they are perfected.

GENETIC ENGINEERING'S MOST IMPORTANT BENEFIT

Probably the most important benefit from genetic engineering's tools--molecular biology and cell and tissue culture--will be their ability to give us an understanding of fundamental biological processes in detail never before possible. Unfortunately, as plant molecular biologists are just now realizing, we do not know very much about the biochemistry and physiology of plants as compared to microbes or animals, and so we are hindered in efforts to put fundamental plant molecular discoveries to work in producing useful whole-plant changes. The millions of dollars now being spent on plant molecular genetics and plant cell and tissue culture will be wasted unless comparable amounts are spent on research in the many intermediate processes between DNA transcription and whole-plant traits. To force funding of research on these all-important intermediate processes may well be the most useful result of the current genetic engineering excitement.

To recapitulate: genetic engineering technology will supplement but not replace traditional plant breeding. In years to come, pieces of technology from genetic engineering, and the added fundamental understanding of plant biology as developed by the practitioners of genetic engineering, will be indispensable to continuing progress in plant breeding.

SUMMARY

Plant breeding and the farm seeds industry affect the nation's food supply to the degree that they succeed in turning out ample quantities of productive, dependable varieties of farm seeds. Plant breeders, public and private, are meeting this challenge successfully, to date. They continue to develop new varieties capable of producing high yields under good growing conditions but also able to withstand disease and insect pests, and environmental stresses such as heat, drought, cool temperatures, and excessive rainfall. A major challenge to plant breeders is to maintain and increase useful genetic diversity. Plant breeders are aware of the need and in recent years have increased their efforts in this direction. However, more genetic diversity is on hand, even now, than is usually realized. Private plant breeding of the major farm crops is becoming increasingly important, relative to public plant breeding. A partnership appears to be developing in which private plant breeders generally are responsible for producing finished cultivars and public plant breeders are responsible for producing broadly based germ plasm pools, which then can be used by those breeders charged with developing finished cultivars. In line with a worldwide tendency for concentration of ownership of maturing industries, farm seeds firms increasingly have been purchased by larger firms, often by agricultural chemical companies, both American and foreign. This trend probably will not have adverse effects upon the breeding efforts of the seed firms, but it bears watching. It should be noted, additionally, that a vigorous and sizeable core of very small seed firms persists and actually is increasing in importance. Finally, genetic engineering will add an important new dimension to plant breeding. Its contributions probably will be essential to maintain and increase the rate of improvements in genetic yield capacity in years to come. However, genetic engineering will not replace plant breeding, nor will it give rise to quantum increases in productivity. Instead it will gradually improve the speed and precision of traditional plant breeding operations. It also will greatly increase basic biological knowledge about farm crops and thus will improve the judgmental and analytical powers of conventional plant breeders.

In summary, plant breeding and the farm seeds industry will continue to produce ample quantities of increasingly productive and dependable varieties of farm seeds. But continuation of these advances will require adequate funding for public plant breeding institutions and an assurance of fair profits for private seed firms.

NOTES

1. Data from private surveys, made by my company, summarized as follows:

The National Seed Corn Market

Year	Total of 7 Major Companies, %	All Others
1971	71.9	28.1
1972	72.3	27.7
1973	71.9	28.1
1974	71.6	28.4
1975	67.8	32.2
1976	71.1	28.9
1977	68.1	31.9
1978	67.1	32.9
1979	64.5	35.5
1980	64.5	34.5
1981	63.5	36.5

Source: Pioneer, DeKalb, Funk, Golden Harvest, PAG, Northrup-King and Trojan.

2. See, for example, applicant listings in the quarterly: Plant Variety Protection Office Official Journal, published by Plant Variety Protection Office, Livestock, Meat, Grain and Seed Division, AMS, USDA, National Agricultural Library Building, Beltsville, MD 20705.

REFERENCES

Bidwell, P. W., and Falconer, J. I. 1941. *History of Agriculture in the Northern United States 1620-1860*. New York: Peter Smith.

Duvick, D. N. 1977. "Genetic Rates of Gain in Hybrid Maize Yields During the Past 40 Years." *Maydica* XXII:187-96.

______. 1981. "Genetic Diversity in Corn Improvement." *Proceedings of the Thirty-sixth Annual Corn and Sorghum Industry Research Conference* 31:48-60.

______. 1982. "Commercial Plant Breeding and Its Relationship to Public Plant Breeding." Proceedings and minutes of the Thirty-first Annual Meeting of the Agricultural Research Institute, October 4-6. Washington, D.C.

______. 1984. "Genetic Diversity in Major Farm Crops on the Farm and in Reserve." *Economic Botany*. Forthcoming.

Large, E. C. 1940. *The Advance of the Fungi*. Reprint. New York: Dover, 1962.

Mooney, P. R. 1979. *Seeds of the Earth: A Private or Public Resource*. Ottawa: Inter Pares for the Canadian Council for International Cooperation.

NAS. 1972. *Genetic Vulnerability of Major Crops*. Committee on Genetic Vulnerability of Major Crops, National Research Council. Washington, D.C.: National Academy of Sciences.

______. 1978. *Conservation of Germplasm Resources, An Imperative*. Committee on Germplasm Resources, National Research Council. Washington, D.C.: National Academy of Sciences.

Russell, W. A. 1974. "Comparative Performance for Maize Hybrids Representing Different Eras of Maize Breeding." *Proceedings of the Twenty-ninth Annual Corn and Sorghum Research Conference* 29:81-101.

Tatum, L. A. 1971. "The Southern Corn Leaf Blight Epidemic." *Science* 171:1113-16.

7
Issues and Perspectives in Plant Breeding

L. J. (Bees) Butler

In attempting to place plant breeding into perspective with food security, it is important to recognize that plant breeding is a technique which potentially increases the productivity of agriculture, thus increasing the farmer's capacity to produce more food from less land. There are two important viewpoints which put increasing agricultural productivity and food security in context.

First, from all available evidence, there is no known physical or technical reason why basic food needs cannot be supplied for all the world's people. The needs are not being met now because of social and political structures, and values, etc., not because of physical scarcity.

Second, the issue of food security (whether we are talking about availability, accessibility or adequacy) involves a direct trade-off between agricultural production and population growth. World population is growing at the rate of about 80 million people every year. This equals the population of Kentucky every two weeks, or a city the size of Lexington, Kentucky twice every day. Ignoring the eventual necessity of institutional arrangements which discourage population growth is akin to mimicking the antics of the ostrich(1).

This chapter, however, does not deal with these important issues. Instead it attempts to examine our current plant breeding system--both the direction it is taking, and the broad issues which appear to be dictating the shape of agriculture in the future. Two major questions are implicit in the ensuing discussion:

(1) Are we headed in a direction which would exacerbate the problems of food security?
(2) Can we decrease the current risks of disastrous food shortages?

Plant breeding is most often identified as a function of the agricultural system which aids in ensuring a sustainable supply of food. One of the biggest threats to this supply of food is massive crop failure due to adverse climate or pathogen attack. Crops which are genetically diverse are not as likely to be uniformly susceptible to adverse conditions or pathogen attack, and we look to our plant breeders and the plant breeding system to provide some of the answers to this issue.

GENETIC DIVERSITY AND VULNERABILITY

For many years, scientists and others familiar with the need to maintain a genetically diverse agriculture have attempted to deal with the problem of genetic erosion within the major food crops of the world.

Why do we need genetic diversity in our food crop? Modern agriculture is dependent upon a relatively small number of commercial crop species, each of which is dominated by a relatively small number of varieties. Our ability to sustain the current output of food to feed the current population relies on our ability to continue to produce these foods successfully. A lack of genetic variability or diversity within these crops means they can be highly susceptible to plant pathogens which could wipe out a significant proportion of the crop in a very short time. With world population increasing at a rapid rate, there is also increasing competition for land, and increasing pressure to produce more food. Thus there is a need to produce more food from less land. This means being able to produce increasing amounts of calories per acre, protein per acre, and other essentials per acre.

But high yield comes at a price. These high quality plants not only need more inputs, but tend to become increasingly uniform in genetic makeup as we continue to breed plants to meet singular needs. This uniformity makes plants increasingly vulnerable to pathogens and therefore threatens our food supplies.

While this cannot be shown with any degree of certainty, many epidemics that have been documented in the past may have occurred because of the lack of genetic diversity of the crop, for example, the coffee rust epidemic in Ceylon in the 1870s and the powdery mildew epidemics in French vineyards in the 1840s and 1850s. Examples closer to home are the red rust epidemics of wheat in the U.S. and Canada in 1916 and 1917, and the 1940s epidemic of Victoria blight in the U.S. oat crop. The most recent epidemic, and one that was almost certainly caused by genetic vulnerability, was the U.S. corn leaf blight of 1969-70. This epidemic resulted from the widespread use of cytoplasmically male sterile hybrid corn which went along with increased susceptibility to corn leaf blight. The disease swept up from Florida and into the Corn belt causing a loss of about 15 percent of the total U.S. corn crop. In some states it created losses of almost 50 percent of the corn crop in that state (NAS, 1972). The problem, however, was able to be corrected the following year, so for consumers the blight meant only higher food prices. It brought home to many, however, the potential seriousness of vulnerability, and the need for genetic diversity. It also brought home the lesson that single gene transfers, regardless of how beneficial, are as much a threat to crops as the more common forms of genetic uniformity.

How much genetic diversity is there? Unfortunately, there are no objective measurements of genetic diversity. One common method used is a sort of crop concentration ratio or crop variety share. Table 7.1 shows some crop variety concentration ratios for a few major crops.

TABLE 7.1
Crop Variety Concentration Ratios of Selected Seed Types: 1979-1980

	Percent of Total U.S. Crop Planted to:			
Crop	One Variety	Four Varieties	Ten Varieties	Twenty Varieties
Soybeans	15.5	33.6	53.6	72.2
Cotton	8.0	30.0	56.0	72.0
Wheat	6.3	21.3	40.2	55.8
Rice	29.3	78.3	99.0	99.9
Barley	20.1	49.0	71.9	83.9
Corn (by company)	36.9	60.5	72.8	79.7

Source: Adapted from Butler and Marion, 1983

However, it is not possible to determine at this stage what is unacceptably high, or what is a desirable level of variety concentration. More importantly, not only do we not know how much genetic diversity there is, we also cannot determine how much diversity we need. A recent survey of leading public and private plant breeders carried out by Duvick (1981) ascertained that

- the percentage of area planted to leading varieties is high, but concentration is less than in 1970;
- turnover of leading varieties is more rapid now than it has been in the past, and is expected to increase in the future;
- breeding pools are more genetically diverse now than in the past, and can be expected to be more

genetically diverse in the future;
- individual varieties will continue to be highly uniform because the costs of diversity outweigh the benefits of uniformity;
- genetic diversity is not enough by itself, i.e., genetic diversity does not prevent epidemics; and
- there is no cause to be complacent about the current levels of genetic diversity.

Let me briefly summarize the situation. Currently, we depend on a few crops to supply the world's food requirements. With increasing world population, and therefore increasing competition for land, there is a widespread belief that we need to increase our food production on less land. But breeding high yielding varieties of food plants can lead to inadvertent narrowing of the genetic base unless diversity is specifically bred into new varieties. We do not know how much genetic diversity there is currently, nor do we know how much we need.

This is essentially the plant breeding issue as it relates to food security. While there is a low probability of immediate threat, it is very difficult not to sound alarmist about these matters. After the 1970 corn blight the National Academy of Sciences did a study on the genetic vulnerability of our major food crops (NAS, 1972). They found that our major crops are highly vulnerable. In 1981, the U.S. Department of State and several other agencies sponsored a strategy conference on biological diversity. The basic message is still very clear--our major crops are highly vulnerable.

The remainder of this chapter examines the major facets of our plant breeding system which affect the genetic vulnerability of our major crops, considering first the availability and maintenance of germ plasm upon which a genetically diverse and secure agriculture is highly dependent.

GERM PLASM AVAILABILITY AND MAINTENANCE

There are two ways of maintaining germ plasm and making it available for innovative plant breeding:

(1) *in situ* - the development and maintenance of rural centers of genetic diversity.
(2) *ex situ* - the development and maintenance of germ plasm in specially controlled artificial environments.

It is generally agreed that long-term maintenance and development of exotic germ plasm urgently requires the conservation of natural tropical communities of plants *in situ*. The most important of these natural centers of genetic diversity are to be found in the twelve tropical and subtropical areas of the world known as Vavilov Centers. They include areas in Central America, South America, Northern and Eastern Africa, the Middle East, Southeast Asia, India, and China. Many of our natural parks and wilderness areas also provide centers of natural diversity. However, in the U.S. germ plasm is mostly developed and maintained ex situ--in gene banks such as seed storage repositories and vegetative propagation centers.

There are several problems associated with these germ plasm maintenance situations with respect to U.S. food security. Although in situ development and maintenance is clearly the more desirable of the two, the U.S. does not have any control over, and in most instances has limited access to, the Vavilov Centers of the world. The Vavilov Centers, in most cases, are tropical and subtropical areas in third world countries with high rates of population growth and extremely fragile environments. Many of these countries are engaged in clearing these areas of natural genetic diversity in order to grow crops to feed their increasing populations. And because of the fragility of their environment these tropical areas require relatively more land on which to grow their food requirements (Chang, 1970). In addition, only an estimated 10 percent of the earth's plant species have been evaluated in any way for potential usefulness to man. Due to the rapid conversion of tropical forests, it is estimated that as many as one-third of the species found in these tropical areas may become extinct during the next thirty to forty years (Dept. of State, 1982).

The ex situ situation, however, is not an ideal alternative. It is obvious that ex situ storage of germ plasm is relatively short term--on the order of 50-100 years. Because ex situ storage is artificial it is also subject to such dangers as failed technology, power-cuts, and accidents. The National Plant Germplasm System has been downgraded in the past and has consequently lacked the resources to develop and maintain gene banks. Only in 1983, for the first time, has the budget been increased to develop and improve the facilities for germ plasm maintenance, but the system is still fraught with problems. The National Seed Storage Laboratory in Fort Collins, Colorado is still understaffed and underfunded. User procedures are extremely poor and there are no formally established procedures to monitor the use of stored genetic resources. Nor is there any information feedback required by users. As a result, this type of system tends to encourage private ownership of germ plasm collections with the result that information and germ plasm exchanges are decreased. An additional problem is that the U.S. is extremely "gene poor." That is, there are very few natural sources of exotic germ plasm in the U.S. and therefore it must rely solely on other countries for its sources of exotic germ plasm. This places additional constraints on the system, particularly in cases where the physical scarcity is exacerbated by political barriers.

In summary, it would appear that the U.S. is not in a good position with respect to germ plasm availability and maintenance. Although the current system houses 500,000 accessions, the difficulties involved in documenting, evaluating, and maintaining these germ plasm resources are a major expense and a large undertaking. No doubt the stored germ plasm resources will expand, but it is unrealistic to expect these resources to provide an unlimited supply of exotic germ plasm in the long run.

I shall now turn to the plant breeding system in the U.S. to examine its adequacy in supplying the genetic requirements of U.S. agriculture.

PLANT BREEDING AND THE U.S. SEED INDUSTRY

The plant breeding system is made up of a large amorphous mass of public and private institutions. Most of the actual breeding work is carried out by a relatively small number of institutions, consisting of the 50-odd State Agricultural Experiment Stations throughout the U.S., the Agricultural Research Center of USDA at Beltsville, Maryland, and approximately 180 private companies which have plant breeding programs. The commercial seed industry, in addition to the private companies, consists of some 3,000-5,000 companies of various sizes which grow, condition, and distribute seed around the U.S.

Until recently, most new varieties of plants were released from public plant breeding institutions such as SAESs and USDA-ARS. However, several recent developments have influenced the incentive structures in the plant breeding system and this has created some institutional changes in the seed industry. For example, a series of technological advances have focused considerable attention on molecular and cell biology, and in particular on genetic engineering. A commodities boom in the early 1970s, and a dramatic increase in the cost of energy and energy-related products, have created a reallocation of resources in factor and product markets resulting in increased interest in cost-effective agricultural inputs such as pest resistant plants. Legislative changes such as the Plant Variety Protection Act and Supreme Court rulings on the Patent Act have added considerably to the protection available to developers of new life forms. These changes have created incentives for private industry to invest heavily in plant breeding R&D, resulting in a modest influx of new firms into the plant breeding arena and a substantial response of older, established firms in the form of a large number of protected variety releases(2).

The degree of merger and takeover activity as a result of these changes in the seed industry in the past decade has been the subject of some concern to the public as well as members of the seed industry. Activities have ranged from simple mergers of small seed companies who wish to expand operations and capture some economies of scale, to the takeover of several of the largest companies in the seed industry by large corporations, some of which are transnational conglomerates, and some of which are petrochemical and/or pharmaceutical concerns.

Acquisitions and mergers may contribute to increased market concentration in two ways. First, if a seed company acquires a direct competitor--that is, a firm that operates in the same relevant geographic and product market--concentration is increased in that particular market. This is particularly troublesome if one of the firms is a market leader, and the market is already relatively concentrated.

Even where the merged firms do not operate in the same market, concentration may be increased if there are benefits from large firm size; for example, the Royal Dutch/Shell group acquired five U.S. seed companies between 1973 and 1980. These companies now have access (at least potentially) to a much larger pool of capital to expand their plant breeding, growing, and marketing activities. Conglomerate companies have available to them competitive options unavailable to the smaller specialized companies. In the long run, the increased conglomerate power of some seed companies may result in increased market concentration.

From available information, however, national concentration levels to date in most seed species are relatively low, with the notable exception of corn. The expectation is that concentration of sales is increasing and will continue to increase in those seed species where plant breeding is largely concentrated in the private sector, and where privately protected varieties have a significant share of the market (Butler and Marion, 1983).

Another change which has occurred in the seed industry is a substantial increase in the price of seed since 1970 (Figure 7.1). This is hardly surprising since patent-like protection provides a firm with greater control over the price and output of its product. Some price enhancement is to be expected. The characteristics of most seed markets suggest that price increases have not been unreasonable or unjustifiable. Pricing discretion of seed companies is constrained by two important checks in the system. First, bin runs or farmer saved seed is a viable alternative for many who use open pollinated seed species such as soybeans, cotton, and most cereals. For other seeds such as clover, alfalfa, and many vegetables, however, bin runs

Figure 7.1

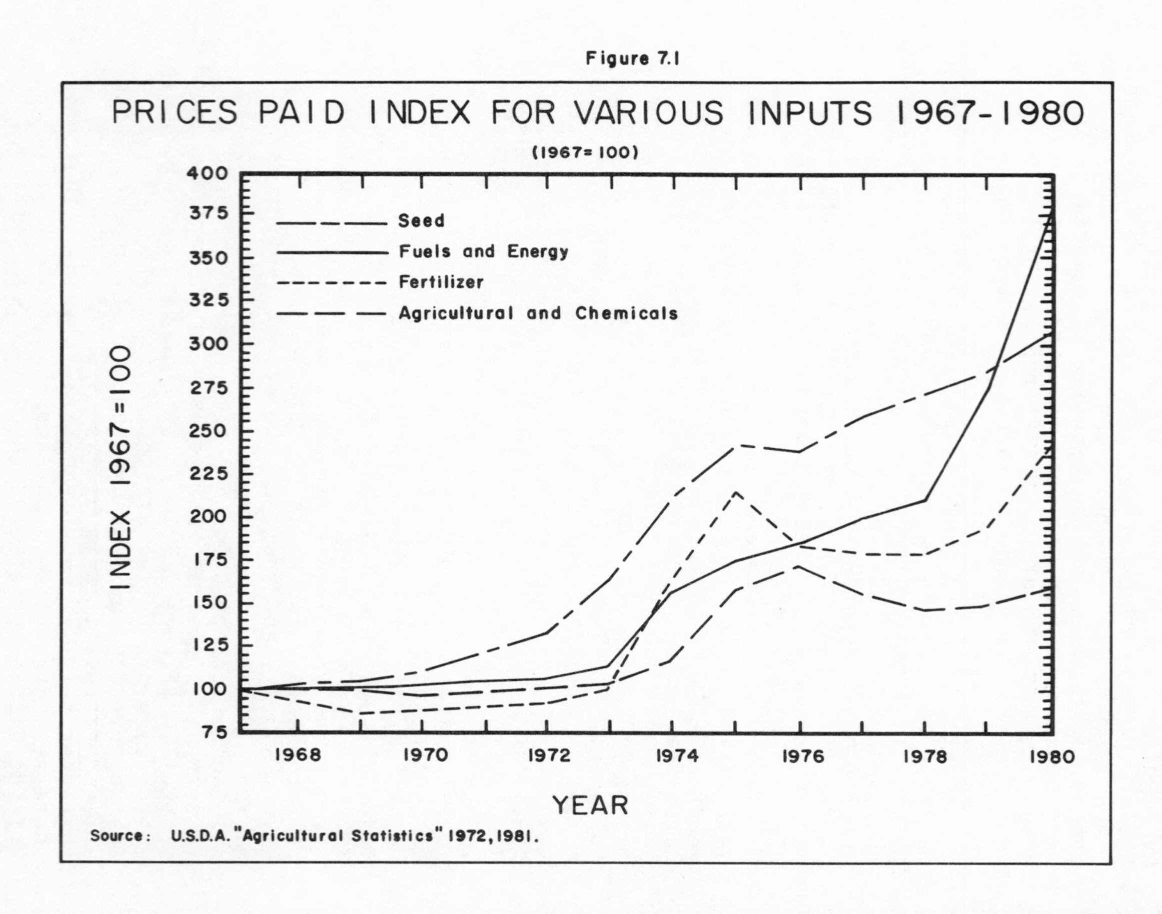

Source: U.S.D.A. "Agricultural Statistics" 1972, 1981.

are not feasible and the pricing discretion for seed companies is higher. The other check on the pricing of many seeds is the availability of publicly developed varieties.

The public plant breeding program in the U.S. serves as the most important competitive factor in the plant breeding system. The changes that have taken place in the seed industry have not had a serious impact on public plant breeding institutions to date, but it should be noted that there is a trend at universities toward more basic research, and most SAESs are currently in a financial pinch. It is imperative that the balance between public and private plant breeding be maintained at least at current levels to ensure the satisfactory delivery of the necessary genetic materials to agriculture. This is particularly true when an increase in genetic diversity is required to increase food security. Although most seedsmen realize the value of genetic diversity and the necessity for it, most private seedsmen cannot afford to breed for genetic diversity because no one will pay them for it. As long as the goals of plant breeding are determined primarily by the commercial feasibility of a variety, there is a strong incentive to breed plants with characteristics which are "saleable." This results in a tendency to inadvertently narrow the genetic makeup of a plant, thereby decreasing, rather than increasing, genetic diversity. Therefore, we must look to the public plant breeding institutions to provide the necessary competition and genetic diversity that may not be provided in a wholly private plant breeding system. Given the incentives created by the structural changes in the industry, the balance between public and private plant breeding may be difficult to maintain.

THE NEW BIOTECHNOLOGIES AND GENETIC ENGINEERING

The potential for the new biotechnologies, and specifically for genetic engineering, is expressed through an as yet undefined potential for increased control and improved techniques in genetically manipulating new plant and animal products. The potential is that it may be possible to bring together populations of genes for which there currently exist natural barriers through sexual reproduction (i.e., crosses between corn and soybeans, wheat and alfalfa, etc. are currently not possible by natural means). Hence it may be possible to engineer plants with such genetic characteristics as nitrogen-fixing abilities, pathogenic resistance, salt and drought tolerance, and perenniality.

The current state-of-the-art in agricultural biotechnologies can be summarized in two main points.

(1) There are new genetic tools and techniques available which may allow scientists to bypass some of the current barriers to increased productivity through plant breeding.
(2) To date, there have been few breakthroughs of any significant economic consequence.

The potential to increase productivity has created immense interest in developing the new biotechnologies by investing large sums of money in research. At the same time, however, there exists a great deal of uncertainty as to the outcome of these investments. What can be realistically expected, and what impact will these changes have on food security in the future?

There are, first of all, a number of technical constraints creating uncertainties for the future of agricultural biotechnologies. For example, the identification of genetic functions linked to agronomic traits is not fully understood because of the complex dynamic interaction between the genetic materials of various crops (Barton and Brill, 1983). The transfer of genetic material from one species to another is not fully understood, and cell tissue culture has yet to prove successful for major crops. These are the challenges which in all probability provide much of the impetus for the research investments which have occurred. Apart from the technical constraints, however, there are a number of economic or institutional constraints which will create repercussions throughout the economy should significant breakthroughs occur. The impetus of biotechnology and the constraints on its development are the forces which shape the development of technological change. It should, however, be recognized that discussion of these changes is essentially speculative. It hasn't happened yet!

I have already discussed the significant changes which have occurred in the structure of the seed industry, in part because of the developments which have occurred in biotechnology. Changes can also be expected to take place at universities, and specifically in the interrelationships between the university and the private sector. In general, industry is supportive of increased cooperation between private and public sectors, but the potential for conflicts of various types has prevented the total consummation of public and private biotechnological R&D. For example, conflicts may occur if private industry demands increased influence over public research agendas, and over the dissemination of research results which have proprietary potential. Currently, the demand for highly trained genetic engineering specialists by the private sector has created a shortage of trained personnel at universities, which could threaten the long-term viability of biotechnological research.

Another change which may take place if genetic engineering breakthroughs are realized is the pressure to change certain types of legislation. Health and safety standards of rDNA technology are strict, and it can be expected that some pressures will be brought to bear for the relaxation of these standards. Similarly, the current arrangements for transferring exclusive licenses on protected materials from universities to private enterprise are unsatisfactory from industry's point of view. The private sector can be expected to lobby strongly for some relaxation of these arrangements. Finally, the level of patent-like protection provided in the laws is apparently insufficient to provide the level of protection desired. This may lead to increasing pressures to change some aspects of the patent/protection laws.

Changes in agriculture and the farm and food system are extremely difficult to predict. Current expectations of new types of plant life which can, for example, biologically fix nitrogen, may have an impact on the fertilizer industry. Plant life which is adapted to pathogen resistance, pesticide immunity or other tolerances will impact on the costs of inputs. Plant life which is adapted to increased yield and disease resistance will have an impact on production, and so forth.

More important, however, is that genetic engineeriing may be used to increase the genetic diversity of our food crops. Unfortunately, this is by no means guaranteed, for several reasons. First, as mentioned previously, there are no objective measures of available or required genetic diversity. Second, since breeding is determined primarily by the "saleability" of genes, it would appear that there will be a commercial bias toward "saleable" genes and therefore away from genetic diversity. Finally, genetic diversity relies on the availability of a wide variety of germplasm. As noted earlier, germplasm may not be widely available in the future unless a considerable amount of public and private funds are spent to upgrade the national and international plant germplasm systems.

And so we have come full circle. What can be said about U.S. food security as it relates to plant breeding? I want to emphasize that it is not my intention to overstate the issues discussed in this article. A more correct interpretation of these issues would be that while the risk of disastrous food shortages due to genetic vulnerability is not at crisis levels, it is perhaps higher than it should be. There is no clear indication that there exist any forces, national or institutional, which create incentives to increase the genetic diversity of our food crops. The availability of exotic germplasm is limited by physical scarcity and institutional barriers. The U.S. germplasm maintenance system does not currently provide a satisfactory long-term alternative to *in situ* storage of germplasm. The plant breeding system, while still dominated by public plant breeding, is currently in a period of transition where the balance between public and private plant breeding is shifting toward the private sector. If the private sector dominates the plant breeding system, it is likely that commercial interests will be put before public interests. Genetic engineering is not expected to replace traditional plant breeding techniques in the foreseeable future, but because of its commercial and proprietary nature it is likely to be used first for commercial purposes.

This does not mean that we should push private enterprise out of plant breeding. It simply means that we must ensure that sufficient public expenditures are allocated to public plant breeding institutions to enable them to increase the genetic diversity of our food crops, to compete vigorously with private enterprise to ensure that farmers have satisfactory alternatives to commercially bred seed, and to perform the necessary role of providing checks and balances in biotechnology research.

NOTES

1. This assumes that the "death rate solution" by war, famine, or pestilence is an unacceptable and destructive solution to the problem of population growth, and that population control is an acceptable humanitarian solution to the problem.

2. Protection is afforded by the Plant Variety Protection Act of 1970 which allows plant breeders exclusive ownership and control of a newly developed variety (see Butler and Marion, 1983).

REFERENCES

Barton, K. A., and Brill, W. J. 1983. "Prospects in Plant Genetic Engineering." Science 219 (February 11):671-76.

Butler, L. J., and Marion, B. W. 1983. Economic Impact of the Plant Variety Protection Act. Madison: North Central 117 Monograph.

Chang, Jen-Hu. 1970. "Potential Photosynthesis and Crop Productivity." Annals of the Association of American Geographers 60(1). March.

Department of State. 1982. Proceedings of the U.S. Strategy Conference on Biological Diversity. November 16-18. Washington, D.C.

Duvick, D. N. 1981. "Genetic Diversity in Major Farm Crops on the Farm and in Reserve." Paper presented to the 13th International Botanical Congress, August 28, in Sydney, Australia.

Mayer, J. 1981. "Genetic Diversity: Strategic Significance and U.S. Opportunity." At Proceedings of U.S. Strategy Conference on Biological Diversity.

NAS. 1972. "Genetic Vulnerability of Major Crops." NAS Report. Washington, D.C.

Labor

8
The Labor Force in U.S. Agriculture

William H. Friedland

INTRODUCTION

An assessment of the agricultural labor force with respect to adequacy, sustainability, nutrition, and equity until the end of the present century is relatively simple, given the fairly extensive history of this labor force. If two fundamental historic policies of the United States are maintained, a hard-nosed assessment, eschewing rhetoric and ideology, will conclude that no problem is likely in the foreseeable future with the labor force with respect to adequacy and sustainability. On the other hand, the issue of equity will continue to be a major problem. Nutrition may become a problem but this issue is more difficult to assess with respect to the factor of labor. The two historical policies of the United States are: (1) the continued destruction of family-based units of production which, until recently, contributed the overwhelming bulk of labor to agricultural production; and (2) the maintenance of an open border with Mexico.

While some readers may disagree with this formulation of outcomes as policies, this chapter will contend that policy is what comes out of a decision-making process rather than the rhetoric of decision makers or even their intentions. This is especially true if the consequences of decisions are consistent irrespective of what is said or done in ostensibly "creating" policy through speeches, legislation, or other activities.

Thus, readers concerned with adequacy and sustainability of the labor supply over the next two decades but who are unconcerned about issues involving equity should proceed to the next chapter since this chapter concludes that, given existing policies, there will be only a few problems which will mainly involve unionization of some segments of the agricultural labor force.

After a brief description of the agricultural labor force based on available data, I turn to a consideration of major anomalies in the data which, I will contend, are artifacts of the methodologies used in defining the various segments of the agricultural labor force as well as of the methodologies used to collect the data. The major critique advanced is that the data on employed farm labor neither describe the characteristics nor their contributions to agricultural production. I center my critique on the process of economic concentration in agriculture and its concomitant effects on the structure of the labor force. I conclude this first section with suggestions for a different methodology to produce more accurate data. In the next section, the security of the agricultural system is examined in terms of the central concepts of this project, i.e., adequacy, sustainability, equity, and nutrition. The chapter concludes with an exploration of the reasons for the unreliability of data on agricultural labor. I argue that this is a product of structural defects in government mechanisms for data collection that occur because the supply of labor has, with few exceptions, historically been more than adequate and, as a consequence, no problematic has been defined. This has had the consequence that data collection is concerned more with myth-making and myth-sustenance than in describing social reality. While the issue of equity has always been--and continues to be--problematic, this issue has been of limited concern to a few individuals and organizations in the policy-making process, including unions of farmworkers, concerned with moral or normative questions. The inadequacies of the data, I conclude, are a function of the nonproblematic created by the continuous availability of a labor supply.

THE PRESENT LABOR FORCE IN AGRICULTURE

The present agricultural labor force consists of two basic segments: (1) nonhired or self-employed labor, consisting primarily of family-farm owners and operators and members of their families; and (2) hired or employed labor.

With the notable exception of the South, family labor has historically provided--and continues to provide, despite its decimation--the overwhelming contribution to production. The employed labor force, however, is contributing an increasing share. These trends are set out briefly in Table 8.1, which shows that "farm operators and their families still account for the largest proportion of labor used in agriculture, but hired farmworkers are providing a greater share of agricultural employment over time" (Smith and Coltrane, 1981:1-2). As the table shows, family and hired employment declined steadily until the early 1970s.

From 1970 on, family employment has continued to decline but hired employment has increased slightly.

TABLE 8.1
Family and Hired Employment on Farms

Year	Annual Average Farm Employment: Total	Family (Thousands)	Hired	Hired labor as a Percentage of Total Farm Employment (Percent)	Total Hired Farm Work Force (Thousands)
1910	13,555	10,174	3,381	25	NA
1920	13,432	10,041	3,391	25	NA
1930	12,497	9,307	3,190	26	NA
1940	10,979	8,300	2,679	24	NA
1950	9,926	7,597	2,329	23	4,342
1955	8,381	6,345	2,036	24	NA
1960	7,057	5,172	1,885	27	3,693
1965	5,610	4,128	1,482	26	3,128
1970	4,523	3,348	1,175	26	2,488
1971	4,436	3,275	1,161	26	2,550
1972	4,373	3,228	1,146	26	2,809
1973	4,337	3,169	1,168	27	2,671
1974	4,389	3,075	1,314	30	2,737
1975	4,342	3,025	1,317	30	2,638
1976	4,374	2,997	1,377	31	2,767
1977	4,170	2,863	1,307	31	2,730
1978	3,957	2,689	1,268	32	NA
1979	3,774	2,501	1,273	34	2,652
1980	3,705	2,402	1,303	35	NA

Source: Smith and Coltrane, 1981:3.

Two important demographic statements can be made about the nonhired segment of the agricultural labor force:

(1) The number of nonhired persons ("farmers") continues to decline. Figure 8.1 shows the process of economic concentration, the "iron cross" of U.S. agriculture: the continual decline in the numbers of agricultural production units and the increase in the average size per unit.

(2) National averages mask important tendencies. A tiny segment of agricultural units has been producing an increasing proportion of agricultural commodities: in

1974, the top one percent of U.S. farms produced 27 percent of food production; by the year 2000, U.S. Department of Agriculture (USDA) experts project that the top one percent of farms will produce 50 percent of production (McDonald and Coffman, 1980:8). Putting this another way, economic concentration in agriculture has begun to approximate that found in other economic sectors although concentration has been much delayed and probably will never be as notable as in some highly concentrated oligopolistic industries.

Economic concentration means that the character of the hired labor force in agriculture is changing. Describing this hired labor force becomes even more complicated than has been the case in the past as production patterns change. It is especially difficult to discuss national patterns when these are made up of variations in and between regions and different commodity systems.

Some of the difficulties are due to the character of and the changes in the agricultural labor market. Historically, much of the labor force was created through biological reproduction and inheritance, i.e., passing agricultural jobs from one generation to the next. Although much of the labor market in agriculture has been relatively stable, it has been so dispersed and individuated that it has been barely conceptualized as a labor market. In contrast, an important segment of the agricultural labor market, seasonal and migratory labor, has been of a secondary character, i.e., unstable and low paid (Doeringer and Piore, 1971:165-67). As agricultural development has occurred, conceptual and analytical processes have not kept pace, particularly in recognizing the enormous social and economic distinctions between different sectors. The class structure within U.S. agriculture has been of little concern, on the whole, to USDA and land-grant demographers, rural sociologists, or agricultural economists. With the exception of an important beginning by T. Lynn Smith (1969), relatively little attention has been given to issues involving class structure, stratification, ethnicity and race, or income distribution in agriculture. An understanding of the character of the agricultural labor force requires the incorporation of such factors.

This broader analytic framework must begin with an historical and structural perspective. Agricultural production involves a number of roles that were once, with some important exceptions, encompassed within a single social unit, the family. These roles include: owner, operator, manager, supervisor, and permanent and temporary worker. In the southern plantation, before and after the Civil War, some of these roles had become differentiated as distinct occupations (Fligstein, 1983); in most other sectors of U.S. agriculture, they

Figure 8.1

NUMBER OF U.S. FARMS AND AVERAGE FARM SIZE

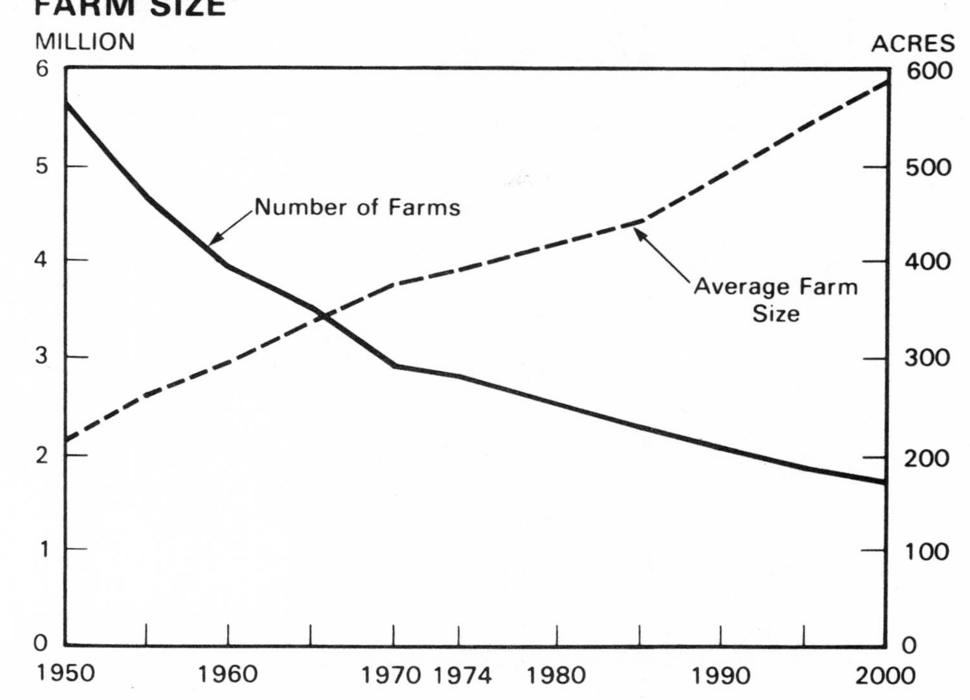

Source: McDonald and Coffman, 1980

had not. The present complexities of the agricultural labor force began to emerge with the turn of the century but trends accelerated with the Great Depression and the New Deal and in the aftermath of the Second World War. As economic concentration increased,

> hired workers accounted for about 23 percent of annual average employment in 1950, but by 1980 the proportion had increased to 35 percent. Furthermore, the rate of substitution accelerated slightly in the last decade (Smith and Coltrane, 1981:4-5).

What is even more important is that, as a small number of agricultural production units has grown beyond family-based operations, the numbers of their employees has increased. Thus, the number of agricultural production units with larger numbers of hired workers is increasing along with the increase in the proportion of hired labor in the total agricultural labor force.

Current methods for categorizing the hired labor force in agriculture are based on the number of days worked annually. The most recent definitions have been set out by Pollack:

> Casual workers (who work less than 25 days per year on farms) and seasonal workers (25-149 days) comprised 71 percent of all farmworkers in 1979, but accounted for only 27 percent of all person-days worked that year. Many casual and seasonal workers are students and housewives who work only a few weeks a year during harvest and other peak labor-demand periods. Others have nonfarm jobs and do farmwork to supplement their income. Regular workers (150-249 days of farmwork) and year-round workers (250 or more days) made up only 29 percent of all farmworkers but performed 73 percent of all person-days worked (Pollack, 1981:4-5).

Figure 8.2 summarizes data on the numbers of hired farmworkers and their respective contributions to agricultural production. Figure 8.2 shows that the secular trend since 1950 in the number of person-days worked and in the number of persons working has declined. The figure also shows that, while the number of casual workers is large in proportion to other categories of workers, their contribution in person-days worked is very small. Figure 8.2 also shows the increased significance, since the early 1970s, in person-days worked of regular and year-round employees.

There is little point in examining these data on demographic trends in the hired agricultural labor force in greater detail. For one thing, the data sources involve considerable anomalies.

Figure 8.2

Hired Farmworkers and Person-Days Worked

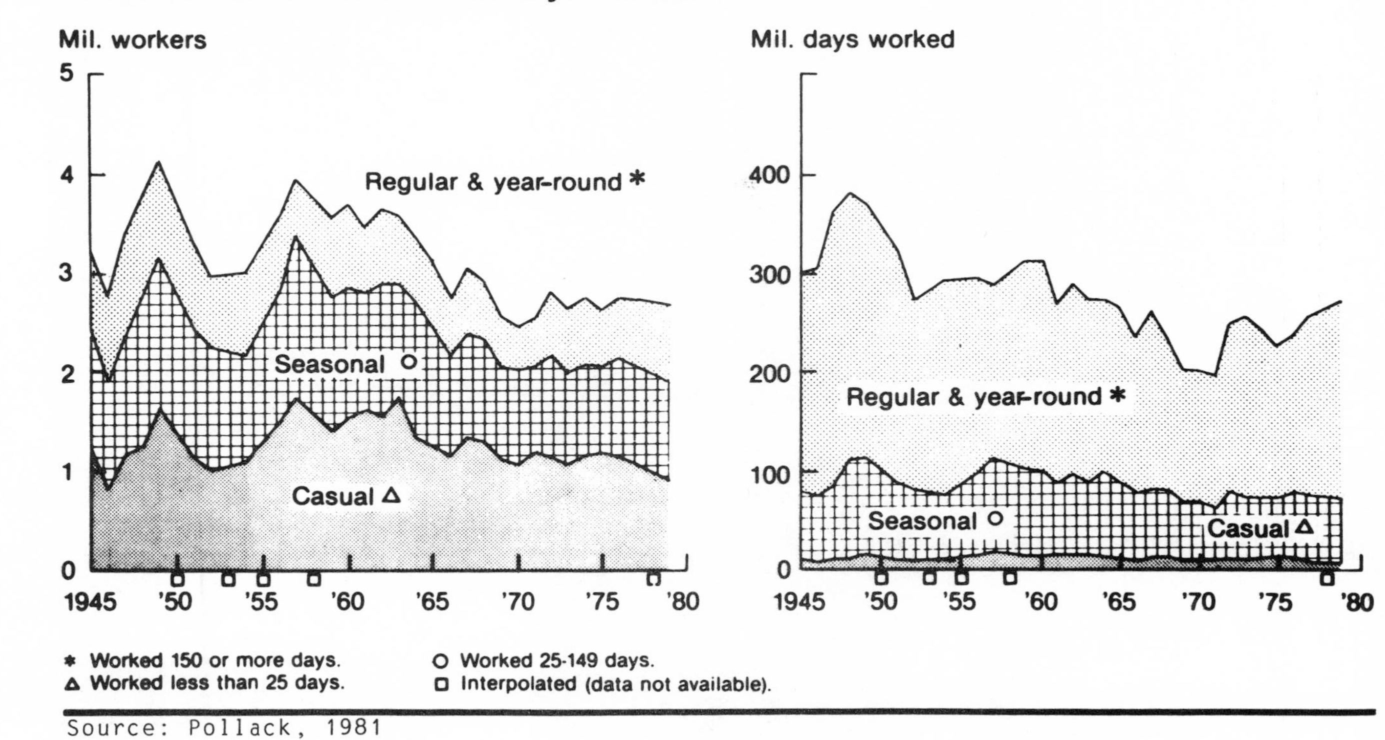

Source: Pollack, 1981

> Annual United States Department of Agriculture estimates exist for the number of hired farm workers in the United States by employment duration.... The [USDA] Statistical Reporting Service...also reports annual data on the number of hired and family farm workers. The USDA-SRS data differ from those of Pollack (Perry, 1982:677-78).

One way that some of the problems about the reliability--and therefore the utility--of the data concerning the hired labor force can be assessed is by examining some of the published reports of the USDA about the characteristics of this labor force. The first by Pollack (1981) is a publication issued regularly by USDA; the second by Smith and Coltrane (1981) is based substantially on Pollack but contains other information as well. Both reports show, for example, that the racial-ethnic background of hired farmworkers and migrants is overwhelmingly white (Pollack, 1981:7; Smith and Coltrane, 1981:10-11). Figure 8.3, taken from Smith and Coltrane, illustrates these data.

There are severe problems that must be resolved, however, at a common sense level when considering such data. This becomes clearest when examining the comparative lack of importance of Hispanic farmworkers. While a great many employed farmworkers in some states are certainly white, seasonally employed labor in many states is overwhelmingly Hispanic, primarily of Mexican origin. This fact fails to be communicated in USDA reports.

A grasp of this problem can be obtained by considering reports of the numbers of Hispanic farmworkers in several regions of the U.S. where it is generally known, by those who study farm labor, that significant numbers of workers of Hispanic--and more particularly Mexican--origin can be found.

Table 8.2, from Pollack, shows the numbers of farmworkers and Hispanics in two regions of the U.S. in which the numbers of Hispanics are common sensically known to be significant but where Pollack shows little or no Hispanic presence. Common sense or impressions, however, are unsatisfactory for assessing such matters. Table 8.3 shows the numbers of Hispanics in a single state, Florida, of Region IV (which includes Alabama, Florida, Georgia, Kentucky, Mississippi, North Carolina, South Carolina, and Tennessee).

Table 8.3 shows more than 19,000 "Tex. Mex." workers in Florida, constituting 37.4 percent of the hired labor force in that state. One important aspect of Table 8.3 is that the time period reported is the same (the end of 1979) as that from which the Pollack data have been derived. Thus we are confronted by an anomaly in which one agency of government, the USDA, basing itself on a survey conducted by another unit of government, the

Figure 8.3

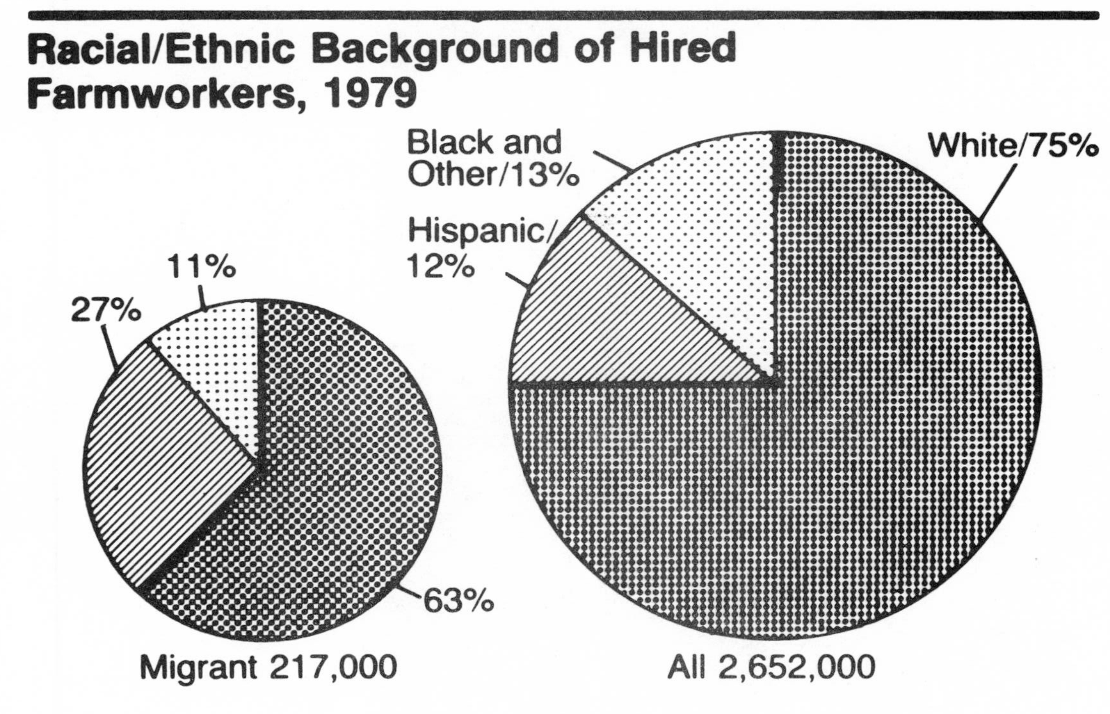

Source: Smith and Coltrane, 1981

TABLE 8.2
Numbers of Hired Farmworkers, by Ethnic/Racial Group, U.S. Regions IV and V, (000s)

Group	Region IV(a)	Region V(b)
White	400	480
Hispanics	3	0
Blacks and others	205	8
Total	608	488

Source: Pollack, 1981: Appendix Table 9:36.

(a) Region IV includes Alabama, Florida, Georgia, Kentucky, Mississippi, North Carolina, South Carolina, and Tennessee
(b) Region V includes Illinois, Indiana, Michigan, Minnesota, Ohio, and Wisconsin

TABLE 8.3
Domestic and Hispanic Seasonal Hired Workers in Florida, Period Ending December 31, 1981

Type of worker	Numbers	Hispanics	Hispanics (as % of Total)
Local	35,930	8,469	23.6%
Intrastate migratory	7,951	5,473	68.9
Interstate migratory	9,056	5,837	64.5
Total	52,937	19,779	37.4

Source: State of Florida, In-Season Farm Labor Report (Form ES-223), period ending Dec. 31, 1981, p. 1. The category named "Hispanics" is described in the original source as "Tex. Mex."

Bureau of the Census, reports a set of data. These data have little relationship to those reported by another unit of government, the State of Florida, functioning under the direction of yet another unit of government, the U.S. Department of Labor.

Similar anomalies, although not originating from government sources, develop when we examine research conducted in several of the states of Region V (which includes Illinois, Indiana, Michigan, Minnesota, Ohio, and Wisconsin). Slesinger (1979), for example, reporting on a study conducted in 1978, states that there were 4,000 migrant farmworkers in Wisconsin of whom 91 percent were of Mexican "heritage." Sutton and Brunner (1983), reporting on a study of migratory farmworkers in Indiana in 1979, provide an estimate of 6,000-7,000 workers of whom 90 percent were Mexican-Americans. Hintz (1976:24) reports on a study conducted by the Ohio Bureau of Unemployment in 1972. This study reported 35,000 migrants in Ohio of whom 76 percent were Hispanic and 71 percent were of Mexican origin. Similarly, a 1971 Michigan report states that "the majority of migratory workers in Michigan are of Mexican-American descent" (Michigan Employment Security Commission, 1971:28). These reports, from different states of Region V and at different time periods, demonstrate the importance of seasonal workers of Mexican origin, a fact consistently underreported in USDA publications.

The systematic underreporting of seasonal farmworkers of Mexican extraction poses serious questions for those who study the agricultural labor force. While it is common knowledge, I would assert, among those of us who research this topic, that farmworkers of Mexican origin are now found pervasively at most locations in seasonal agricultural employment, the adherence to outmoded and improper methodologies fails to reflect this social reality. How can this be accounted for? Why is it that the main survey method used remains focused on households (a reporting unit that creates exceptional methodological problems because of the character of seasonal agricultural employment) with the survey being conducted during December (a period when agricultural employment is at a minimum)? Facile accusations about institutional racism might be levied; a more likely case is that government bureaucracies have become remote and indifferent to segments of the agricultural labor force and that, because there is an endemic oversupply of labor, there exists no problematic for policy formation. These are matters which need not preoccupy us here; I will return to them in the conclusion of this chapter.

Some idea of the contribution which Hispanic farmworkers make to agricultural production can be obtained by examining Figure 8.4. This shows that 50 percent of Hispanics work 150 days or more in agriculture in contrast to lower percentages of white or Black workers.

Data collection on farmworkers is unquestionably difficult. They constitute a difficult social category to "capture" since many farmworkers are socially hidden or invisible (Nelkin, 1970:51-70). But data collection and publication appear to be almost deliberately obtuse on this subject. Perhaps the most systematic analysis of procedures and methods has been undertaken by Lillesand, Kravitz, and McClellan (1977) in an

Figure 8.4

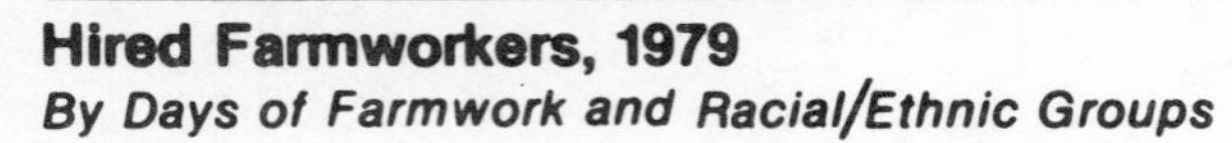

Hired Farmworkers, 1979

By Days of Farmwork and Racial/Ethnic Groups

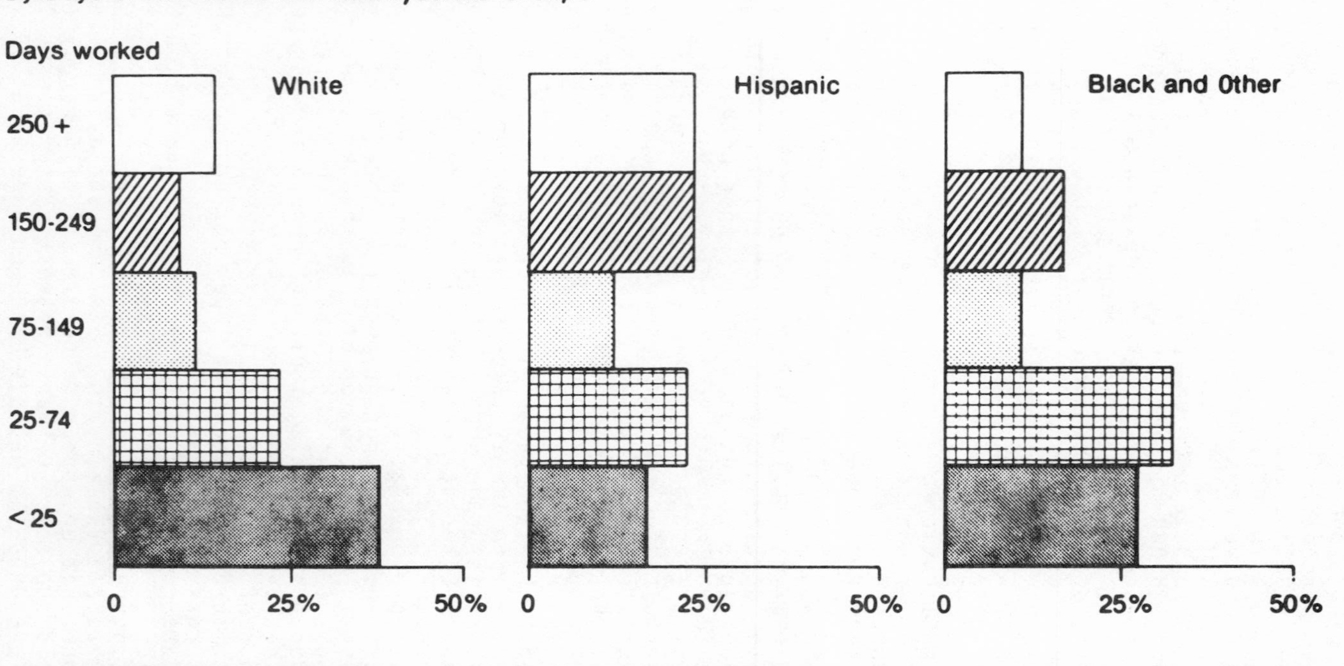

Source: Pollack, 1981

TABLE 8.4
Existing Sources of Farmworker Data, 1977

A. U.S. Department of Commerce, Bureau of the Census
   1. Decennial census
   2. Agricultural census

B. U.S. Department of Agriculture
   1. Current population survey (conducted by the Census Bureau)
      ---- report: The Hired Farm Working Force of 19xx
   2. Farm labor
      ---- quarterly report of the USDA Statistical Reporting Service

C. U.S. Department of Labor
   1. The In-Season Farm Labor Reports
   2. The Employment Security Automated Reporting System
   3. CETA 303 Program Reports
   4. Certified numbers of foreign workers

D. U.S. Department of Health, Education, and Welfare
   1. Office of Education, Migrant Programs Branch
      ---- reports numbers of school-age migrant children
   2. Migrant Health Program
      ---- report: Migrant Health Program Target Population Estimates

E. Ad hoc studies, mostly by states

Source: Lillesand, Kravitz, and McClellan, 1977.

unpublished study. This study shows that four departments of government (Commerce; Agriculture; Labor; and Health, Education, and Welfare), including various agencies within each, as well as various ad hoc agencies, provided data about farmworkers.

Space limits preclude a detailed recital of the methodological inadequacies found by Lillesand, *et al*. If we consider, however, sources of error in the USDA *The Hired Farm Working Force* reports, we can obtain some idea about the compounding of errors in data collection about the agricultural labor force. The authors point out, for example, that

> ...less than 300 migrant households provide data on what USDA estimates to be a population of around 200,000 workers not counting dependents. In 1974 it was 209,000; in 1975 it was 188,000. The difference was accounted for by a USDA researcher as due to "statistical variability".... There are several

> causes for the low reliability of USDA-CPS data. First, the small population yields a very high standard error. Second, all of the problems associated with the decennial census in finding migrant housing units apply here. Third, the survey is done in December when a large number of migrants, including U.S. nationals, are living in Mexico.... Fourth, the geographical distribution of migrants is irregular. During December migrants are concentrated in a few identifiable areas of the country. The CPS survey, however, is designed so that theoretically any housing unit in the United States whether in Maine, Indiana, or Arizona, has an equal chance of being included in the survey.... Fifth, the USDA definition excludes migrants employed in food processing... (Lillesand, _et al_., 1977:16-17, 19).

Similar criticism of equal trenchancy and appropriateness is provided by Lillesand, _et al_., for the other systems of data collection on migrant and seasonal farmworkers.

This brief and incomplete treatment of the data problems suggests some of the problems that will have to be overcome in developing data that reflect the social realities of the hired labor force in agriculture.

Before turning to an assessment of adequacy, sustainability, and equity, I want to suggest some issues that must be dealt with as part of a serious approach to policy formation in understanding the hired labor force in agriculture. Two arguments underlie the analysis that will be offered. First, the agricultural labor force has been conceptualized incorrectly. Second, data collection methods are so flawed that it is currently impossible to develop an understanding of the agricultural labor force that will lead to adequate policy formulation. These two issues, I contend, are interrelated. By failing to conceptualize agricultural labor appropriately, erroneous data get collected.

These two arguments will be elaborated in four topics: the importance of corporate units in agriculture and the increased size of production units; the attendant increase in the numbers of year-round employees and the increase in the numbers of production units employing ten or more workers; the development of a significant new category within agriculture of managers and supervisors; and finally, the need for new categories to describe farmworkers.

The Importance of Corporate Units and Increased Size

The _1974 Census of Agriculture_ summarizes the importance of corporate growth in agriculture:

> In 1974, corporations operated 1.7 percent of the farms and accounted for 18 percent of the value of agricultural sales. The average value of sales for corporation farms was $515,030 compared with an average of $39,571 for the three other types of organizations (individual, partnership, and "other"). Corporations tended to specialize in the production of certain commodities. They sold 60 percent of the nursery products; 37 percent of the vegetables; 33 percent of the cattle and calves; 32 percent of fruits, nuts, and berries; and 28 percent of poultry and poultry products in 1974.... The significance of corporate farm commodity sales is dependent both upon the percentage of a given commodity produced by corporations and upon the relative proportion of total sales accounted for by that commodity. For example, corporate sales of cattle and calves are of particular importance not only because they constitute 33 percent of total cattle sales, but also because cattle sales account for 22 percent of the value of all agricultural products sold. As a consequence, the value of cattle sold by corporations account for 7.4 percent of total agricultural sales by all farms (Bureau of the Census, 1978:4).

Although much of the concern with corporate agriculture is with "publicly held" in contrast to "privately held" corporations(1) (Bureau of the Census, 1978:3), increasing concern should be focused on the latter as well. As far as government agencies are concerned, there is apparently little to worry about with respect to the latter: "The incorporation of family farms to facilitate farm transfer from one generation to the next or to provide greater management flexibility is usually not viewed as a threat to the 'family farm'" (Bureau of the Census, 1978:1).

There should be increased concern about this latter phenomenon, however. There are indications that privately held family corporate farms are becoming increasingly important in numbers and in their contribution to total agricultural production (Time, Nov. 6, 1978:92-102). Evidence from a variety of sources indicates that some privately held corporate units are of significant size. Thus, for example, the original farm of the Garst family of Iowa was only 200 acres and "run down"(2) but now consists of 8,000 acres raising seed corn and cattle and a business that has been integrated backward and forward into grain-elevators and storage, machinery manufacturing, agricultural chemicals, and five banks and an insurance company (Time, Nov. 6, 1978:101). This Horatio Alger success story has been repeated by J. R. Simplot, who made his fortune of an estimated $500 million selling frozen french fries from his extensive potato acreages in the northwest (U.S. News and World Report, May 31, 1982:58-59). While some originally family-based units have become publicly held corporations, others have undoubtedly remained private.

The mushrooming during the 1970s of center-pivot irrigation systems in states such as Nebraska also indicates the presence of another type of corporate entity in agriculture (Center for Rural Affairs, n.d.). In these cases, agricultural entrepreneurs and promoters created tax shelters for urban investors. The enterprises buy land with capital invested by nonagricultural investors, put it into production, hiring managers and employees as necessary. While little is known systematically about the phenomenon as a whole in the U.S., the Nebraska study indicates that this corporate phenomenon deserves analysis, if only to determine how widespread it is in other circumstances(3).

These examples indicate that, beyond tendencies toward reducing the numbers of farms and the increase in their size, corporate organization is becoming manifest in a variety of ways. There are vertically integrated or conglomerated agricultural subsidiaries of large corporate enterprises as, for example, Tenneco's agricultural subsidiary Tenneco West. There are corporate ventures under the direction of entrepreneurs or large growers to utilize surplus capital available from outside of agriculture, as manifested in many of the Nebraskan center-pivot corporations or the land-owning dummy corporations formed by Giffin in California to conform to the Reclamation Act limitations. And finally, there are at least two types of family corporations, i.e., those that have already grown substantially beyond the capacity of a single family to handle all activities (with the concomitant necessity of employing significant numbers of managers, supervisors, and employees), and those in which legal incorporation is undertaken to resolve issues of intergenerational transmission, particularly with respect to the effects of taxation on inheritance. The effects of incorporation of family-based units on the reproduction of the "family-farm" stratum represents an area badly in need of research (Friedland, 1982:600).

Systematic data do not exist about the various types of larger and corporate units, but the illustrative material suggests the growing importance of large employers of farm labor. It also suggests that increased attention must be given to the analysis of larger farm units, if only to understand the changing character of the employed labor force in agriculture.

The Increased Importance of Larger Employing Units

Impressionistic data on larger agricultural

production units were noted in the previous section. Evidence on agricultural employment available in the 1969 and 1974 Censuses of Agriculture support these examples quantitatively. Table 8.5, derived from the 1974 Census of Agriculture, summarizes the argument. The units reported in the table are, sequentially, the three states which have the largest numbers of agricultural hired workers working 150 days or more annually in California, Texas, and Florida, and the largest employing state within each of the regions included in the Census of Agriculture. Table 8.5 shows that, in every state reported except one, the average number of employees per farm increased in the 5-year reporting interval. The single exception is Connecticut. Table 8.5 also shows that the greatest increase in the percentage change for the U.S. and for each of the states shown has been in the categories of 5-9 workers and 10 or more workers. Individual exceptions are scattered through the table but the trend is clear.

Table 8.5 must be treated with caution because it shows the changes occurring over a 5-year period. Comparability for earlier periods is not feasible because of inconsistency in data reporting. Data from the last Census of Agriculture in 1978 were not available when this chapter was being completed since the appropriate report had not yet been published.

The Need for New Census Reporting

The changing structure of agriculture provides some indication of the need for (1) new reporting categories of agricultural employees as well as (2) different methods for data collection. New categories are necessary since present categories fail to grasp the social realities of U.S. agriculture. Different methods of data collection are necessary if an adequate statistical picture is to be constructed of the hired labor force in agriculture. As pointed out earlier, the major categories presently utilized to characterize the hired labor force are based on the number of days employed in agriculture. This method yields four categories of workers: "year-round" (250+ days per year), "regular" (150-249 days), "seasonal" (25-149 days), and "casual" (less than 25 days).

Data collection and the method of reporting for farm units (i.e., farm owner/operators) appear, to this writer, to suffer only "normal" problems of data collection. Information on this category of the labor force in agriculture seems to most scholars to be adequate; hence, no suggestions are made for change.

For the hired labor force, four new categories are suggested:

TABLE 8.5
Farms, Hired Workers, Average Number of Hired Workers Per Farm, and Percentage Change in Hired Workers for Selected States, 1969-74. (Hired Workers Working 150 Days or More.)

	Number of Farms		Number of Workers		Average Workers/Farm		Percentage of Change, 1969 to 1974						
									No. of Workers Employed/Farm				
State	1969	1974	1969	1974	1969	1974	No. Farms	No. Workers	1	2	3-4	5-9	10+
U.S.	247,642	233,093	654,370	712,715	2.64	3.05	-5.9	+8.9	-20.1	-2.5	+11.7	+23.4	+23.0
CA	17,436	15,856	105,381	136,216	6.04	8.74	-9.1	+29.3	-28.1	-14.1	+1.0	+19.5	+26.5
TX	26,385	20,249	55,015	53,106	2.09	2.62	-23.5	-3.5	-33.8	-14.6	-3.4	+22.7	+25.4
FL	5,197	4,530	42,804	47,443	8.24	10.47	-12.8	+10.8	-25.8	-10.8	-0.4	-12.5	+4.2
CT	1,004	772	7,251	4,976	7.22	6.54	-23.1	-31.4	-36.8	-19.3	-9.7	-3.6	+2.9
PA	8,064	6,876	18,151	20,654	2.25	3.00	-14.7	+13.8	-25.3	-7.8	+18.3	+22.9	+40.2
WI	8,942	10,112	14,588	19,259	1.63	1.90	+13.1	+32.0	-5.0	+43.4	+115.3	+101.0	+32.9
IA	8,847	10,053	12,622	16,849	1.43	1.68	+13.6	+33.5	+0.3	+49.2	+100.0	+139.8	+55.6
MS	6,313	4,717	20,603	18,641	3.26	3.95	-25.3	-9.5	-35.4	-28.8	-24.0	+3.6	+5.3
AZ	1,981	1,682	14,676	12,755	7.40	7.58	-15.1	-13.1	+34.7	-19.1	-9.4	+16.5	+6.5

Source: Calculated from Bureau of the Census, 1974 Census of Agriculture, Vol. II, Part IV, Table 12, page II-9.

Managers and Supervisors. The emergence of a new category of managers and supervisors is indicated not only by impressionistic evidence but by the increase in the size and scope of production units and in the numbers of employed workers. As the numbers of workers increase per unit, it becomes necessary for farm owner/operators, as well as any corporate units involved, to focus attention on farm management activities and the supervision of the employees. The category of managers and supervisors is presently not reported; it simply does not exist in the present data collection system.

There will be some conceptual and legal difficulties in delineating the supervisor category. In California, there have been some problems with the delineation of those eligible to be included in a bargaining unit under the state Agricultural Labor Relations Act. While, in many cases, the delineation of managerial employees is easy, in others, the question of the appropriate definition of supervisory employees has created difficulties. This is because some workers function as year-round employees for much of the year but are given supervisory duties over seasonal workers during harvest periods.

Permanent, Year-round Employees. As farm production units increase in size, the number of workers employed as regular, year-round employees increases. The employment characteristics of these workers are distinctly different from those of managers and supervisors, although some of them may become supervisors on a temporary basis during the harvest peak-season employment of temporary workers.

Career Farmworkers. These are farmworkers who derive the majority of their income, including that of members of their families, from agriculture and agriculturally-related employment such as food packing and processing. My research, as well as that of others, indicates that these workers are preeminently Hispanic workers presently reported as "seasonal" workers or as "casual" workers in USDA reports. Since these workers are defined either by a census of households or by periods of employment with a single employer, the present reporting system fails to capture the fact that they may work for many employers, whether or not they are migratory, during the annual cycle. Yet they, and members of their families, contribute significantly not only to person-days in agriculture but also to seasonal employment in agriculturally related employment such as food packing and processing.

Casual Farmworkers. These are workers who are presently employed for relatively short periods in agriculture, i.e., those workers who work fewer than 25 days total per year in all agricultural or related employment. These workers contribute only a small segment of the total number of person-days required for overall production but are important in flash-peak harvests and in a great many localized production circumstances. Casual workers can be divided into two subcategories: regular casuals, those who work in agricultural or related labor for extended periods of time, say more than four years; and temporary casuals, those who work for one season or several but do not stay more than three years.

The Need for New Methodology

I have indicated the need for new methodologies of data collection if a realistic picture of the employed labor force is to be developed. The present methodology of the Census of Agriculture which uses the household as the unit for sampling can continue to be utilized to develop information on managers and supervisors and permanent workers, since these workers usually have stable residences which are not systematically underenumerated. However, utilizing households or employing units as units of analysis to develop data on what the USDA calls "seasonal" and "casual" workers and what I have named as "career farmworkers" and "casuals" is unsatisfactory. If data are to be developed on these workers it is essential that a new methodology be developed which makes the worker the key unit reporting.

This can be accomplished in several ways but only one method is possible as a practical and inexpensive solution, although it, too, has methodological problems. While it would be feasible to develop data on career farmworkers by direct interview, as a practical matter this has methodological difficulties and is very expensive. Its major advantage is that the social reality of the family unit could potentially be captured as a data source which otherwise will be lost.

The available potential solution is to utilize social security reportage as the means of locating agricultural workers in their employment and to determine the number of days they are employed in agriculture and related occupations. With social security information, it is feasible to determine the number of days workers are employed in agriculture and related employment as against nonagricultural employment. It thus becomes possible to determine the category of career farmworker and to distinguish this category from casuals. Through this method, better data can be collected on the two categories of farmworkers. The career workers are of critical importance because of the number of person-days they contribute to agriculture; the casuals are important because of their sheer numbers and their significance in local harvest situations(4).

THE AGRICULTURAL LABOR FORCE AND FOOD SECURITY

Turning to the issue of food scarcity, utilizing the stratified categories of the hired labor force that have been suggested, it should be recognized that each stratum makes distinctive contributions and receives benefits from the agricultural production system differentially and unequally. Translating factors such as adequacy, sustainability, or equity, into criteria relating to the five categories of labor, it should be noted that adequacy and sustainability involve the degree to which requisite supply and skills are available for each category. The issue of equity translates into income distribution, perhaps, better than any other single criterion.

Let me begin by asking the questions: To sustain the food system, what factors influence the flow of labor to each of the five categories, to family based farms as well as to the four suggested types of hired labor? What systems exist for the reproduction of each category? I argue that adequacy and sustainability of family farmers poses no problems other than the continuing historical one, what to do with the numbers of such farmers _unwanted_ in agriculture? In contrast, there are various problems for each of the other four categories of workers, especially concerning equity, and especially for the permanent, career, and casual workers.

Adequacy and Sustainability

Family farmers are present in sufficient numbers to continue to sustain the food system. Although their numbers will continue to decline (McDonald and Coffman, 1980), there will be more than enough to sustain the declining sector of family farming _and_ to provide numbers for the manager-supervisor category which can be expected to grow in significance. While technical and managerial skills necessary for agriculture continue to increase, biological reproduction within the family and social reproduction in the land grant colleges of agriculture are more than adequate to train future family agriculturists. In addition, with the decline in the numbers of production units, there will continue to be a surfeit of this type of labor and it will continue to move into part-time agriculture (with the main base of employment off the farm) or out of agriculture.

Managers may pose a motivational problem. Many managers are currently trained by colleges of agriculture where an important part of training is not appropriately organized for the fact that students will be managers rather than farm owner-operators. Agriculture colleges are oriented ideologically to family farm ownership and operatorship even if this has been a declining pattern. The problem of motivation--which includes but is not limited to financial rewards--of managers trained through land grant education may surface sometime in the future. Most likely, this will not be a serious problem since those entering managerial positions will learn, within relatively short periods of time, present economies of ownership and managership and accommodate to their employee-managerial status.

Permanent employees present a complex of problems. Here I report on the basis of personal observation, discussion with knowledgeable observers, but with no basis in any systematic analyses which, to my knowledge, do not exist. At present, recruitment and training of such workers are handled unsystematically and few systems of formal training exist. The pool from which permanent workers are drawn, i.e., the supply of such labor, remains substantial and, in this sense, employers (and managers) have a large potential supply of such workers. While it is theoretically feasible that some of the permanent workers might come from family farmers who can no longer "make it" in agriculture, a greater probability is that such farmers will "skid" out of agriculture rather than remain within it as employees--although some may do so. More likely, the supply of permanent workers will be drawn from career farmworkers for whom such employment constitutes upward mobility.

A major problem with permanent workers may develop with respect to training since the capacity for educating them for a range of agricultural activities is presently limited. Permanent workers require a variety of technical skills ranging from driving abilities and tractor maintenance to pruning or other technical procedures. They must also have some interpersonal and supervisory skills to work with other people including their own supervisors and seasonal workers whom they supervise in peak employment periods. At present, few organizations provide technical skills for this category of worker. In some places in California but also recently at the national level, recognition of the changing character of labor and the importance of employees as a factor in agricultural production has given rise to an emphasis on the need for _personnel management_ training. I shall return to this topic below. This more systematic focus on personnel management stands in sharp contrast to the ad hoc arrangements for other training of permanent workers in agriculture. No systematic assessment of this process exists, to my knowledge, either nationally or for the Sunbelt, where such employment is most important. Many ad hoc arrangements have developed. Some community colleges or high schools offer training programs for tractor drivers or longer classes on tractor maintenance and repair or other subjects. Few programs have been initiated to deal with supervisory or interpersonal skills of permanent workers. My impression is that this area remains largely untouched.

Thus, sustainability and adequacy of personnel and skills for permanent employees remains in question. The character of training at the community level indicates the strengths and the weaknesses of the present system. On the one hand, local educational systems (including high schools and community colleges) are often responsive to the needs of the local labor market and will mount classes accordingly. On the other hand, there have to exist some coherent local forces recognizing the need for such training. Since agricultural organizations and individuals historically have not understood the significance of labor as a factor of production(5), the lag time in creating training programs for permanent employees may be substantial.

To summarize, for the moment and for the foreseeable future, the supply of permanent workers does not appear to constitute a significant problem. The development and training of such labor, the upgrading of technical abilities to confront new skill requirements, however, remains uncertain.

Career farmworkers and casual workers manifest some similarities and some differences. There is little likelihood of a deficiency in the supply of seasonal labor to agriculture. In this respect, the U.S. and Mexico constitute a single labor market(6). Although there is a border and legal limits on the movement of persons, in a sociological sense the border is fictional. Workers fitting the requirements for seasonal labor are available in unlimited quantities in U.S. agriculture (although local constraints occasionally impede their flow).

Assessing the issues involved with one segment of the career farmworkers--immigrants from Mexico--poses considerable difficulties, as the debate over immigration policy and the adoption of the Simpson-Mazzoli bill indicate. At present, different segments of the U.S. population benefit from the massive infusion of Mexican workers, whether to farm labor or to other employment and whether such labor is present "legally" or "illegally." As Thomas (1981) has demonstrated, citizenship status impinges on all aspects of the employment relationship, making workers vulnerable to their employers. This has not only benefitted agricultural employers; as those employers continually emphasize, reduced labor costs benefit consumers (although the extent of consumer vs. grower benefit has not been demonstrated). Agricultural employers have shown mixed feelings about the proposed legislation. On the one hand, they are enthusiastic about a revived bracero program; on the other, they do not want to be held responsible for the employment of undocumented workers. While the complexities of the Simpson-Mazzoli bill and its consequences cannot be treated here, neither the adoption of the bill _nor_ its defeat promises to produce any significant change in the _de facto_ character of the integrated labor market between the U.S. and Mexico. Whether present as legally established noncitizens ("greencarders"), temporary legal workers (as in the present H-2 program or in a resuscitated bracero program), or as undocumented workers is the main consideration, _not_ whether there is an adequate supply with adequate skills. No significant change can be envisioned in supply until Mexico's economy has caught up with its birth rate or political relations between Mexico and the U.S. change.

While the skills required by career farmworkers, whether stable or migrant, will continue to remain relatively low, some workers will require increased skill levels. As capital substitution for labor continues, the skill requirements of seasonal labor will continue to decompose, to be reconstituted as a small number of relatively easily learned motions requiring little exercise of judgment. As has occurred elsewhere in industry with increased capitalization, skill levels for those at the bottom will decrease (Braverman, 1974). The situation in agriculture seems certain to follow suit (Perry, 1982). Problems of equity for these categories of workers will assuredly continue. For a small number of workers, however, the operation of complex equipment will require an upgrading of their skills.

Casual workers will undoubtedly continue to be produced locally. The availability of such workers depends on the character of the labor market. Some regular casuals will have high investments in real or pseudo-seniority or in personal relationships with farmers. Far more important, especially to attract temporary casuals, is the question of rates of pay, availability of nonagricultural employment, dangers and hazards involved, the pleasantness and unpleasantness of work, etc. These issues are difficult to assess on a national basis.

Equity

Equity issues vary greatly among the strata of the agricultural labor force. Put simply, family agricultural units that survive and grow will not experience "equity problems" whereas those that fail will experience crisis. Of the strata within the hired labor force, managers will benefit most (other than agricultural capitalists) from economic concentration. Permanent employees will find themselves in constant struggle over issues of equity and will probably begin organizing unions or other protectionist organizations. Career farmworkers will probably continue to suffer most from inequities of income distribution although they will probably participate in attempts at organizing.

Family-based units present a most difficult issue with respect to equity considerations. Agriculture continues to experience endemic crises during which protests get reported in the press. Farmers for whom the system has proven to be "equitable" ("fair") are those who continue to grow and get larger whereas the system is regarded as inequitable ("unfair") by those who are being put out of agriculture.

There are complex and difficult problems. If the issue is simply survival (i.e., the ability to continue in agriculture), many individual family farmers are currently threatened. But many farmers are threatened less by survival than from beliefs encouraged by the USDA, the land grant colleges, and the political system, that they are entitled to high rates of return for their management abilities and capital investments. Since agriculture is no longer a way of life as well as a basis for economic sustenance but has become a business, low returns to farmers for management and capital have become an increasing source of dissatisfaction.

The problem of equity will continue to represent a sore on the body politic. The number of farmers involved, however, has now dropped to such a low level that those that will be affected in the future are relatively small in number. The problem will remain, in other words, but it will be less intense nationally. It will continue to have important localized effects in some states, particularly in the midwest, still the bastion of U.S. family farming.

Managers and supervisors will experience the greatest benefits (other than agribusiness employers) in the developing system of economically concentrated agriculture. Because the cost of managerial labor is unlikely to constitute a significant segment of total costs of production, managerial employees will be relatively well compensated and will experience benefits approximating managerial levels outside of agriculture. While management employees cannot be expected to reach remunerative levels equivalent to those found in industry, there will be some relationship with other managerial categories(7).

Permanent workers will, in some respects, be similar to managers in that the costs of their labor as part of the total costs of production will not be very high. However, permanent workers cannot expect to be adequately remunerated until they generate labor unions or other protectionist organizations. Permanent workers remain insufficiently understood or appreciated either by agricultural capitalists or managers. Experience indicates that they will win appreciation--and equity--only through organizing. Since they play a pivotal role in production, they can be expected to experiment with organizational forms of one type or another, including (but not limited to) unionism.

Career farmworkers can be expected to experience continual inequity in the distribution of income in agriculture. Although some limited farmworker unionism has been successfully established in several places (California and Hawaii), career farmworkers, whether stable or migratory, will remain in oversupply and their skills and abilities will remain inadequately rewarded(8). From time to time, as has occurred in the past, changes in political climate may produce legislation or welfare benefits for seasonal workers but the inequities of the present system can be expected to continue. Until some fundamental change occurs in the labor supply, cutting off Mexico (or some future location) as a continuing source of agricultural labor, the system will continue to be inequitable for career farmworkers. Only union organization of permanent and career farmworkers will help. But seasonal workers have historically suffered in the U.S. and there is little indication of changes about which one can be sanguine without such organization.

Costs and Benefits

The costs and benefits of the developing agricultural system for the various categories of labor pose difficult evaluative questions. The system can be expected to reward managers, a social category that will become increasingly important in agricultural production.

In contrast, costs and benefits to family farmers will be very differentiated. Farmers growing in size will benefit considerably; those experiencing economic difficulties will follow the strategies of farmers since time immemorial and increase self- and family-exploitation to remain in business except when they organize protest activities, "raising less corn and more hell."

For permanent, career, or casual farmworkers, the developing system of agricultural production has already created new risks whose consequences are only just beginning to be understood. These risks will be experienced as high personal costs at some future date but may, in some cases, also be experienced as social costs. The risks are explicit in industrialized agriculture which accelerates exposure to chemicals and pesticides, on the one hand, and the effects on the body of highly repetitive activities, often involving strenuous physical activity, on the other. Thus, workers are more exposed to carcinogens now than they were previously. Although this is becoming increasingly clear for workers generally in U.S. industry, agricultural workers have been exposed to more chemicals, as Coye points out in this volume, than many other workers. In addition, where workers paced themselves earlier in agricultural activities, the shift to capital-intensive agriculture produces situations in which workers are exposed to noise, continuous bending, etc., such as they never experienced before. For example, workers on tomato harvesting machines are subject to continuous noise of powerful engines; lettuce workers must bend continuously; grape harvester operators get noise, dust, and other exposures, etc. Tools such as the short-handled hoe, originally used during only part of the work day, became used continuously, thereby producing high rates of back injury (Murray, 1982).

Among career farmworkers, the development of consequences often occurs years after exposure, e.g., in increased respiratory problems, back injuries or weaknesses, cancer, or other occupational diseases. These are not experienced "socially" except as costs in a medical system when a worker becomes indigent. The costs are borne more by workers than by any other social category involved in agriculture. There is nothing unusual in this since the same phenomenon is found in industry, business, and elsewhere and, in this sense, agriculture joins other industrial sectors in producing the inequities in the distributions of costs and benefits.

A few words are probably appropriate on the organization of labor in agriculture. The issue here has to do with the fact that the organization of workers in agriculture has been much delayed and remains, until the present, very uneven(9). The major locus of organization has occurred in California in the United Farm Workers Union. Other unions of farm workers have been begun in Arizona, Texas, and Ohio, and to a lesser degree in Florida and several other states. With the exception of Hawaii, where farmworker organization is very powerful (primarily because it is linked to longshore unionism), farmworker organization and unionism remain quite tenuous. Some sectors of California agriculture have been substantially organized and some developments have occurred in Arizona and Texas where interesting experiments in cross-border organizing have been begun by the Arizona Farm Workers Union and the Texas Farm Workers Union. The organization of Ohio tomato harvesters in the Farm Labor Organizing Committee (FLOC) has been temporarily important and the demands by FLOC for three-party bargaining (farmworkers, farmers, and processors) represent a significant and interesting idea but, up to the time of writing, no significant continuous organization has been put in place.

What can be expected over the long run as agriculture becomes increasingly concentrated and capital intensive is that organization will spread and unionism will become a more permanent part of the system of food production (Friedland and Nelkin, 1972). The development of unionism will contribute to the adequacy of a flow of labor into agriculture as wages and benefits rise. Costs of production will also rise although this factor will in part be mitigated by the effects of the world market (and the general surplus in agricultural commodities) and the degree to which trade barriers are established to prevent importation of foods. Increases in the costs of production will have the greatest effect on low income populations including those inside and outside of agriculture.

Nutrition

As far as nutritional aspects of the food system are concerned, the issue with respect to agricultural labor is simple. All strata in the agricultural labor force will benefit from nutritional qualities based on income levels (i.e., equity) and the nutritional information available to them. Income usually correlates with education and education frequently (but certainly not always) correlates with nutritional information. Therefore, the higher the income, usually the better the nutrition. For workers with low income and little nutritional knowledge, the inadequacy of nutritional information may lead to serious problems. With some workers, as the pace of industrialized agriculture increases, there will be tendencies (such as have been noted among lettuce workers) toward drug dependency (to permit maintenance of a very difficult work pace) coupled with poor nutrition.

More important will be the growing alienation and detachment of all participants in the system from the quality of the food produced. As agriculture becomes increasingly industrialized, each social category will be concerned with specific and limited goals rather than with nutritional quality. Thus, investors will be concerned with the "bottom line," profit on investment. Managers will be concerned with the bottom line but also with specific factors, such as costs of production. Since nutritional quality is not measurable as far as they are concerned, and since there will be no financial benefit for either improving it or ignoring it, managers can be expected to focus their energies on other issues such as output and productivity.

The same will hold for other employees. Permanent workers, career farmworkers and casual workers are going to worry about issues affecting employment, remuneration, and benefits. Until nutritional concerns are built into the reward system, the continued acceleration of the division of labor can be expected to reduce concerns for those factors that remain unremunerated.

The consequences of decline in nutritional quality of food can be anticipated for consumers. Since this subject is being handled elsewhere in this volume, I will not discuss it here. It should be noted, however, that at some stage in the future, the development of such a system may produce what could be called the "Japan effect," exemplified in the current situation in which automobiles, electronic equipment, and other commodities are produced more effectively in Japan with higher quality than in the U.S. It is not inconceivable, were a country such as Mexico to pay more attention to nutritional quality factors than the U.S. (although there has been no indication that this is the case), that competition could undermine U.S. production. There has been one case where this has been partially true, with winter tomatoes harvested relatively ripe in Mexico but harvested "mature green" in Florida. Even in this case, however, there have been pressures on Mexican producers to harvest tomatoes "mature green" so that distribution can take place throughout the U.S. and Canada.

KNOWLEDGE PRODUCTION ON AGRICULTURAL LABOR

The earlier discussion of some of the problems with data on agricultural labor inevitably raises the question: Why are the data about the hired labor force so uncertain and dubious?

This section undertakes a partial answer to this question. The argument can be summarized as follows: Knowledge about hired agricultural labor is bad because agricultural labor has not been a significant priority in national policy, except as a moral question, for the past century except in relatively brief and unusual periods such as wars(10). Knowledge production has masked social inequities rather than revealing them.

The Nonproblematic of Agricultural Labor

Agricultural labor has been nonproblematic because supply was resolved historically by an essentially unrestrained flow of labor to agriculture over the past 100 years and because adequate numbers of some categories of workers have been available as a result of the continuous destruction of family farming.

There is little point in expanding on this latter point even though it describes a monumental demographic transition in the United States, the shift from a rural to urban population. As what has been referred to as "the iron cross of agriculture" has unfolded in U.S. policy, "unsuccessful" farmers became available for employment in agriculture and industry. Most of these workers experienced a process of proletarianization that was personally traumatic. They moved out of agriculture and rural society and ended up in urban, industrial employment. More than enough were always available to work as permanent "farm hands." During the 1930s, many displaced farmers became farm laborers but most left seasonal employment during the Second World War.

The nonproblematic of the seasonal workers is illustrated by the bracero program initiated during the Second World War and the maintenance of an "open border" since 1964 when the bracero program was terminated. Earlier, an extensive and open labor supply also existed but we need not dwell on the situation of agricultural labor from 1880s onward. The bracero program delivered Mexican workers to agricultural employers throughout the U.S. at wages sufficiently low to discourage attraction to agricultural employment by domestic workers.

With the end of the bracero program, the "open border" policy came into effect. This policy has involved two approaches, fairly open access to legal status for some workers, giving them "green-cards" permitting legal residence in the U.S.; and effectively permitting large numbers of workers to cross the border without documentation(11). Although there is a border, immigration controls, and an Immigration and Naturalization Service that deports thousands of undocumented workers from the U.S., there are an estimated 3-6 million undocumented workers in this country (Massey and Schnabel, 1983:8). The maintenance of this open border that permits the continuation of Mexican migration is evidence of the policy of guaranteeing a cheap and controllable labor force.

Many Mexican indocumentados can be found in agriculture although it now appears that the bulk of these workers are located in nonagricultural employment. The critical point, as far as the agricultural labor force is concerned, is that there has been a continuing flow such that a labor shortage has not constituted a significant problem for U.S. growers. This has meant that serious research has not been necessary to understand the problem and what has been transpiring. The main reasons for what research has been done originate with those persons of conscience who worry about the equitability of the agricultural system. Thus, during the War on Poverty we had a Subcommittee on Migratory Labor in Congress, long since disbanded, and various agencies that developed data and held hearings. The 1960s saw the first significant coverage of agricultural workers by protective labor legislation, something from which they had been historically excluded. Since the 1960s and early 1970s, legislative and public interest has largely ended except where farmworker unionism has been a matter of continuing concern, i.e., in states such as California, Arizona, and Texas. What legislation has been adopted since the mid-1970s has occurred more at the behest of agricultural employers than of workers, with the exception of California'a Agricultural Labor Relations Act.

Media presentations have occasionally stirred interest and the effects of Ed Murrow's "Harvest of Shame" of 1960 are still remembered, usually with horror in agricultural employment circles. Occasional human barbecues in substandard housing fires and bus catastrophes provide (usually) localized reminders of the plight of farmworkers. On the whole, however, because there has been continuity of supply, there has been little concern to develop adequate knowledge about the problem of hired employment in agriculture.

Masking Reality

Because there has been an adequate agricultural labor supply, the important organized and politicized constituencies in agriculture have been more concerned with masking reality in knowledge production than in revealing social reality.

I do not intend to enter into a discussion of whether this has been deliberate or not. Not believing much in conspiracies, I doubt that there is, in fact, any conspiracy to mask reality. Rather, unreality is probably a product of the reification of data collection in a contextual vacuum in which there is no "real" problem. What is meant is that, because pressures get generated from time to time to produce information about farm employment, government agencies either with substantive interest in the problem (the Bureau of the Census) or with a grower-oriented bias (the USDA) utilize standard techniques of data collection that fail to grapple with social reality. Because there is no problematic defined by politically-relevant client groups, there is no need to develop new methods of data collection. Instead, standard methods such as surveys of households or farms serve. The methods become reified when publications emerge, year after year, census after census, and are "accepted" as pictures of reality.

Knowledge Production as a Class-biased Activity

The class bias of knowledge production on agricultural labor is indicated by the fact that the agency bearing the primary responsibility for such knowledge is not contained within the U.S. Department of Labor (USDL), that agency that bears the "normal" responsibilities for research on issues involving labor. Instead, responsibility rests with the U.S. Department of Agriculture, the agency whose mandate involves agriculture. There are historical bases for this responsibility since, as has been reported earlier, family farm labor made the overwhelming contribution to agricultural production in the past. But if this has been historically the case, the fact of its continuation in a period in which family farm labor decreases and hired labor increases indicates the need to place data collection responsibilities within an agency with less of a constituency attachment to agricultural employers than the USDA.

That the USDA is a "friendlier" agency to agricultural employers than other government units is well understood in the agricultural employing community. Some years ago, when the United Farmworkers Union began to gain success in California, many state agricultural employers and organizations favored the establishment of an Agricultural Labor Relations Board to be administered by the USDA. They knew that they could expect more "understanding" treatment from USDA than from the National Labor Relations Board, despite the fact that the NLRB is no longer very sympathetic to organized labor.

An indication of how agricultural circles approach the issue of employment can be seen by examining the situation in California and the manner in which USDA has been moving, although still not in a formalized way, into this arena. In California, with the emergence of farmworker unionism and with growing complaints about the anti-union and anti-worker bias of agriculture, funding was obtained through the State Legislature to encourage the Agricultural Division of the University of California to "enter" this area. Five positions were created within Agricultural Extension of the University to deal with personnel management issues. These positions, it should be noted, are concerned with problems of the _management_ of labor and not of labor in itself, or workers, or workers' organizations(12).

The involvement of USDA has come through the establishment of an Agricultural Employment Work Group (AEWG) consisting of persons concerned with agricultural employment and representative, I believe, of employer interests. In the listing of persons involved in the AEWG no representation of unions can be found(13). Significantly, the major policy thrust of the AEWG in their first report was to recommend:

> A major effort to educate employers, farm labor contractors, workers, and their organizations in the need for and benefits of improving personnel management practices in agriculture. [This would include] ...The identification of innovative and progressive practices in farm labor management and farm labor market operations across the country... The development of teaching and training in modern farm labor management practices... The hiring of staff specialists in agricultural personnel management in the Extension Services of those states with a high usage of hired labor... The professionalization of agricultural labor

> management... The incorporation of industry leaders and experience into the education process... The involvement of farmworker organizations in developing improved personnel management practices and a more skilled work force for agriculture (Agricultural Employment Work Group, 1982:2).

Personnel management is viewed by the AEWG as a neutral activity, somehow unrelated to the role that personnel managers play with respect to their employers, on the one hand, or to the workers that they manage, on the other. In my view, personnel management cannot be regarded as class-neutral but as an instrument representing the interests of employers.

The California developments indicate the class biased character of this approach. Having received funding for Extension specialists to work on the subject of labor, the Extension segment of the University of California's Agricultural Division has employed five specialists in personnel management. Most of the activities of these specialists involve concerns about personnel management. The degree to which approaches have been made, for example, to workers' organizations is not presently known and, since no reports have yet been made public, cannot be estimated. A solid guess based on experience and the historic client relationships between Agricultural Extension and California growers, however, would lead one to suspect that most of the activity will be aimed at growers, at personnel management, and that little activity will be aimed at workers' collective interests other than at managing them.

One further note might be made of the potential distinction that exists between the various kinds of agricultural employers, on the one hand, and managers and supervisors, on the other. The present developments aimed at personnel management are not yet clearly defined with respect to which constituency is intended to be served, i.e., family farmers, different types of agricultural investors, managers, supervisors, temporary supervision, etc. Developing knowledge and skills about personnel management can be generally applicable to all employers of labor but, as has been noted, there are different types of employers and differential levels of management and supervision. While it is not yet a problem, the issue may arise in the future as to which segment of the ownership-managerial strata such knowledge production and training programs are aimed.

CONCLUSION

The issue of labor in agriculture poses important normative or moral questions. Labor supply, given present existing policies of the U.S. and Mexican governments, can be expected to continue to provide more than enough for present labor requirements. The continual decline in family-based farming, as represented by the "iron cross," will continue, providing more than enough skilled farm managers. The elaborate mechanism of agricultural knowledge production and transmission embodied in the USDA and the publicly supported colleges of agriculture in almost all of the states will continue to handle most problems of training and education of growers, managers, and supervisors.

The only present "fly in the ointment" on the labor scene, from the point of view of present arrangements, is the spread and expansion of unionism. Difficulties are likely since many agricultural employers are reluctant to "live" with unionism. Despite this, a continuous struggle for organization can be expected.

Whether there will be a resolution of the present uncertainties and unevenness in coverage of farmworkers through legislation will probably depend, in all likelihood, on the degree of success of such organization. It might be noted, for example, that some California growers are, on the whole, dedicated believers in national coverage of farmworkers under the National Labor Relations Act. This is because they experience a "comparative disadvantage" with respect to Arizona, Texas, and Florida growers and therefore prefer a single, uniform system. The spread of farmworker organization can be expected to produce eventual coverage of farmworkers under federal legislation, perhaps even its expansion, as well as the provision of protection for farmworkers equivalent to that found for other workers.

NOTES

*I am indebted to the following readers and commentators on this chapter: Robert R. Alford, Molly Joel Coye, Marshall Ganz, Robert W. Glover, Phillip LeVeen, Robert Marotto, James O'Connor, Howard R. Rosenberg, Robert J. Thomas, and Leslie A. Whitener. Although I am certain some of these readers will not agree with many of the arguments, I am grateful to all of them.

1. Privately held corporations are thus defined: "All or almost all of the corporation stock is owned by the few persons who formed the business firm or by their successors," where in publicly held corporations "the stock... is bought and sold on recognized stock exchanges..." (Bureau of the Census, 1978:3).
2. See entry for Roswell Garst in the <u>Current Biography Yearbook, 1964</u>:139-41.
3. The phenomenon has also been documented in California where Russell Giffin, a major landowner and

grower on the west side of the San Joaquin Valley, has legally divested himself of land ownership to conform to Reclamation Act limitations by selling off legal ownership of the land while maintaining managerial control over it. For documentation, see Ballis, 1975: Part 1, 57-69; Part 1C, III A 1, 1632-1800.

4. Utilization of social security numbers has administrative and methodological problems. Administratively, it will be necessary to obtain the cooperation of the Social Security Administration to obtain data on earnings (using scrambled identification numbers). On the whole, this has proven to be clumsy and difficult but, assumably, agencies of government might be more successful at this than nongovernmental researchers. Methodologically, there are problems since some undocumented workers have no, stolen, or fictional social security numbers. In some cases, several people (or families) report earnings on a single social security number. Despite these problems, social security reporting promises to be a more accurate methodology of reporting than the present methods.

5. Agricultural employers have been most sensitive to issues of supply of labor (keeping it high) and control. Volume in supply has not only kept wage levels low but has permitted substitutability of workers which has, in turn, been important in the maintenance of control. Control has involved those situations in which workers have organized unions despite the existence of labor surplus. For an historical discussion, see Jamieson (1946). For a more recent and relevant discussion see Jenkins (1978).

6. Cf., Bach (1978). While emphasis has been placed here on the integration of the U.S. and Mexican economies and populations, some localities draw their labor supply of seasonal workers from off-shore sources such as Puerto Rico and the Caribbean islands and, in a few instances, Canadian labor is important.

7. Managerial employees will have to be compared to equivalent organizations outside of agriculture. Very few agricultural organizations are presently comparable to most industrial enterprises. Indeed, in directories such as <u>Fortune's</u> 500, there are no agricultural units. Most large agribusiness organizations are heavily involved in nonagricultural activities such as transportation, storage, processing, sales, etc.

8. With a few notable exceptions, agricultural employers, despite their rhetoric about the importance of career farmworkers, have failed to provide the material returns appropriate to hard and difficult labor. Were the labor supply from Mexico to dry up, the operations of a national labor market would make agricultural wages properly competitive with industrial labor.

9. This subject has been discussed at length by various students of agricultural worker unionism. I

have summarized this discussion in Friedland (1981).

10. There are two important exceptions to this statement. During World War II, the shortage of all labor contributed to the establishment of national policies favoring the importation of labor from Mexico for U.S. agriculture, i.e., the bracero program. During the war on poverty in the 1960s, when social programs aimed at migrant and seasonal farmworkers were established, funding of such programs depended on formulas that ultimately related to data collected and published by agencies such as USDA. Thus there were real consequences to these programs of the inadequacies of the data base.

11. It is not my intention to argue for or against the legalization of such migration and/or the stringent enforcement of legal controls. I do not know where I stand on this subject personally. What is crucial to recognize, irrespective of individual attitudes or formal "policy," is that there is an open border.

12. Several readers of an earlier draft of this chapter understand personnel management to be more "neutral" than I do. While personnel management as a profession undoubtedly emphasizes the formal delineation of worker rights as well as adherence to laws affecting the employment relationship, personnel managers operate at the behest of employers. They may feel that they represent the interests of workers but experience reveals that such interests are best represented by workers themselves, not by external representatives.

13. I am informed that such representation was solicited but not forthcoming. Several of the AEWG participants, in personal communications, contended that variously 9 or 10 of the 25-27 members of the AEWG represented worker interests. None, however, represented unions and the question of who "represents worker interests" is certainly open to considerable controversy.

REFERENCES

Agricultural Employment Work Group, U.S. Department of Agriculture. 1982. Agricultural Labor in the 1980's: A Survey With Recommendations. Berkeley: Division of Agricultural Sciences, Univ. of California.

Bach, R. S. 1978. "Mexican Immigration and the American State." International Migration Review 12(4):536-58.

Ballis, G. 1975. "Statement of George Ballis, Executive Director, National Land for People." In Will the Family Farm Survive in America? Joint hearings before the Select Committee on Small Business, Committee on Interior and Insular Affairs, U.S. Senate, 94th Congress.

Braverman, H. 1974. Labor and Monopoly Capital: The Degradation of Work in the Twentieth Century. New York: Monthly Review Press.

Bureau of the Census, U.S. Department of Commerce. 1978. "Corporations in Agricultural Production." 1974 Census of Agriculture, vol. 4, part 5. Washington, D.C.: U.S. Government Printing Office.

Center for Rural Affairs. n.d. Wheels of Fortune: A Report of the Impact of Center Pivot Irrigation on the Ownership of Land in Nebraska. Walthill, Nebraska: Center for Rural Affairs.

Davidson, J. 1982. "Modern Tycoons - How They Made it, How They Live." U.S. News and World Report (May 31):58-59.

Doeringer, P. B., and Piore, M. J. 1971. Internal Labor Markets and Manpower Analysis. Lexington: D.C. Heath.

Fligstein, N. 1983. "The Transformation of Southern Agriculture and the Migration of Blacks and Whites, 1930-1940." International Migration Review 17(2): 268-90.

Friedland, W. H. 1981. "Seasonal Farm Laborers and Worker Consciousness." In Research in the Sociology of Work: A Research Annual, ed. Ida Harper Simpson, pp. 351-80. Greenwich: JAI Press.

______. 1982. "The End of Rural Society and the Future of Rural Sociology." Rural Sociology 47(4):589-608.
Friedland, W. H., and Nelkin, D. 1972. "Technological Trends and the Organization of Migrant Farm Workers." Social Problems 19(4):509-21.
Hintz, J. 1976. Seven Families. Tiffin, Ohio: Joy Hintz.
Jamieson, S. 1946. Labor Unionism in American Agriculture. U.S. Department of Labor Statistics. Bulletin No. 836. Washington, D.C.: Government Printing Office.
Jenkins, J. C. 1978. "The Demand for Immigrant Workers: Labor Scarcity or Social Control." International Migration Review 12(4):514-35.
Lillesand, D.; Kravitz, L.; and McClellan, J. 1977. "An Estimate of the Number of Migrant and Seasonal Farmworkers in the United States and the Commonwealth of Puerto Rico." A report prepared for the Legal Services Corporation.
Massey, D. S., and Schnabel, K. M. 1983. "Background and Characteristics of Undocumented Hispanic Migrants to the United States: A Review of Recent Research." Migration Today 11(1):6-12.
McDonald, T., and Coffman, G. 1980. Fewer, Larger U.S. Farms by Year 2000 - and Some Consequences. Agricultural Information Bulletin 439. Washington, D.C.: Economics and Statistics Service, U.S. Department of Agriculture.
Michigan Employment Security Commission. 1971. 1971 Post Season Rural Manpower Report. Detroit: Michigan Employment Security Commission.
Murray, D. L. 1982. "The Abolition of El Cortito, the Short-Handled Hoe: A Case Study in Social Conflict and State Policy in California Agriculture." Social Problems 30(1):26-39.
Nelkin, D. 1970. On The Season: Aspects of the Migrant Labor System. Ithaca: New York State School of Industrial and Labor Relations.
Perry, C. S. 1982. "The Rationalization of U.S. Farm Labor: Trends Between 1956 and 1979." Rural Sociology 47(4):670-91.
Pollack, S. L. 1981. The Hired Farm Working Force of 1979. Agricultural Economic Report No. 473. Washington, D.C.: U.S. Department of Agriculture, Economic Research Service.
Slesinger, D. P. 1979. "Migrant Agricultural Workers in Wisconsin." Population Notes 8:1-4.
Smith, L. W., and Coltrane, R. 1981. Hired Farmworkers: Background and Trends for the Eighties. Rural Development Research Report No. 32. Washington, D.C.: Economic Development Division, Economic Research Service, U.S. Department of Agriculture.
Smith, T. L. 1969. "A Study of Social Stratification in the Agricultural Sections of the U.S.: Nature, Data, Procedures, and Preliminary Results." Rural Sociology 34(4):496-509.

Sutton, S. B., and Brunner, T. 1983. "Life on the Road: Midwestern Migrant Farmworker Survival Skills." Migration Today 11(1):25-31.

Thomas, R. J. 1981. Citizenship and Labor Supply: The Social Organization of Industrial Agriculture. Ph.D. dissertation. Evanston, Illinois: Northwestern University, Department of Sociology.

Time Magazine. 1978. "The New American Farmer." (November 6):92-102.

9
Occupational Health and Safety of Agricultural Workers in the United States

Molly Joel Coye

Agriculture has long been recognized as one of the most dangerous of occupations; it has consistently ranked with coal mining and construction among the three most hazardous jobs in the United States. I will be discussing the impact of occupational health problems in agriculture on domestic food security in the United States, defined briefly as the long-term sustainability of agricultural practices and the extent of risk imposed by these practices on agricultural producers.

The significance of occupational hazards for domestic food security has mainly to do with the second aspect, that of risks imposed on the agricultural producers. In this chapter, I will present and discuss the implications of several aspects of these risks: (1) the shift in occupational hazards from safety to health, and within health from acute to chronic problems; (2) the limitations of current knowledge about the health effects of many occupational hazards, particularly agricultural chemicals, and the paucity of good diagnostic and surveillance methods; (3) the relative lack of protection afforded agricultural workers, compared to those in industry, in health and safety legislation, enforcement and compensation coverage; and (4) the growing concern of both agricultural workers and the public regarding the long-term health effects of pesticide exposure.

This concern for the potential consequences of pesticides(1) exposure has focused on directly occupational exposure to pesticide applicators and farmworkers, on communities incidentally exposed by spray drift during serial application or by contamination of water supplies, and on consumers exposed to pesticide residues on marketed produce. It has already resulted in direct intervention to change agricultural practices, limiting or preventing certain usages. Increasing awareness of the potential health risks imposed by other agricultural practices, such as some forms of mechanization, may also lead to similar interventions, and increasing recognition of the economic costs of agricultural occupational disease may force growers to internalize a greater share of these costs through improved compensation benefits. The review of occupational health and safety in agriculture presented here should help to clarify some of these issues and encourage a consideration of potential health effects in future discussions of research and the development of agricultural technology.

During the past three decades, as control over farmland has become increasingly concentrated and the total agricultural workforce has steadily declined, the structure of agricultural employment has also changed, becoming more "industrial" in character. An increasing proportion of agricultural workers--approximately one-half of farmworkers in the western third of the United States--are nonfamily hired employees. As in most other occupations, injury rates for agricultural workers have been declining. Occupational disease rates, however, have not shown a significant decline, and represent a rising proportion of all job-related morbidity for agricultural workers. In public health terms, this resembles the broader shift in developed industrial societies to predominant patterns of chronic disease.

The predominant forms of presently acknowledged occupational disease in agriculture are dermatitis, osteoarthritis, and pesticide-related illnesses. Concern about the latter has increased dramatically during the past decade, both among farmworkers and among the general public, and has already resulted in regulatory changes directly affecting agricultural practices. Partly as a result of increasing worker and public concern, physicians and other health providers are also becoming more aware of pesticide-related and other occupational diseases, and are seeking information with which to evaluate difficult issues such as the reproductive effects of chronic low-level pesticide exposure.

In addition to the direct health effects outlined above, the economic impact of this injury and disease on the agricultural workers themselves, on their families, and on the nation as a whole should not be underestimated. In all industries, including agriculture, the economic cost of lost work days due to occupational hazards has risen with the relative productivity of labor. The cost of lost workdays to the agricultural employer is somewhat mitigated, however, because the majority of farm work is not highly skilled, and the employer can often replace injured or ill workers without incurring additional direct costs.

Furthermore, workers compensation does not cover agricultural workers at all in 24 states, and only partially in 13 others. In states where agricultural workers are covered, studies have demonstrated that few farmworkers actually receive workers compensation for job-related illnesses. Thus economic costs to the employer in lost production time and in compensation for injury and illness--factors which motivate industrial employers to protect workers under some conditions--are significantly reduced in agriculture.

I will begin with a description of the agricultural workforce in the United States, and a review of available information on occupational injury and illness among that workforce. Because patterns of occupational injury are more generally known, and better understood, than issues of occupational disease, I will discuss the latter in more detail. In particular I would like to spend some time on the chronic health effects of low-level pesticide exposure, because that appears to be--with occupational skin disease--the most widespread form of occupational illness among farmworkers, and the form which is most seldom recognized. I will conclude by outlining regulatory and compensation issues, and by identifying areas in which agricultural specialists and policy makers could ameliorate some problems, thereby improving the current status of domestic food security.

DESCRIPTION OF THE AGRICULTURAL WORKFORCE IN THE UNITED STATES

The most reliable estimates of the agricultural labor force suggest that 4-5 million persons work in agriculture as their primary means of earning a living. Hired, nonfamily farmworkers constitute between one-third and one-half of those primarily employed in agriculture (hired workers are probably underreported, because employers have economic incentives not to report seasonal and migrant workers). As many as 5 million more persons are engaged in commercial agriculture at some time during the year (Bureau of Community Health Services, 1979).

Seasonal and migrant workers are probably at greatest risk because they are concentrated in particular crops and activities which result in increased occupational exposure. More than 50 percent of seasonal workers are hired for harvesting, predominantly in fruits, vegetables, and tobacco. Most of the crop-activities utilizing seasonal hired labor--especially harvesting--involve contact with foliage during or immediately after pesticide application periods, and therefore represent a proportionally greater risk of exposure to pesticide residues. Of the 27 percent hired for cultivation, almost one-third work in cotton, a crop which uses a very high rate of pesticide applications (Task Group on Occupational Exposure to Pesticides, 1975). In addition, acreage under irrigation has been steadily increasing, and irrigated crops tend to be those most likely to require intensive hand labor in preharvest and harvest operations, with concomitant pesticide residue exposure.

The general health status of many agricultural workers--not solely seasonal and migrant workers--is already sigificantly compromised by factors which are indirectly job-related but nevertheless derivable from occupation. Low income and relative geographic and social isolation contribute to deficiencies in nutrition, housing, sanitation, education, and access to preventive and medical care services. Some of these factors exacerbate occupational health risks: housing in or near fields exposes workers and their families to pesticide spray drift; lack of potable water in the fields and some housing forces workers to drink irrigation water, frequently contaminated with pesticides; lack of education makes it difficult or impossible for workers to read pesticide labels; protein deficiency increases the toxic effects of many pesticides (Shakman, 1974); and the same economic and geographic factors which limit access to medical care for farmworkers also delay or prevent appropriate treatment for job-related injuries and illnesses. For some ethnic groups, such as Mexicans or Mexican-Americans, Haitians, Puerto Ricans, Filipinos, and more recently Southeast Asians, language differences further exacerbate these problems.

OCCUPATIONAL SAFETY AND HEALTH OF THE AGRICULTURAL WORKFORCE

Agriculture, as I noted in the introduction, ranks with mining and construction as one of the three most hazardous occupations, and reported injury and illness rates in agriculture are probably more underestimated than are those for mining and construction; far more agricultural workers are self-employed, and therefore do not report work-related injuries for compensation purposes.

In discussing the hazards of any occupation, injury and illness rates must be considered separately, because the causes and temporal trends for each are usually quite disparate. As California has the most complete and detailed data available for agricultural worker health and safety--and represents a large part of U.S. agricultural production, and an even larger share of agricultural employment--I will rely primarily on statistics reported by the California Department of Food and Agriculture (CDFA) and Department of Industrial Relations in this review.

Occupational Injuries

Rates for occupational injuries have declined for many sectors of agriculture, due to mechanization (particularly in harvesting processes), changes in cultivation practices which have reduced the risks of sprain-and-strain injuries, and new agricultural chemical uses which reduce the need for labor in certain activities. Between 1965 and 1970, for example, injuries involving ladders (used primarily for orchard harvesting) decreased by 40 percent (Whiting, 1975). Reductions in the available labor force caused by mechanization in one crop have stimulated efforts to mechanize others: mechanization of cotton harvesting reduced the number of seasonal/migrant workers available for grape harvesting, and spurred the partial mechanization of grape harvesting (Runsten and LeVeen, 1981). The ending of the bracero program supply of laborers also gave impetus to the mechanization of some labor-intensive crops.

In a few cases pressure to change agricultural practices has come from the farmworkers themselves. For five decades, farmworkers in California agriculture sporadically protested the enforced use of "el cortito," the short-handled hoe used for "stoop labor" in the cultivation of lettuce, tomatoes and other crops. In the late 1960s and early 1970s, the California Rural Legal Assistance program responded to this sentiment among its farmworker constituency and developed a political, administrative, and legal campaign against the short-handled hoe that was finally successful.

Changes in cultivation practices such as the use of herbicides to eliminate the need for stoop labor have reduced labor requirements to a certain extent, although some of these changes have also increased potential chemical exposure. Due to these factors, among others, injury rates have declined.

Workers in California field crops which traditionally utilized stoop labor had a 34 percent decrease in sprain-and-strain type injuries, including back strain, during the late 1960s. Workers on fruit and nut tree farms, where there is less bend-over work but much lifting, experienced an intermediate decrease in such injuries of 19 percent. Workers on dairy, livestock, and poultry farms, and in nursery and greenhouse work, showed almost no decline in sprain-and-strain injuries for the same period; mechanization of these farms had taken place much earlier (Whiting, 1975) and there was less room for innovation in cultivation practices or in agricultural chemical uses.

For the workers who operate harvesters and other new machinery, mechanization conversely may represent potentially increased risks. Pesticide residue exposure is now both dermal, as before, and also by inhalation since the harvesters shake the foliage and dislodge pesticide residues. The severity of injuries also increases because of the machines themselves. Tractors alone are responsible for an estimated 40-60 percent of farm accidents and fatalities, and equipment operators on larger, more complex machines find the consequences of heat stress and pesticide effects more difficult to deal with.

Agricultural aviation, or pesticide spray planes, are a special case within the general category of mechanization. Manual and ground rig applications have been rapidly replaced by aerial application. A 1980 review of FAA statistics found that of more than 6,000 registered agricultural aircraft in the U.S., approximately 1,300--or 20 percent--experience difficulties requiring an unscheduled landing each year, with an average of 40 pilot fatalities per year (Pechan and Jansson, 1980). Contributing factors identified in another study of cropduster accidents were long work hours, weather conditions, low-altitude runs, fixed obstacles, noise, vibration, heat stress, and pesticide exposure (Richter et al., 1981).

Occupational Illnesses

Occupational illness rates for agricultural workers are also high relative to other occupations; in fact, in California they are the highest in the state (Table 9.1).

TABLE 9.1
Occupational Illnesses for 100 Full-time Workers
California, 1979

Industry	Rate
All industries	0.3
Agriculture, Forestry, Fishing	0.6
Agricultural production	0.7
Agricultural services	0.5
Manufacturing	0.5
Mining	0.3
Construction	0.3

Source: Adapted from Division of Labor Statistics and Research, 1981.

Occupational morbidity and mortality. Very few studies have been done of overall patterns of morbidity and mortality among those employed in agriculture. As early as 1963, DHEW reported an elevated mortality rate for leukemia among male farmers and farm laborers (DHEW Vital Statistics Special Report, 1963). Further studies by Milham and Peterson found elevated rates of skin cancer and leukemias (Milham, 1976; Peterson, 1980). Death certificate studies in Iowa and Nebraska also found an increased risk of leukemias, variously associated with the production of certain crops and livestock and with the use of herbicides and insecticides (Blair and Thomas, 1979; Burmeister, et al., 1982, 1983). A New Jersey study recently reported an increased risk of liver cancer mortality associated with agricultural work (Stemhagen, 1983). Blair, in a study of cancer mortality among pesticide applicators, found elevated rates for leukemia and for cancers of the brain and lung (Blair, 1983).

A study of mortality among agricultural workers in California found that the mortality rate from respiratory diseases in farm laborer groups was triple the rate for the farm management group (resident owners and managers) (Carlson and Peterson, 1978). A later study comparing mortality rates for Nebraska farmers with all other occupations did not find greater mortality from respiratory diseases, nor a difference between farmers (farm owners) and farm laborers (Burmeister and Morgan, 1980). There are, of course, significant differences between the two states in the nature of the agricultural work force, in agricultural production and pesticide usage, and in other factors which account for the differences between the findings of the two studies.

While definitive conclusions cannot be drawn from these limited and disparate reports, they suggest the possibility of a higher rate for some health problems among agricultural in comparison to non-agricultural populations. The report of a class difference in mortality among agricultural workers in California is particularly interesting.

Occupational skin disease. The most frequently reported occupational disease, for all industries and for agriculture, is dermatitis. The rate of occupational skin disease for all California industries combined was 2.1 per 1,000 workers in 1977. The rate for agriculture was 8.6, for manufacturing 4.1, for construction 2.5, and for mining 2.0. While agriculture represented only 3 percent of state employment, it accounted for more than 13 percent of all occupational dermatoses (Division of Labor Statistics and Research, 1982). The majority of these cases are attributed to plant exposures such as poison ivy, but a growing proportion is now attributed to pesticides.

The economic consequences of pesticide-related dermatoses are significant. In 26 percent of all dermatoses caused by agricultural chemicals--that is, by pesticides--the reporting physician expected a period of work disability, and therefore of income loss. Furthermore, a large number of pesticides in common use have been reported to cause sensitization (an allergic reaction) as well as direct irritation; farm workers who develop sensitization are forced to permanently abandon work in crops on which that pesticide is utilized.

Pesticide-related Illnesses.

(1) Use patterns and exposure groups. Approximately 900 million pounds of pesticides are produced and used in the United States each year--one-third of the total world consumption--with another 800 million pounds produced for export. Roughly 80 percent of pesticides sold are used in commercial agriculture, the rest being divided almost equally between household/garden use and industrial/structural use. Thus agriculture accounts for the bulk of potential occupational exposures to pesticides.

California Food and Agriculture data allow us to be even more specific in identifying high risk occupational groups within agriculture. Ground applicators, mixers, loaders, and field workers represent the greatest proportion, approximately one third, of all reported cases of pesticide-related illnesses. Greenhouse and nursery workers also have relatively high rates, presumably because their workplaces are largely enclosed and pesticide applications are geographically concentrated. Truck drivers, who handle highly concentrated pesticide formulations, and firemen, who are exposed in uncontrolled emergency situations, have rates only slightly lower than the greenhouse and nursery workers.

(2) Reporting of pesticide illnesses. Statistics on pesticide-related occupational illnesses should be regarded with even more caution than most statistics in occupational health. Based on several studies of pesticide illness reporting patterns, the director of the Pesticide Program for the California State Department of Health estimated in 1976 that probably only a very small fraction of pesticide-induced illnesses among farmworkers are reported, possibly no more than 1 or 2 percent (Kahn, 1976). This is despite the fact that California is the only state in the nation which requires physicians to report pesticide illnesses (and may levy a $250 fine against physicians failing to file such a report). Statistics for other states which do not have reporting requirements--and have either inadequate or no workers compensation coverage for agricultural workers--are even less likely to accurately reflect the extent of pesticide-related occupational illness.

There are many factors which contribute to this reporting problem, of course, including the nonspecific nature of early and mild symptoms of pesticide exposure, the sociology and political economy of agriculture and its field labor force, and the lack of physician knowledge regarding occupational disease in general and pesticide-related disease in particular. With respect to this last factor, a 1978 survey of medical school curricula showed that only 30 percent required coursework in occupational medicine, and the mean time devoted to this was 4 hours in the 4 years of medical school (Levy, 1981). Increased education in the health effects of pesticide exposure would benefit the general population as well as agricultural workers, for more than 90 percent of households in the U.S. use one or more pesticides, and non-occupational pesticide illnesses (particularly accidental ingestion by children) are distressingly common.

(3) Changes in patterns of pesticide illnesses. While the number of reported cases of acute pesticide poisoning of all kinds has held constant over the past fifteen years in California, the type of cases reported has changed substantially. Until the mid-1970s, a large proportion of reported cases involved entire crews of field workers, frequently harvest crews in citrus orchards, who entered the orchard or field while high concentrations of pesticide residues were still present on the foliage. By the early 1970s, California had established a series of re-entry periods requiring growers to restrict fieldworker entry into fields for specified periods after certain pesticides had been applied. As a result of these re-entry periods, and a relatively strong pesticide monitoring and enforcement program, the number of severe mass poisonings declined abruptly.

Cases of pesticide-related systemic illness reported in the last 5-7 years (as opposed to skin or eye irritation) have usually been workers who had a single exposure to a relatively large quantity of pesticide, and immediately developed specific symptoms of acute poisoning referable to that pesticide. For example, a pesticide applicator who is accidentally sprayed with methyl parathion and rapidly develops blurred vision, pinpoint pupils, nausea, vomiting, excessive salivation, chest tightness and difficulty breathing, weakness, and tremors is easily diagnosed and treated, and is very likely to be reported as a case of pesticide-related occupational illness.

In other words, the systemic illness cases reported in recent years are still cases of acute severe poisoning, although they involve small numbers of individuals in each incident rather than entire crews. Recalling the estimate that as little as 1 or 2 percent of all pesticide illnesses in farmworkers are reported, what do the un-reported cases look like? Are literally thousands of other acute cases similar to the one just described above "silently" occurring--either undiagnosed or unreported by a physician? This is highly unlikely.

Evidence has accumulated to suggest that the vast majority of pesticide induced illnesses among agricultural workers are in fact chronic rather than severe or acute--that is, continuous or intermittent symptoms in response to continuous or intermittent low-level exposures. The difference between acute and chronic pesticide exposures and illnesses can be represented in a chart (Table 9.2).

TABLE 9.2
Occupational Disease Chart

Exposure	Disease: Acute	Disease: Chronic
Acute single episode large quantity	marked symptoms temporal relation to exposure clear	persistent non specific symptoms after acute episode
Chronic/ Intermittent multiple exposures small quantities	mild or non-specific symptoms temporal relation to exposure may be clear if exposure is intermittent	
Acute and/or chronic	both acute and chronic/intermittent exposures MAY cause reproductive effects or delayed-onset disease such as cancer; little known yet	

Acute exposure to large enough amounts of any pesticide will cause acute illness. In the case of many pesticides, however, recovery from moderate acute illness is usually complete, and does not leave any residual chronic disability. Similarly, many chemical groups of pesticides do not cause clinical symptoms at very low levels of dermal or inhalation exposure (the predominant forms of occupational exposure to pesticides).

The major exceptions to both of these generalizations are the organophosphate pesticides and, to a lesser extent, the carbamates. These are among the most widely-used pesticides, of course, and exposure to low levels is associated with a broad range of relatively nonspecific symptoms. (Solvents, which are widely used as vehicles for the active pesticides, may also play a role in chronic mild "pesticide-related" illnesses independently of or in synergism with the active pesticide.)

How prevalent is such chronic pesticide-related illness among agricultural workers? In the next section I will review existing evidence regarding the prevalence of both symptoms and cholinesterase depression among a variety of agricultural workers.

PREVALENCE OF CHRONIC PESTICIDE EXPOSURE AND EFFECTS

Low-level exposure to organophosphate pesticides produces nonspecific central nervous system (CNS) symptoms, that is, symptoms which also occur with influenza and many other common nonoccupational diseases. The symptoms which have been found to be associated with low-level exposure to organophosphates are:

- headache, fatigue, drowsiness
- insomnia and sleep disturbances with excessive dreaming
- mental confusion and difficulty in maintaining alertness and in concentrating
- memory disturbances
- restlessness and anxiety
- tremulousness, irritability, and emotional liability.

Behavioral effects of organophosphate exposure have also been studied in both humans and animals. Significant impairment of performance at low levels of exposure has been found, including impairment of repetition of learned behaviors, the inability to maintain alertness and attention and to concentrate, memory impairment, slowing of information processing and performance of simple tasks, expressive language defects, and slowed responses with inappropriate indifference (Levin, 1976). These effects suggest that such performance decrements may seriously compromise the safety and efficiency of agricultural workers excessively exposed to pesticide residues on a chronic basis.

A recent study examined workers exposed to low levels of organophosphates who did not have clincal signs of cholinergic toxicity (the more widely known symptoms of "acute" poisoning such as pin-point pupils, nausea, salivation, tremors, and so on). Personality tests, a structured interview, and cholinesterase levels were used to compare the exposed and control populations.

As Table 9.3 shows, the exposed workers produced significantly higher anxiety scores than the controls; when the exposed workers were separated into commercial spray applicators and farmworkers, both had higher scores than the controls, and the difference between applicators and controls was even more significant, $P < .005$.

Although plasma cholinesterase levels were significantly depressed among the exposed workers in comparison to those of the controls, there was no correlation between anxiety scores and plasma cholinesterase values for individuals. This is an important finding and is confirmed in many other studies: the nonspecific CNS symptoms associated with

TABLE 9.3
Organophosphate Exposure and Psychiatric Symptoms

	Controls (N=24) Mean Values	Exposed (N=24) Mean Values
Age, years	38.8	39.2
Education, years	12.4	12.4
Anxiety	9.9	14.5(a)
Depression	3.3	3.9
Symptom severity	1.8	2.9
RBC ChE	13.2	13.6
Plasma ChE	4.8	3.8(b)

Source: Adapted from Levin, et al., 1976.

(a) P < .05, (b) P < .01

organophosphate exposure do not correlate with cholinesterase inhibition. This fact is not known to most medical practitioners, who frequently will not diagnose an organophosphate-related illness without laboratory confirmation of cholinesterase inhibition.

Unfortunately, in many cases these CNS effects persist for weeks to months after the patient is removed from exposure. With time, however, these symptoms are completely reversible, a fact which underscores the importance of protecting workers from chronic pesticide exposure. Four relevant empirical studies reflect the prevalence of nonspecific symptoms and cholinesterase inhibition among farmworkers.

(1) A 1976 investigation compared migrant farmworkers in New Jersey to nonagricultural controls, and found a significant difference between the migrant workers and controls for symptoms of headache and nausea and for plasma cholinesterase activities. The cholinesterase differential persisted when Puerto Rican controls were included to match the ethnic origins of the farmworkers. Again there was no correlation between symptoms and exposure for individuals. (That is, the individuals who reported symptoms were not necessarily those who manifested lower cholinesterase activity) (Quinones et al., 1976).

(It should be pointed out that, in these studies, plasma cholinesterase values are actually a more useful measure than red blood cell values, because the cholinesterase activity is being used as a surrogate indication of exposure to and absorption of the pesticide, rather than as an index of health effect. Plasma cholinesterase is, of course, more sensitive to most organophosphate exposures than the red blood cell enzyme.)

(2) Even when red blood cell cholinesterase is used as the index of exposure, as in a study of farmers in a vegetable farming region in Ontario, Canada, seasonal differences are found which correlate with periods of organophosphate pesticide application. Cholinesterase activities were determined for each farmer at the beginning of the season, at the end of the pesticide application period, and several months later. The majority of individuals demonstrated a decrease in red blood cell cholinesterase activity during the pesticide application period, and an increase during the post-spray season (Brown et al., 1976).

(3) In another study conducted in Nebraska, 98 farmers and commercial applicators were tested for organophosphate and carbamate exposure. Baseline cholinesterase values were drawn 11 months after the previous year's applications, and the second value was drawn immediately after the spring planting and pesticide application period. All participants had attended a pesticide training program and were certified by the state to use pesticides. Significant reductions in plasma cholinesterase were found in 30 percent of individuals tested. Twenty-two percent of the study participants reported mild symptoms consistent with organophosphate exposure. The correlation between symptoms and cholinesterase inhibition was not analyzed (Spigiel et al., 1981).

(4) The prevalence of potentially pesticide-related symptoms was investigated by the California Department of Public Health in household interviews of more than 1,100 nonmigrant farmworkers. Only symptoms which were of sufficient severity to cause the worker to seek medical treatment were recorded. Because the symptoms are not specific to pesticide exposure, a group of controls were interviewed at the same time. The interviewers were "blinded" to the study's real purpose, and believed they were conducting a general health survey. The farmworkers were found to have experienced the potentially pesticide-related symptoms at 15 times the rate for the control group, although almost none of the cases had been dignosed or reported by physicians in the area. The study report commented that "there is apparently something peculiar to the agricultural working environment which is causally related to the nausea, eye and skin irritation, chronic headaches, sleeplessness and other symptoms which have been reported in large numbers," and concluded that the only credible hypothesis offered as explanation for these findings had been that of chronic low-dose pesticide exposure (Kahn, 1976).

These studies indicate that cholinesterase inhibition, taken as an index of exposure, is widely prevalent among a variety of agricultural work groups, and that certain nonspecific but potentially pesticide-related symptoms are also more prevalent among farmworkers than the general population. Although the extent of cholinesterase inhibition reflects organophosphate (and carbamate) exposure only, we must assume that low-level chronic exposures to other pesticide residues also occur. Most of the nonspecific neurological symptoms considered as pesticide-related have been associated with low-level exposure to organophosphates, in studies where the exposure is more specifically identified than in the populations discussed above. For this reason we attribute such symptoms, when found in general farmworker populations, to organophosphate residues; but it should be pointed out that solvent vehicles, nonorganophosphate pesticides, and other exposures may also contribute to these findings.

WORKERS COMPENSATION COVERAGE FOR AGRICULTURAL WORKERS

Workers compensation coverage for farmworkers is nonexistent or extremely inadequate in almost every state. In the few states with coverage somewhat approximating that for industrial workers, such as California, farmworkers are largely unaware of their eligibility or even of the existence of the compensation system.

In the California State Department of Public Health study just described, the investigators also inquired about the source of payment for medical care sought for the potentially pesticide-related symptoms. Only 6 percent of cases were treated under workers compensation, although technically and legally every one was entitled to be treated as an occupational illness. None of these cases had actually been reported as occupational by the treating physician. This supports the Department's estimate of underreporting of pesticide-related illnesses among farmworkers, as the major impetus for workers or physicians to report an illness is to seek reimbursement under workers compensation. In a separate study by the Department, only 8 percent of farmworkers surveyed (and only 6 percent of Mexican-American farmworkers) had any idea of what workers compensation actually is (Division of Labor Statistics and Research, 1982).

Farmworkers suffering from the chronic health effects of pesticide exposure therefore have little choice but to continue working, for a number of reasons: (1) the physician is not aware of and does not seek environmental/occupational causes for illness; (2) the worker is unaware of his/her right under workers compensation; (3) the physician will rarely diagnose symptoms as pesticide-related in the absence of cholinesterase depression; (4) even if the physician will make the diagnosis, some symptoms require weeks to months of removal from exposure to resolve; (5) all workers compensation programs require employment for minimum time periods before workers are eligible, and many farmworkers do not stay long enough with a single employer to earn eligibility; and perhaps most importantly, (6) farmworkers are not organized to the extent industrial workers are, and most have no means of defense against retaliation by employers for questioning the health effects of pesticide exposure on the job.

REGULATORY AGENCIES AND POLICY ISSUES

A welter of agencies have overlapping responsibilities for agricultural worker health and safety. Some, like the Environmental Protection Agency (EPA), have a large number of regulations pertinent to occupational health, but have almost no enforcement staff. Others, like the Occupational Safety and Health Administration (OSHA), have a relatively larger pool of inspectors available for enforcement purposes, but have ceded jurisdiction over major aspects of agricultural work to other agencies and have developed almost no regulations to enforce. The following is a brief and much simplified overview of agency regulatory and enforcement efforts; although some of the generalizations may seem sweeping, I assure you that this is a reasonably good representation of what these agency programs look like from the perspective of field and other agricultural workers.

OSHA negotiated an agreement regarding jurisdiction over agricultural workers with the Environmental Health Agency, under which OSHA has primary responsibility for manufacturers and formulators of pesticides, and responsibility for fieldworker safety and sanitation. The EPA is responsible for registering pesticides, for developing and promulgating regulations for commercial application of pesticides (both agricultural and structural/institutional), and for enforcement of these regulations.

The National Institute for Occupational Safety and Health (NIOSH)(2) proposed a generic standard for manufacturers and formulators of pesticides in 1978 which would have grouped all pesticides into three levels of potential hazard for regulation of work practices, environmental controls, labeling, medical surveillance, and so on. OSHA has not yet acted on this proposed standard, and current OSHA standards simply set a Permissible Exposure Limit for industrial exposures to a small number of the most toxic pesticides.

OSHA has regulations regarding equipment operation in agriculture, although in many states the OSHA enforcement staff have industrial rather than agricultural safety training and are poorly prepared to evaluate farm machinery safety. OSHA has not promulgated any standards for field safety and sanitation; after more than 5 years of litigation, in 1981 the Supreme Court ordered OSHA to develop and promulgate such a standard, and OSHA staff are currently working on this. The major aspects of field work which will be considered for regulation are requirements to provide potable water, nonpotable water for washing purposes, and latrines.

The EPA, as stated above, has primary responsibility for registration of pesticides. A large number of pesticides were "grandfathered" in when the EPA was first established; since then, requirements for registration have become more stringent. Public and scientific access to the data submitted to the EPA by manufacturers in support of petitions for registration has been blocked by court suits by the manufacturers, so there is little opportunity for independent review of EPA staff decisions; this ruling is currently under appeal.

The EPA is also responsible for regulating the labeling and use restrictions for registered pesticides. The primary difficulty with regard to the use restrictions is the extremely limited number of enforcement personnel. The EPA has contracted with state agencies, usually with the Departments of Agriculture in each state, to enforce usage restrictions such as re-entry periods in the field; in some cases this is carried out by fewer than ten officers for an entire state. Labeling regulations require only that acute health effects be listed; dermatologic and chronic effects are not included.

There are no national requirements for medical surveillance of agricultural workers, and only one state, California, requires periodic screening of applicators, the group with the greatest potential exposure. California is also the only state to require physicians specifically to report pesticide poisoning.

Other agencies have smaller pieces of the jurisdictional pie: the Department of Transportation regulates transportation of pesticides, the FAA regulates agricultural aviation, and state and local highway, police, and fire departments regulate worker exposure under emergency conditions such as warehouse fires and highway spills. State and local air and water control agencies regulate permissible levels of pesticide contamination in community exposures, and the EPA establishes food residue tolerances.

This combination of agencies and regulations can result in wide variations in safety and health conditions for agricultural workers. The following scenarios, all based on actual cases, illustrate the range of these conditions and potential health effects: an agricultural pilot is reasonably well protected, while the workers who mix and load the pesticides into his plane have no personal protective equipment and suffer periodic acute symptoms of poisoning as a result. A seasonal farmworker employed for ten years in harvesting a certain crop returns in the eleventh year to find he or she is suddenly developing headaches, sleep disturbances and a skin rash--the pre-harvest pesticides have been changed, and there was no notification. A group of farmworkers harvesting onions become sensitized to the fungicide Captan, and goes to local physicians; none of them is correctly diagnosed, none of the cases is reported, the workers are not able to return to work, but receive no compensation. Ownership of a large citrus orchard changes hands, and harvest dates are now determined by agriculturalists and market specialists far removed from that specific orchard--when the foreman thinks there might still be too much pesticide residue on the foliage, his opinion is overridden, and the pickers are ordered into the grove.

What specific policy changes in agriculture would affect this rather dismal picture? I will approach the question in terms of our basic public health strategy, that is, to emphasize first the prevention of disease and injury itself, and second the prevention or limitation of physical, psychological, and economic damage in cases where disease and injury occur.

To prevent or limit occupational disease and injury among agricultural workers, four elements must be considered: education, design of the workplace (equipment and materials), monitoring and medical surveillance, and compensation.

Education. Most agricultural workers have limited knowledge of the risks of job-related injury, and no knowledge of the risks of job-related disease. Particularly in an economic sector such as agriculture, where the workforce is geographically dispersed, improvement of work conditions cannot be expected to result primarily from enforcement actions. Education should include information on potential health risks, appropriate work practices and protective clothing and equipment, when to seek medical care, and sources of compensation for job-related injuries and illnesses. (I would like to point out that in my experience it has not been only the hired agricultural workers, but many family farmers as well, who want and need education about these issues.) Because most agricultural workers are not unionized, the industrial model of union health and safety department responsibility for education of the workforce is not available. No private institution or governmental agency has undertaken even a modest version of this educational effort; without such an effort, however, much of the rest of preventive efforts will have limited value.

It is not only the worker or farmowner, of course, who needs education regarding the occupational hazards of agriculture. Medical providers, rural clinic administrators and outreach or social service workers, public health professionals such as environmental sanitarians, and community health educators should all be provided with continuing education courses on the evaluation and prevention of these hazards, or the recognition and treatment of the injuries and illnesses they produce. Our NIOSH Regional Office has conducted more than 25 training sessions of this type in cooperation with other agencies over the past 4 years, for more than 800 health professionals; the result has been a significant increase both in requests for information and consultation and in the interest of attending practitioners in these problems.

Certification of farm equipment and pesticide health and safety. Safety experts have been involved in the development of new equipment, and changes such as the installation of roll bars on tractors have led to reductions in injuries; currently existing equipment needs to be further revised according to the patterns of injury among workers operating such equipment. In addition, however, the long-term health effects of poorly designed equipment such as arthritis and other musculoskeletal disease should be considered. It does not help the worker a great deal to save him from broken bones only to condemn him to crippling back pain or arthritis.

Prevention of pesticide-related illnesses requires policy developments in two areas. With the exception of worker education programs as described above, the greatest improvement in worker protection would result from improved registration procedures: (1) submission of more data on chronic and low-level health effects should be required for registration; (2) all data submitted in support of registration applications, past and present, should be available for public and scientific scrutiny outside of EPA; (3) alternatives for pest control should be more systematically considered before granting registration; (4) more rapid mechanisms of response to new evidence of health effects for currently registered pesticides should be developed; and (5) labeling requirements should include information on chronic health effects, dermatitis, and more detail on protective equipment and clothing.

In addition to improvements in pesticide registration procedures and policies, enforcement of regulations regarding pesticide use must also be strengthened. Enforcement by state agencies at present is fraught with contradictions, because in most states enforcement authority is vested in Departments of Agriculture, which are charged with the primary task of promoting agriculture; health and safety is usually not a major priority for these agencies, they have few personnel trained in health, and they have extraordinarily small budgets with which to protect worker health and safety.

Transfer of enforcement responsibility to OSHA has been proposed, because OSHA has greater numbers of personnel trained in health and safety, and to avoid the conflict of interest within some Departments of Agriculture. While such a renegotiation of the present interagency agreement between EPA and OSHA might relieve some of the most pressing inadequacies in enforcement, we must recognize that inspectors from either agency will never reach more than a very small percentage of workplaces in agriculure.

The primary responsibility for enforcement of pesticide use regulations lies with the growers and farmworkers themselves; to facilitate this task, however, the following changes are necesary: education for both the growers and the farmworkers; stiffer economic sanctions for noncompliance; and legal protection against grower retaliation for farmworkers who request information or investigation of pesticide use in their own workplace.

Monitoring and medical surveillance. The only way to know whether policy changes are improving worker health and safety is to collect information on the health status of workers in a systematic and useful way. This is not done in any state at present, and is part of the reason that we have to patch together our estimation of farmworker health indices as roughly as I have demonstrated earlier. Simple approaches, such as requiring the inclusion of occupation on death certificates, or conducting periodic health surveys on sample populations, have yielded a tremendous amount of very useful information in industrial populations, and there is no reason why this should not be undertaken with respect to agricultural populations. I will not discuss this in greater detail here, but would like to point out that unless agricultural researchers and leaders manifest some interest in the health of their workforce, we in public health will continue to have difficulty in pressing for the implementation of such a data collection system.

Medical surveillance, which means the periodic screening of workers for health effects of their known occupational exposures, is currently required for workers directly handling pesticides only in California. Based on the statistics on acute pesticide poisoning which I reviewed earlier, extension of this program to other states would have the most immediate and direct impact on overall pesticide-related illness.

Compensation for occupational injury and illness. Farmworkers are not covered by workers compensation in half the states of the country, and do not have parity with industrial workers in 13 more states. This gives us a good place to start in making policy recommendations. The reasons for extending compensation coverage, for improving its adequacy, and for educating workers about their rights to compensation, however, are more than simple humanity and decency. In addition, workers compensation is the only existing mechanism whereby industries are required to partially internalize the economic costs of the disease and injury created by the working conditions which they maintain. The U.S. Department of Labor estimated in 1980 that less than 5 percent of the direct economic costs of all occupational disease in industry is reimbursed under workers compensation; this figure, although unknown for agriculture, would certainly be lower. Improvement and national standardization of workers compensation for agricultural workers--particularly in view of the migrant labor force--would contribute significantly to the improvement of occupational safety and health in this sector.

CONCLUSION

Most workers involved in U.S. agricultural production face hazards to their health as a price of employment. These hazards are preventable, and agricultural worker health and safety programs and legislation are in fact significantly better in some European countries, particularly Scandinavia. Changes in agricultural technology are creating new problems as well as solving old ones, and the health effects of these changes usually occur without being recorded, commented upon, or considered in the further evolution of these technologies.

As long as there is an ample supply of poor and unorganized workers in agriculture, it will be difficult for the workers themselves to challenge these undue risks. Improvements in their working conditions come largely as incidental effects of changes demanded by other occupational groups, or by consumers. In this chapter I have described the actual extent (so far as it can be measured) of occupational health and safety risks for agricultural workers, and outlined those areas in which we, as scientists involved in the development of technology in agricultural production and in the development of agricultural policies, can incorporate these concerns into our own work and policy recommendations.

Over the past 15 years industrial workers have increasingly challenged the exclusive right of scientists and corporate owners and managers to make decisions regarding production technology and occupational health. This process is weaker in agriculture, because, as I have noted before, the labor force is significantly less organized. Even if the agricultural workforce were organized, of course, the wide range of occupational health risks in agriculture profiled here would not disappear. In the absence of such organization, however, the need is even greater for research, policy development, legislation and enforcement which will improve the health and safety of workers in agriculture, and strengthen the sustainability and security of agricultural production.

NOTES

1. Although agricultural chemicals are used as pesticides, herbicides, fungicides, fertilizers, growth stimulants and for other purposes, the generic term "pesticides" is widely used to include all of the above. For example, Environmental Protection Agency programs involving agricultural chemicals are housed in the Office of Pesticide Programs.

2. OSHA is an agency of the Department of Labor, charged with promulgation of standards to protect worker health and safety and with the enforcement of these standards. NIOSH is an agency of the Centers for Disease Control, Public Health Service, Department of Health and Human Services, charged with the development and recommendation of standards, research, and the training of occupational health professionals. The Environmental Protection Agency is an independent agency within the executive branch of the federal government. Approximately half of the states have assumed responsibility for the administration of OSHA programs; in the other half, OSHA is directly administered by federal representatives.

REFERENCES

Blair, A. 1983. "Lung Cancer and Other Causes of Death Among Licensed Pesticide Applicators." Journal of the National Cancer Institute 71(1):31-37.

Blair, A., and Thomas, T. L. 1979. "Leukemia Among Nebraska Farmers: A Death Certificate Study." American Journal of Epidemiology 110:264-73.

Brown, J. R.; Chai, F. C.; Chow, L. Y.; et al. 1978. "Human Blood Cholinesterase Activity--Holland Marsh, Ontario, 1976." Bulletin of Environmental Contamination and Toxicology 19:617-23.

Bureau of Community Health Services. 1979. "Farmworkers in the U.S." Internal data. Public Health Service. DHEW.

Burmeister, L. F., and Morgan, D. P. 1980. "Mortality in Iowa Farmers and Farm Laborers, 1971-1978." Office of Pesticide Programs Report, Environmental Protection Agency.

Burmeister, L. F.; Van Lier, S. F.; and Isacson, P. 1982. "Leukemia and Farm Practices in Iowa." American Journal of Epidemiology 115(5):720-28.

Burmeister, L.F.; Everrett, G. D.; Van Lier, S. F.; and Isacson, P. 1983. "Selected Cancer Mortality and Farm Practices in Iowa." American Journal of Epidemiology 118(1):72-77.

Carlson, M. L., and Peterson, G. 1978. "Mortality of California Agricultural Workers." Journal of Occupational Medicine 20(1):30-32.

DHEW. "Vital Statistics Special Report." 1963. U.S. Dept. of Health, Education and Welfare, National Center for Health Statistics. Washington, D.C.: U.S. Government Printing Office.

Division of Labor Statistics and Research. 1982. "Occupational Skin Disease in California." San Francisco: California Department of Industrial Relations, January.

______. 1981. Occupational Injuries and Illnesses Survey: California, 1979. San Francisco: California Department of Industrial Relations, May.

Kahn, E. 1976. "Pesticide Related Illness in California

Farm Workers." Journal of Occupational Medicine 18(10):693-96.
Levin, H. S., and Rodnitzky, R. L. 1976. "Behavioral Effects of Organophosphate Pesticides in Man." Clinical Toxicology 9(3):391-405.
Levin, H. S.; Rodnitzky, R. L.; and Mick, D. L. 1976. "Anxiety Associated with Exposure to Organophosphate Compounds." Archives of General Psychiatry 33:225-28.
Levy, B. S. 1981. "The Teaching of Occupational Health in American Medical Schools." Annals of Internal Medicine 95:774-76.
Metcalf, D. R.; and Holmes, J. K. 1969. "EEG, Psychological, and Neurological Alterations in Humans with Organophosphorus Exposure." Annals. New York Academy of Sciences 160:357-65.
Milham, S. 1976. "Occupational Mortality in Washington State 1959-1971." DHEW/NIOSH Research Report, Pub. No. 76-175-C.
Pechan, D., and Jansson, E. 1980. "What Can Be Done to Reduce Cropdusting Aircraft Crashes and Pilot Injuries in the United States." Report from Friends of the Earth, July 1980, Washington, D.C. Figures cited based on FAA data.
Peterson, G. R., and Milham, S. 1980. "Occupational Mortality in the State of California 1959-61." DHEW/NIOSH Research Report, Pub. No. 80-104.
Quinones, M.A.; Bodgden, J. D.; Louria, D. B.; and Nakah, A. E. 1976. "Depressed Cholinesterase Activities Among Farm Workers in New Jersey." Science of the Total Environment 6:155-59.
Richter, E. D.; Gordon, M.; Halamish, M.; et al. 1981. "Death and Injury in Aerial Spraying: Pre-crash, Crash, and Post-crash Prevention Strategies." Aviation, Space and Environmental Medicine (January)53-56.
Runsten, D., and LeVeen, P. 1981. "Mechanization and Mexican Labor in California Agriculture." Monographs in U.S.-Mexican Studies, No. 6, 1981. Program in U.S.-Mexican Studies, Univ. of California, San Diego, La Jolla, California.
Shakman, H. E. 1974. "Nutritional Influences on the Toxicity of Environmental Pollutants." Archives of Environmental Health 28:105-13.
Spigiel, R. W.; Gourley, D. R.; Holcslaw, T. L.; et al. 1981. "Organophosphate Pesticide Exposure in Farmers and Commercial Applicators." Clinical Toxicology Consultant 3:45-50.
Stemhagen, A.; Slade, J.; Altman, R.; and Bill, J. 1983. "Occupational Risk Factors and Liver Cancer." American Journal of Epidemiology 117(4):443-54.
Task Group on Occupational Exposure to Pesticides. 1975. Occupational Exposure to Pesticides. Report to the Federal Working Group on Pest Management. U.S. Government Printing Office No. 0-551-026.

Whiting, W. B. 1975. "Occupational Illnesses and Injuries of California Agricultural Workers." _Journal of Occupational Medicine_ 17(3):177-80.

Nutrition and Food Processing

10
Food Security in the United States: A Nutritionist's Viewpoint

Joan Dye Gussow

An analysis of food security from the standpoint of nutrition must begin, of course, with the simple question of whether there will be enough food. But we must then step back and ask what is meant by food. In this article, food will refer to substances, whether plant or animal, that are commonly judged as edible in the United States (thus excluding dogs, cats, insects and a variety of other substances which--while nutritious--are not acceptable as food to most people in this society). The term food will not, however, be assumed to refer to the precise mix of edible substances presently consumed in different parts of the country--as reflected, for example, in the most recent USDA survey of household food consumption (USDA, 1982). In other words, the possibility that food security may in the long term involve a changed dietary pattern is not excluded.

From a nutritional standpoint, to ask whether there will be enough food is to ask on the most basic level whether there will be enough calories. In order to be nourished humans need to consume enough calories to meet the growth, maintenance, and activity needs of their bodies. What "enough" means varies by age, size, sex--and by activity level (National Research Council, 1980). Adequacy of calorie intake is most easily measured by looking at body weight. The fact that overweight is considered a significant health problem in the United States while underweight is not suggests what is true, that the U.S. population as a whole now has access to many more calories than it needs.

There are a few imaginable scenarios that might result in the reduction of total calories available in the U.S. to below subsistence levels. A not necessarily complete list of these not necessarily in order of likelihood would include: nuclear war, other ecological catastrophes, and economic concentration.

The effects of a nuclear exchange are so unpredictable as to make further discussion of its potential effect on food availability pointless, since nuclear war would obviously affect the demand for food as well as its supply in unanticipatable ways (Ambio, 1982). A variety of other ecological catastrophes are also possible--e.g., the levels of carbon dioxide in the atmosphere might get high enough to raise the mean global temperature so as to melt the polar icecaps and flood the coastlines; or alternatively, air pollution might block enough solar radiation to lower the mean temperature and bring on a new ice age. Or forest destruction in the tropics might so alter rainfall patterns as to turn present croplands--even in the U.S.--into deserts. None of these scenarios is impossible though none of them is certain. All would affect future food availability (which is already being affected by such ongoing though less dramatic events as soil erosion, ground water mining, and acid rain).

Another eventuality that could affect national food availability is continued economic concentration. The nation might allow food producing resources (e.g., farmland, fertilizer, phosphates, crop germ plasm, water) or the major commodities themselves to fall under the control of those who would use them up wastefully, withhold them entirely in time of need, or charge us more than we could pay (see George, 1977; Hightower, 1975; Kramer, 1977; Morgan, 1979; Perelman, 1977).

On a regional level, it is possible to imagine the urban areas running out of food if there were a prolonged truckers' strike, a breakdown of the country's social structure, or regional warfare of the kind that began to emerge during the first energy crisis when bumper stickers in the oil and gas-producing states carried the unfriendly suggestion: "Let the bastards freeze in the dark."

For present purposes, however, it will be assumed that these potential problems will all be dealt with elsewhere in this volume, and that those concerned with such crop inputs as water and soil, and with such other aspects of the food supply system as marketing and transportation, will take responsibility for seeing to it that the population has enough food, and hence enough calories.

From a nutritional standpoint, then, the next major concern would be protein, a substance whose etymology (from the Greek proteios, meaning of "prime importance") reflects its centrality in nutrition. The U.S. diet has long put heavy emphasis on meat and other animal products. Although various reasons for this have been advanced, it is clear that availability has much to do with it. The U.S. meat emphasis is obviously related, at least in part, to the fact that the country began with such an abundance of animal protein. Some of it was in the form of fish and game; anything that moved--from cod and lobster to buffalo and passenger pigeons--was consumed in astonishing numbers (Root and deRochemont, 1976). Another source of meat was the ubiquitous hog, whose meat was readily preserved by curing and who could be left to forage in the country's extensive woodlands or to feed on its native corn (Cummings, 1970; Hess and Hess, 1976). In more recent times, this dietary emphasis on meat has been encouraged by a powerful livestock industry which is quick to complain to its Washington representatives whenever the suggestion is made that reduced consumption of certain animal products might be beneficial to health(1) (see, e.g. Hausman, 1981).

The national overconcern with protein--especially animal protein--helped contribute to the heavy emphasis once put on this nutrient in world hunger circles. In the 1950s the United Nations formed a Protein Advisory Group (PAG, 1974) to work on increasing protein production around the world. A great deal of interest was generated in such "solutions" to the world food problem as fish protein concentrate, amino-acid fortification of various grain products, and so on (Altschul, 1974; Pariser, *et al*., 1978). Reality ultimately obtruded, however, when it began to be recognized that protein *per se* was not a problem; most people in the world could meet their protein needs if they could only get enough of whatever they were eating as a major staple (F.A.O., 1975). That is, if they could get enough calories, they could get enough protein. The Protein Advisory Group was duly transmuted into the Protein-Calorie Advisory Group (PAG, 1974).

There are a few exceptions to the rule that calorie adequacy ensures protein adequacy: babies and young children need a higher percentage of protein in their diets than can normally be provided without including some foods relatively higher in protein than--for example--wheat. And it is difficult to get enough protein on a diet made up largely of fruit, plantains, sweet potatoes, or cassava. These exceptions are essentially irrelevant to the present discussion, however, because North Americans are not heavy consumers of sweet potatoes or cassava and, what is more important, the U.S. midwest shares with Canada the greatest grain and soybean growing region in the world. In planning for any imaginable dietary future, crops from this area would obviously be a primary resource.

Moreover, most of that prime farmland now grows grains and soybeans that are either fed to animals or exported so that there is an enormous margin of safety, even assuming that some crops would continue to be exported for food emergencies elsewhere. Finally, it must be assumed that any rational food security system would include plans to use, for grazing animals, land in various parts of the country unsuitable for plant crops. Thus animals could be raised even if we stopped entirely the practice of feeding them grain and soybeans. It is likely, therefore, that the U.S. population would almost inevitably have not only an abundance of protein but ample animal protein, assuming--as was noted earlier--that those concerned with creating a sustainable agricultural system have seen to it that we produce enough food and make certain it is controlled by friends.

What about vitamins and minerals? For the most part they occur in surprising abundance in this protein-calorie package. Although iodine may be low in areas where soils are iodine deficient(2), consumption of enough seeds (whole grains and legumes) to meet calorie and protein needs would go a long way toward meeting trace mineral and magnesium requiremments, especially if care is taken in fertilizing soils so as to replace minerals removed by erosion or previous crops, or to add minerals (such as zinc) in areas where soils may be naturally low. Iodine (extractable from the ocean) could be provided, where necessary, by iodizing salt. These same plant foods would supply many of the needed vitamins as well--the B-complex especially--with significant additional quantities coming from even relatively small amounts of animal products (see Table 10.1). As was earlier indicated, these are likely to be available in any future U.S. food system. With animal foods available, B-12 would also be amply supplied; in any case, it could be provided by products made from fermented soybeans, a crop the U.S. grows in abundance.

What is there left to worry about? Vitamins A, C, and D. Leaving aside the possibility of providing vitamin D synthetically, which is how most of it is now provided, it is not unrealistic to imagine a lifestyle that would take people out of doors so that Vitamin D could once again be treated as a hormone. Vitamins, of course, are substances that must be provided by the diet because the body is unable to make enough of them fast enough. Hormones are substances made in the body. Vitamin D was a hormone until the indoor work and air pollution associated with industrialization prevented people from getting enough sunshine to make enough of their own vitamin D.

As for vitamins A and C, they are abundant in plants--that is, vitamin C is abundant. Plants have no vitamin A but rather carotenes, substances which animals can convert to vitamin A. These nutrients are found in green leaves and fruiting parts of plants so that it is hard to imagine--absent a catastrophe--any kind of situation in which the population of the U.S. would suffer an absolute shortage of them.

It _is_ imaginable, however, that many parts of the country _might_ need to derive their ascorbic acid from something other than orange juice. That is not a source of concern, for not only is there much more variety in vegetables than is popularly recognized, but they are popularly under-rated as sources of many nutrients, including protein. As Table 10.1 shows, for example, broccoli has as much protein per hundred calories as beef. It is also high in available calcium, high in potassium (but appropriately low in sodium), a good source of iron, thiamin, niacin and riboflavin, and an excellent source of carotene and vitamin C. And it grows well even in the northeast (where citrus fruit is not produced). This is merely a single illustration of the ease with which we could provide enough vitamins and minerals to the populace in quite a palatable form--so long as there is enough food. And so long as we do not lose sight of the fact that food is not limited to the grains and oilseeds used as world bargaining chips and thus most attended to in discussions of the "world food crisis."

Suppose, however, that agriculture were re-regionalized--because of energy shortages, system failures, bioregional warfare or other catastrophes--so that the diet was necessarily more seasonal and local than it is now. Regional diets, it is obvious, can be nutritionally adequate; if they could not be the country would not now be populated--at least not by former Europeans. The settlers would have perished, even with the help of the Indians (see, e.g., Haughton, 1982). Moreover, given present methods of food preservation not available to the early residents, there should be no real barrier to maintaining adequate regional food supplies all year, even in Maine--even perhaps in Alaska--although food habits would have to change. The spartan diet whose nutrient value is presented in Table 10.1 serves merely as an illustration of the ease with which adequate nutrients can be obtained from a diet made up largely of grains, beans, and hardy vegetables with a small amount of added meat and milk. In less than 1300 calories, a diet consisting of just over 2 1/2 cups cooked bulgar wheat, a cup of soybeans, one stalk of broccoli, one cup of kale, 1 cup of milk, and a small hamburger (1/4 pound before broiling) meets the Recommended Dietary Allowances for an adult male of every nutrient listed except (perhaps) vitamin B-6 and magnesium(3).

The intention here is not to recommend a diet like that in Table 10.1, but to illustrate the ease with which nutritional adequacy can be achieved with whole foods, even within the confines of a highly restricted diet. Thus in one sense, the food security question is very easy to answer. Is there in the foreseeable future any obvious technical or biological obstacle to assuring everyone in the U.S. access to domestically-produced and safe foods from which to compose a nutritionally adequate diet? The answer to that question is "no." Once having established that, however, a more difficult question arises, namely: are there problems in the way of actually achieving such national food security? And that question must be answered in the affirmative. Many of these problems fall outside the scope of this chapter. But one of them

TABLE 10.1
Nutritive Value of Six Common Foods Compared to the Recommended Dietary Allowances

Foods	Calories (kcal)	Protein (gm)	Fat Total (gm)	Fat Satd (gm)	Fat Poly (gm)	Carbohydrate (gm)	Crude fiber (gm)	Vitamin A (IU)
2.7 cups cooked bulgar wheat	602.	19.0	2.6	0.0	0.0	128.7	2.9	0.
1 cup cooked soybeans	234.	19.8	10.3	1.5	5.3	19.4	2.9	54.
1 lg. stalk boiled broccoli	73.	8.7	.8	0.0	0.0	12.6	4.2	7000.
1 cup boiled kale	43.	5.0	.8	0.0	0.0	6.7	1.2	9130.
1 cup skim milk	88.	8.8	.2	0.0	0.0	12.5	0.0	10.
2.9 oz. hamburger-Med. well	235.	19.8	16.6	8.0	.3	0.0	0.0	30.
Total	1275.	81.1	31.3	9.5	5.7	179.9	11.2	16224.
R.D.A. adult male		56.						5000.

TABLE 10.1 (continued)
Nutritive Value of Six Common Foods Compared to the Recommended Dietary Allowances

Foods	Thiamine (mg)	Riboflavin (mg)	Niacin (mg)	Vitamin C (mg)	Calcium (mg)	Phosphorus (mg)	Iron (mg)	Sodium (mg)	Potassium (mg)
(bulgar wheat)	.5	.2	7.7	0.0	49.3	575.	6.3	.007	.389
(soybeans)	.4	.2	1.1	0.0	131.4	322.	4.9	.004	.972
(broccoli)	.3	.6	2.2	252.0	246.4	174.	2.2	.028	.748
(boiled kale)	.1	.2	1.8	102.3	205.7	64.	1.8	.047	.243
(skim milk)	.1	.4	.2	2.5	296.5	233.	0.0	.127	.355
(hamburger)	.1	.2	4.4	0.0	9.0	159.	2.6	.049	.222
Total	1.4	1.8	17.3	356.8	938.3	1526.	17.8	.262	2.929
R.D.A. adult male	1.4	1.6	18.	60.	800.	800.	10.		

	Folate	B6	Zinc	Magnesium
Total	.416	.7	14.7	307.8
R.D.A. adult male	.400	2.2	15.	350.

Computer Analysis by: U. Mass. Nutrient Data Bank 6/83

requires brief discussion here. Will the people and their governments--whether local, state or federal--have the political will to protect the resources--the soil, water, farmer skills, and so on--necessary to produce the food?

In thinking about such a question it is important to keep in mind that 97 percent of the population of the United States is not on farms and that increasing numbers of Americans are more than one generation removed from agriculture as a livelihood. Even granting a level of agricultural sophistication, however, it would be difficult for the New York City purchaser of Mexican asparagus, Chilean grapes, Guatemalan broccoli ("grown by Indian communities") or even California tomatoes to be knowledgeable about the sustainability of the agricultural systems which have produced these foods (Haughton, 1982:61; Kinley, 1982). As I have argued elsewhere (Gussow, 1983), the price of the product, however high it might be in a case such as South African nectarines, is unlikely to reflect the true environmental and human costs involved.

Given human nature, then, it seems clear that food security in the long run can only be achieved by greater regional self-reliance in food production and distribution--in the United States as well as in the rest of the world. Self-reliance does not mean self-sufficiency. The nation's security will not be increased by a society of homesteaders with guns prepared to take care of their own. But if people are to be aware of the impact their food demands have on the resources necessary to sustain food production, at least some of that production needs to be brought closer to home.

There is another food security problem, one which comes acutely into consciousness in the face of lengthening bread lines: How is food security to be provided for everyone, not just those who are affluent? Food, unlike most other goods our economic system allocates, is a need rather than a want. Air and water, which are the only other <u>essentials</u> for survival, have for the most part been treated as free goods in our economic system (inappropriately perhaps from an environmental standpoint). Clothing and shelter are also essential for survival in other than tropical climates, but they differ from food in that they are not in the literal sense consumed and thus do not have to be re-acquired every day. They can, as a consequence, often be improvised by the poor from other people's cast offs.

While a full discussion of the issue of economic justice is obviously outside the scope of this chapter, a few points need to be made in the present context. The first is that food security ought to include everyone and that it is obviously unacceptable in a country that produces so much more food than it can consume (and that purports to be civilized) that people should suffer from hunger or malnutrition. The second point is that even if national security could be achieved by some combination of missiles and battleships it will clearly not be achieved while those are being purchased at the expense of the hungry. Third, food security for all will not be achieved by punitively administered, reluctantly funded food programs whose level of adequacy is at the mercy of each new class of congresspersons or each new economic theory. Nor is food security in the U.S. to be achieved by giving the poor humiliating access to food pantries which industry is encouraged to fill--for a handsome tax benefit--with products which have been unsuccessful in the frantically competitive food marketplace. Outdated canned egg salad does not represent food security. Food security, in short, requires that some way be found to give everyone in the population fair access to the common food supply.

But if food security at any given moment depends on access to the food supply that exists at any given moment, then it is necessary to examine whether nutritional problems are likely to emerge if the food system we presently have in place, and the economic system that drives it, continue to develop in their present direction. It is this issue which the remainder of this chapter will address.

The task of nutrition science is to "formulate a diet over the lifetime of an individual that will optimize health, well being and longevity," a task that "calls for providing the necessary chemical components in the right proportion and avoiding or minimizing toxic substances" (Morowitz, 1980). In practice, what is needed is a food supply from which a reasonably informed populace can select the mix of necessary chemicals while avoiding substances that may prove dangerous over the short, or long, runs. Given the fact that individual nutritional requirements as well as individual tolerances for toxins may vary markedly (see Williams, 1956; Yew, 1975), and given the sheer numbers of animal and vegetable substances from which choices must be made, the fact that humans have survived at all begins to seem quite remarkable.

Eaters who are specialists have an easier task. Herbivores are programmed to like only certain vegetable substances and they meet their nutrient requirements by following their desires (Yudkin, 1978). Carnivores can be wired with a nutritional program which says simply "pursue and consume warm moving things"--so that a lion will only become vitamin deficient by consuming a vitamin-deficient zebra (Rozin, 1976). But humans are, like rats, generalists in their food selection. They therefore have the terrifyingly difficult task of getting enough of the right kinds of nutrients within rather strict caloric limits without ingesting poisons. It is not known how they do this--though there is some helpful internal wiring that assists the tissues in maintaining appropriate sodium and water levels and there are sensors that attract the human tongue to sweet things and warn it away from bitter poisons (see Steiner, 1977).

But what is clear is that as far as humans are concerned most of the choice of what will be viewed as tasty food is determined by the culture they were born into. Geographic and cultural constraints, in other words, determine which substances will be available to eat and which of the available substances will be interpreted by a given culture as edible and satisfying. That is why it was earlier pointed out that housepets and insects would not be considered food here, although they are considered so elsewhere. These cultural differences account for the shock that occurs when people from other cultures bring their definitions of food into our society--as when Indochinese refugees in California are found hunting squirrels, stray dogs, and pigeons in Golden Gate Park (New York Times, 1980).

Cultural wisdom acquired over the millenia--presumably sometimes painfully--was the only available mechanism for keeping civilized humans alive and even relatively healthy before nutrition science was discovered (4). What new information has that science provided? Here it will be useful to quote from one of the more perceptive outside observers of the field: "The information available to diet planners consists of a small body of universally accepted results such as the pathways of intermediate metabolism; a set of direct minimum requirements to avoid dietary deficiencies; data on toxic substances and level of acute toxicity, and a very large body of results--many of which do not measure up to the minimum standards of statistical acceptability" (Morowitz, 1980).

Without going into detail it can be said, then, that although we know enough about human nutrient requirements to keep sick people alive through intravenous alimentation (at rather formidable cost--as much as $55,000 a year by one recent estimate), we are nowhere near the point, as is sometimes optimistically asserted, where we will be able to prescribe perfect individual diets for optimum health and longevity (New York Times, 1981). Much more important, however, than our present lack of knowledge is the fact that attempting to feed the population properly on such prescribed--and therefore doubtless largely synthetic--diets would raise insoluble political and social questions. This, of course, omits entirely the issue of whether such diets would be ecologically, energetically, or economically rational--or moral--in a society that has not yet figured out how to get everyone access to ordinary foods presently in abundant supply. Finally, to imagine a population living on prescribed nutrient mixes is to imagine a degree of regimentation of the food supply--Soylent Green springs to mind--that is, to put it mildly, improbable in a country that cannot even bring itself to limit the number of colored and sugared breakfast cereals marketed to its children.

Therefore, assuming that we would not choose to be fed on the Compleat Pellet (Wolff, 1973) even if we could be, it is necessary to assume that the food supply will continue to develop along much the same lines as it has during the past decades. What does this imply for our nutritional security?

Based on present health indicators, there would seem to be no cause for concern. Although the U.S. has higher infant mortality rates than many countries, that fact has more to do with food and care not provided for pregnant and pre-pregnant women than with any identifiable problems inherent in the food supply. Moreover, life expectancy continues to rise, deaths from cardiovascular disease are going down, and the nation is not suffering an epidemic of cancer; indeed, rates of cancer at most sites have been relatively constant over the last 40 years (National Research Council, 1982; Stamler, 1982; Walker, 1977). Thus, contrary to frequent alarms in the media, there is no convincing evidence that anything that has happened to our food over the last few decades has had a negative effect on our national health--at least to date. Everything that follows must be read in light of that observation.

Within the last half century, the U.S. marketplace has drastically altered the food selection process by which humans stayed alive over the millenia. The number of items in the typical supermarket, for example, has gone from 800 to 12,000 (Molitor, 1980). What are the nutritional implications of this? Looking strictly at the measured nutrient content of the food supply, it turns out that despite extensive product proliferation, there has actually been little visible change in available nutrients (Table 10.2). The food supply changes are much more visible when per capita consumption is expressed in terms of specific food types; for example, the consumption of flour has been effectively halved since 1909 (Figure 10.1). The changes become even more visible when specific food products are considered. For example, Americans consume per capita five times as much commercially processed tomato as they did 50 years ago in the form of catsup, tomato sauce, barbecue sauce, and so on (Brewster and Jacobson, 1978). To assume that the nutritional worth of tomatoes in tomato catsup can be equated with the nutritional worth of a vine-ripened tomato simply by comparing their levels of vitamin C and carotene is to ignore two significant facts: (1) that foods consist of hundreds of chemicals put together in combinations that are capable of reproduction and respiration--that is, are alive, and (2) that we have no reason to assume, without evidence, that the only part of this complexity

Figure 10.1 Contribution of various food groups to per capita supply of food energy (calories).

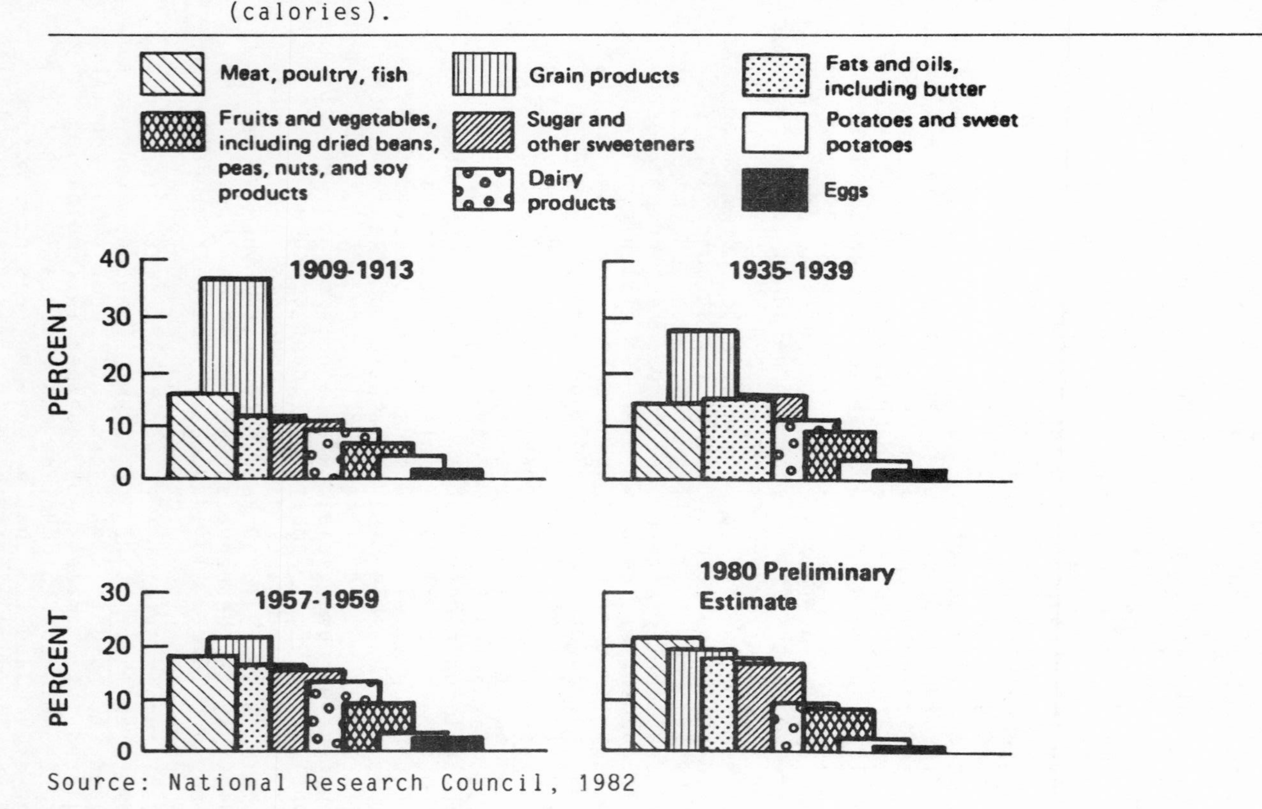

Source: National Research Council, 1982

TABLE 10.2
Daily Per Capita Intake of Nutrients in the United States

		Energy Sources			Minerals			
Year	Calo-ries	Protein (gm)	Fat (gm)	CHO(a) (gm)	Ca (mg)	P (mg)	Fe (mg)	Mg (mg)
1909	3,480	102	125	492	820	1,560	15.2	408
1927	3,460	95	134	476	860	1,510	14.1	388
1948	3,230	95	140	403	1,000	1,550	18.6	369
1965	3,150	96	144	372	960	1,520	16.7	339
1976	3,300	101	157	376	930	1,550	18.6	344

	Vitamins						
Year	B1 (mg)	B2 (mg)	Niacin (mg)	B6 (mg)	B12 (μg)	A (μg)	C (mg)
1909	1.64	1.86	19.2	2.26	8.4	2,280	104
1927	1.55	1.87	18.0	2.05	8.1	2,400	106
1948	1.91	2.30	21.4	1.99	9.0	2,640	114
1965	1.81	2.30	21.9	2.02	9.1	2,310	97
1976	2.04	2.46	25.2	2.26	9.6	2,430	118

Source: Adapted from Page and Friend, 1978.

(a) Carbohydrate.

relevant to human health is a few chemicals that happen to have been identified as nutrients.

When we deal with fabricated foods (powdered orange juice substitutes, for example) we make a further assumption--that the isolated substances we have identified as related to deficiency diseases can be combined into foods that will sustain humans. Such an assumption, in the words of a Sufi sage, is based on the notion that because we know what _one_ is and because we know that _one_ and _one_ equal _two_, therefore we know what _two_ is. In fact, we don't, because we don't know what _and_ is (Meadows, 1982). That we do not know what _and_ represents where the human diet is concerned is splendidly illustrated by the recent NAS report on _Diet, Nutrition and Cancer_ (National Research Council, 1982). Here it becomes evident that certain substances which occur quite widely in foods are powerful mutagens (therefore putative carcinogens) even though consumption of some of the foods that contain them appears to be associated with lower rates of cancer in human populations. Such a finding suggests that where food is concerned, the whole is not only _more_ than but may actually be very different from the sum of the parts.

There are other problems as well in dealing with nutrients as if they were separate substances--which in food they never are. Foods in the same family as broccoli, consumption of which has been associated with lower rates of cancer at some sites--have a variety of "nutritional" characteristics--high fiber and water content, low fat content, high levels of vitamin C and carotene, any or all of which might account for their apparent cancer-inhibiting characteristics. The brassicas contain, in addition however, several "non-nutritive" chemicals which have been shown to inhibit chemical carcinogenesis in the laboratory. It would be hard, therefore, to construct a simple "broccoli-equivalent" since it is not at all clear which of the vegetable's components, in which sorts of biochemical matrices, and in which sorts of actual human diets, are actually helping prevent cancer. As the Committee on Diet, Nutrition and Cancer concluded, therefore, the most rational diet based on present knowledge would be one rich in fruits and vegetables and whole grains, restrained in its use of smoke-and salt-cured foods, and relatively low in fat--a diet of whole foods, not of food equivalents. Yet the marketplace continues to offer new products in which food components are manipulated and rearranged so as to make them more durable, attractive, palatable--and profitable.

But the task of nutrition science as it was earlier presented is not limited to merely "providing the necessary chemical components in the right proportion" but also "avoiding or minimizing toxic substances." Toxic substances get into foods in several possible ways. There are naturally occurring toxicants, many of which humans have learned by experience to avoid--deadly nightshade berries, rhubarb leaves--or to eliminate by traditional methods of processing--like the water treatment of manioc roots to remove prussic acid in making cassava flour.

There are also contaminants. Some of these get into food intentionally but remain there unintentionally--pesticides and herbicides for example--and some of them may be accidentally introduced. Among these introductions are the PCBs spilled in North Dakota packing-house wastewater which contaminated the food supply of 17 states before their source was discovered, or the PBBs that traveled through the Michigan food chain after being accidentally added to cattle feed (Comptroller General, 1979, 1980; Office of Technology Assessment, 1979).

Other non-food substances are intentionally added to the food supply and are intended to stay there. The subject of these intentional additives is much too complex to discuss here in detail. It is enough to note that such substances are actually much less regulated than is commonly believed; of the approximately 2,700 substances intentionally added to foods only about 400 are technically "additives" and therefore subject to the rigorous testing about which food safety "reformers" complain. Not included in the term "are approximately 500 food ingredients termed GRAS (Generally Recognized as Safe) substances; about 100 other unpublished GRAS substances; approximately 1,650 flavoring agents, most of which are classified as GRAS; prior sanctioned food ingredients, consisting of about 100 substances approved by the U.S. Department of Agriculture (USDA) or the FDA prior to 1958; and approximately 30 color additives" (National Research Council, 1982). Most of these have not been thoroughly tested for carcinogenicity.

But if we leave all these "added" substances aside, it is obvious from what has been discussed earlier that no one has any idea what has been happening to the chemical composition of the food supply--additives aside--just as a consequence of the processing foods undergo. That is, even if all the additives were proven safe, how do we know whether the novel sorts of compounds produced by novel sorts of processing are equally safe.

Several years ago, Dr. William Castelli, Director of the Framingham Heart Study, testified at a hearing in Boston about the effects of processing on the fat composition of the diet (Commonwealth, 1974). Dr. Castelli was complaining about the use of totally saturated fats (listed on labels as vegetable oils) in the manufacture of numbers of food products, and he noted that although he knew these fats improved the shelf life of the product, they were unhelpful from the standpoint of those trying to combat heart disease. "The shelf life of the product is fantastic," Dr. Castelli said. "The shelf life of the guy that eats it; boom!" Unfortunately that was an exaggeration. If the consumer went "boom!" the product would fail because the next consumer would know enough not to buy it. Moreover, in this case, Dr. Castelli was talking about a substance, a fat, whose presence was known.

What may require somewhat more attention are the long-term health effects of continuous and wholly unmonitored changes in food composition. Neither consumers nor our epidemiologic methods are sophisticated enough to evaluate these. Food technologists--those persons actually involved in the molecular rearranging that brings new products to the shelves--might be expected to be least sympathetic to the notion that there is anything to worry about in what they do. It is of some interest, therefore, that in his W.O. Atwater Memorial Lecture to the 1982 Annual Meeting of the Institute of Food Technologists, Foster added to a list of hazards associated with food, "reaction products formed during processing or preparation for eating" (Foster, 1982). However, in keeping with his general theme that concern over food safety was largely unwarranted, he limited his specific examples of "reaction products" to those associated with home or restaurant cooking--the hydrocarbons produced by charcoal broiling or grilling of meat or fish. Such circumspection is to be expected from a food technologist, but more than broiled hamburger is involved. The extent of the actual information gap is better illustrated by reference to a document put out in 1980 by the Food and Drug Administration--the agency most responsible for assuring the safety of the food supply. The document is called The Bureau of Foods Research Plan, and it is, in a sense, the research wish list of the Head of the Bureau of Foods. It lays out the research which the Bureau scientists believe to be necessary to assure that the nation's food is safe and wholesome.

The document lists 719 research needs, grouping them into a series of overall goals such as "Nutritional Requirements" and "Toxicological and Epidemiological Testing." Goal C, "to isolate, purify and identify potentially hazardous food and cosmetic constituents and adulterants," includes research needs number 117 through 286 inclusive. Item 133, "Survey natural plant extracts for mutagenic/estrogenic properties; isolate and identify the responsible toxic agents," reminds us again that plants contain certain natural toxicants.

This is item 124 under Goal C: "Determine products formed as a result of chemical modification of foods, e.g. bleaching of flour with chlorine dioxide, chlorine, peroxides, and oxides of nitrogen." Flour is simply used as an example--and a relatively uncomplicated one. What is suggested here is what was pointed out earlier: we know almost nothing about the chemical compounds formed from natural foods in the course of manufacturing the extraordinary array of products in the supermarket.

Finally, item 128, "Determine from literature on production, volume and uses, and laboratory research in chemical and physical properties, which chemicals of the more that 43,000 commercially produced have the greatest potential to enter the food chain and present a human health hazard." Those are two of the 719 things the Bureau of Foods thinks it needs to know, and--more importantly--<u>does not currently know</u> in order to assure us that our food supply is safe. This is the "universe of knowledge needed by the Bureau," in the words of its director, "to fulfill its responsibilities to protect the national food supply" (USDHEW/PHS, 1980).

Many people seem to believe that someone is already on top of all this. No one is. And, what is more, no one will be. Even given the propensity of scientists to exaggerate the amount of research that needs to be done, what is outlined here is mind-boggling. The research plan contains 719 items, a large number of which are, like those quoted, each a giant research plan of its own. Clearly the FDA is desperately behind in its work, yet in August of 1982 alone 100 new edible products were introduced (not counting 34 new gourmet or health food items), contributing to an all-time record in new introductions to the supermarket (New York Times, 1982).

What is going on? In a recent article in the Ecologist, Kauber (1982) makes it clear why we cannot get the knowledge we need. Contemporary technology, he observes, is "dominated by the process of innovation....The appearance of wholly new substances" testifies "to the increasingly swift introduction, diffusion and turnover of things and ways of doing. Increasingly too 'unnatural synthetic' substances are being injected into the environment...compounds of all sorts previously unknown in nature." This rapid innovation, he points out, "undermines the experimental nature of empirical evaluation by radically increasing the number of variables required to be taken into account (43,000 chemicals, for example) as a result of prior innovations." Trouble arises "at the very start of the evaluative process. The data which ordinarily set the parameters for experimentation must arise out of prior experience with the elements from which the object or process in question is composed. But under conditions of extreme innovation, we often do not possess that kind of knowledge. Thus under conditions of rapid innovation we find ourselves encountering new substances whose combinations are poorly understood."

Ideally, therefore, one would extend the testing process, but marketplace forces urge haste; market forces also work against a slow introduction of the product, although inadequate background data and "hurried and incomplete" testing would make slow diffusion prudent. As a consequence, persons responsible at the later post-testing stages (the Food and Drug Administration, for example) "are faced with an awesome task ... to assess on a continuing basis, the long-term effects of an innovation which, we recall, has been injected into an environment already overloaded with novelties. Such testing, were it responsibly carried out, could swallow up a quantity of resources which would dwarf that of the original implementations, while the 'correction of undesired conditions' could conceivably become the sole occupation of an entire society of technicians" (Kauber, 1982). Or the entire staff of a score of FDAs.

To return then to the issue of food security, let us assume that individuals (or a nation of individuals) can only have food security by knowing how to select wisely from among the foods available to them; and assume further that they are faced--in the future as they are now--with a marketplace containing some 12,000 items, most of them bearing little resemblance to anything their neolithic ancestors might have eaten, many of them unfamiliar even to immediately preceding generations. (Many of them unfamiliar even to this generation; between January and June of 1982--a not atypical period--manufacturers introduced, among other new products, 103 new frozen foods, 48 new snacks, crackers, and nuts, 33 new beverages, and 27 new cakes, cookies, and breads.)

Each shopper is expected, in the words of one consumer publication (White, 1982) to be "aware, knowledgeable, decisive and honest"--presumably about all 12,000 items in the store. As early as 1929, when there were approximately 800 items in foodstores, made largely from familiar raw materials, home economists worried that the problem of making choices was a burdensome one (Richardson, 1929). Indeed, an even earlier champion of the informed and educated homemaker, Ellen Swallow Richards, thought that the only way housewives might protect themselves was to set up to test products (Hunt, 1912). Nowadays, of course, such testing is beyond even the capacity of the FDA.

To select among items like Tang, Start or Count Chocula, the modern consumer cannot use what one observer has called "craft skills"--the knowledge of how spoiled meat smells or how a ripe fruit feels (Leiss, 1976). Therefore, the consumer cannot make personal judgments about whether these foods are safe and nutritious, but must trust someone else--presumably the food manufacturers, the nutritionists, or the Bureau of Foods of the FDA. The fact is that increasing numbers of shoppers appear not to trust any of these people; and, as the FDA research plan suggests, they are probably wise not to do so.

What, then, can consumers do? They can't make independent judgments and they aren't sure they can trust anyone. All they can really do is avoid. And that is precisely what some of them are doing. Research done on food labeling for the FDA (Heimbach and Stokes, 1979) shows that of all persons who pay attention to the list of ingredients on the label (that is, 54 percent of <u>all</u> food shoppers), 70 percent do so to <u>avoid</u> one or more substances. They are avoiding sugar and salt; they are avoiding "preservatives/chemicals." They are turning to foods labeled "natural" and "organic" even though the government has persistently refused to define these terms in such a way as to ensure that they mean anything. The authors of the FDA study call the information about label readers "disturbing." Nutrition professionals tend to denounce such behavior as superstitious, faddist, irrational, and so on, terms that are difficult to defend when no rational counterstrategy is readily available.

What can one make of all this in relation to national food security? It is important to recall that, as was spelled out at the beginning, there is no unavoidable technical or biological reason why the U.S. cannot have a safe and nutritious food supply into the foreseeable future. In examining the problem of getting adequate nutrients it became clear that it was really not at all difficult to put together a nutritious diet from grains, beans, fruits, vegetables, and animal products (although it may be extremely hard to figure out how to put it all together when there are 12,000 products to choose from). It is not true, as a nutritionist recently commented to a group of attendees at a Short Course on Ingredient Technology, that "the American diet, good as it is, can always be improved. Increasing the number and kinds of foods available on the market shelf can help" (Leveille, 1982). More food products will not help; they will only make things worse.

Therefore: (1) If product proliferation for profit continues unabated, if manufacturers continue treating food substances like playdough, which can be molded into increasingly realistic and increasingly synthetic approximations of familiar foods without reference to the bio-chemical changes they are thereby creating; and (2) if we continue to pollute the environment--and thus, inevitably, the food producing environment--with chemicals of high toxicity and persistence, then there is no way of guaranteeing that we will continue to have the necessary, let alone optimal, levels of the good things and the minimal levels of the bad things--in short, good nutrition.

Thus if we want to ensure nutritional security, our concern ought to be to bar the introduction of and encourage the phasing out of chemicals that are both highly toxic and persistent, and to discourage the excessive processing of foods and the proliferation of highly engineered food products. These solutions, of course, raise difficult economic and political questions.

NOTES

1. As a member of the National Academy of Sciences Committee on Diet, Nutrition and Cancer, the author had access--as did other panel members--to clippings from around the country which made reference to the *Diet, Nutrition and Cancer* report, as well as to a number of letters sent to the Academy. The repetitiveness of the arguments used by opponents and the obvious access of some of the reporters and letter writers to the same "inside" information made it appear that there was a well-orchestrated campaign to discredit the report's recommendations regarding the consumption of fat and products (largely meat products in the U.S.) preserved by smoking or salt curing.

2. Seawater (and hence seafood and seaweed) contains abundant iodine as do soils which have once been seafloor and are not heavily eroded. Present levels of iodine in the U.S. diet are surprisingly high.

Per capita intake is officially estimated to range from 64 to 677 micrograms per day (RDA 150 micrograms), much of it adventitious, e.g., from iodine added to cattle feed or used to sterilize milking equipment.

3. Nutrients shown only as totals in the table are likely to be underestimated since values are not available for each of the foods reported.

4. Yudkin has argued (1981) that humans would have the same instinctive ability as other animals to select nutritious foods if they ate only substances their neolithic ancestors would have recognized--and were subjected to no outside influences. There is some evidence to support this notion in the only study ever done of dietary self-selection among young infants (Davis, 1928, 1934). The children in the study were given a choice of only unmixed, unsalted, unsugared foods and were left free to choose as much or as little of each as they wished. They selected "balanced" diets. However, they also put everything within reach in their mouths for the first few days, so they would have poisoned themselves if anything tasty but toxic had been available. Moreover, as Davis has pointed out, all the choices offered them were nutritious.

REFERENCES

Adams, C. 1975. "Nutritive Value of American Foods in Common Units." Agriculture Handbook No. 456 ARS/USDA. Washington, D.C.: Supt. of Documents, November.

Altschul, A. M. 1974. "Fortification of Foods With Amino Acids." Nature 248:643-46.

Ambio 11(February-March). 1982. "Nuclear War: The Aftermath."

Brewster, L., and Jacobson, M. 1978. The Changing American Diet. Washington: Center for Science in the Public Interest.

Commonwealth of Massachusetts Consumers' Council. 1974. Hearing on Health, Nutrition and Dietary Problems. Testimony of Dr. William P. Castelli, Director of Laboratories at Framingham Heart Study. Boston, Massachusetts. July 15. Unpublished mimeograph.

Comptroller General. 1980. Further federal action needed to detect and control environmental contamination of food. Comptroller General's report to the chairman, Committee on Appropriations, United States Senate. No. CED-81-19, Washington, D.C.: General Accounting Office, December 31.

Cummings, R. O. 1940. The American and His Food. Reprint. New York: Arno Press and New York Times. 1970.

Davis, C. 1928. "Self-selection of Diet by Newly Weaned Infants." American Journal of Diseases of Children 36:651-79.

______. 1934. "Studies in the Self-selection of Diet by Young Children." Journal, American Dental Association 21:636-40.

Food and Agricultural Organization of the United Nations. 1975. The State of Food and Agriculture, 1974. Rome: F.A.O.

Foster, E. M. 1982. "Is There a Food Safety Crisis?" Food Technology (August):82ff.

George, S. 1977. How the Other Half Dies. Montclair: Allanheld, Osmun & Co.

Gussow, J. D. 1983. "Food: Wanting & Needing &

Providing." Food Monitor (July/August):12-15.
Haughton, E. 1982. "The Cosmopolitan Radish: Procedures for Constructing a Food Guide for New York City and State in the Year 2020." Unpublished doctoral dissertation, Teachers College, Columbia University, New York.
Hausman, P. 1981. Jack Sprat's Legacy: The Science and Politics of Fat and Cholesterol. New York: Richard Marek Publishers.
Heimbach, J. T. and Stokes, R.C. 1979. FDA 1978 Consumer Food Labeling Survey. U.S. Dept. of Health, Education and Welfare, FDA Bureau of Foods, Division of Consumer Studies. Washington, D.C.: FDA, October.
Hess, J. L. and Hess, K. 1976. The Taste of America. New York: Viking.
Hightower, J. 1975. Eat Your Heart Out. New York: Crown Publishers, Inc.
Hunt, C. L. 1912. The Life of Ellen H. Richards. Boston: Whitcomb and Barrows.
Kauber, P. 1982. "Technology and the Quest for Rational Control." The Ecologist 12(2):87-92.
Kinley, D. 1982. "Case study questions the value of Reagan's 'Caribbean basin initiative.'" Multinational Monitor 3(5):17-19.
Kramer, M. 1977. Three Farms: Making Milk, Meat and Money From the American Soil. Boston/Toronto: Atlantic-Little Brown.
Langman, M. 1981. "Traditions of 40 Centuries Still Survive in China." The Soil Association (September): 9-11.
Leiss, W. 1976. The Limits to Satisfaction: An Essay on the Problem of Needs and Commodities. Toronto and Buffalo: Univ. of Toronto Press.
Leveille, G. A. 1982. "Food Fortification--Opportunities and Pitfalls." Lecture to Institute of Food Technologists Short Course on Ingredient Technology. Mimeo.
Meadows, D. H. 1982. "Whole Earth Models and Systems." Co-evolution Quarterly (Summer):98-108.
Molitor, G. T. T. 1980. "The Food System in the 1980's." Journal of Nutrition Education 12 (Suppl. 1):103-111.
Morgan, D. 1979. Merchants of Grain: The Power and Profits of the Five Giant Companies at the Center of the World's Food Supply. New York: Viking Press.
Morowitz, H. 1980. "Food For Thought." In The Wine of Life and Other Essays on Societies, Energy and Living Things, H. Morowitz,ed. New York: Bantam.
National Research Council. 1980. Recommended Dietary Allowances. 9th ed. Committee on dietary allowances, food and nutrition board, division of biological sciences, assembly of life sciences. National Academy of Science. Washington, D.C.: National Academy Press.
National Research Council. 1982. Diet, Nutrition and Cancer. Committee on diet, nutrition and cancer,

Commission on life sciences. Washington, D.C.: National Academy Press.
New York Times (December 27). 1981. "Chemical Diet Replaces Food For Patients Unable to Digest."
______. (August 14). 1980. "Indochina Refugees Accused of Poaching on Coast."
______. (July 25). 1982. "What's New: 186 More Items on the Shelf, But No Soup."
Office of Technology Assessment. 1979. "Environmental Contaminants in Foods." Washington, D.C.: Congress of the United States, Office of Technology Assessment, December.
PAG Bulletin 4:3(1-2). 1974. "PAG's Name Changed, Scope Widened."
Page, L. and Friend, B. 1978. "The Changing United States Diet." Bioscience 28:192-97.
Pariser, E. R., Wallerstein, M. B., Corkey, C. J., and Brown, N. L. 1978. Fish Protein Concentrate: Panacea for Protein Malnutrition? Cambridge: MIT Press.
Perelman, M. 1977. Farming for Profit in a Hungry World. Montclair: Allanheld, Osmun and Co.
Richardson, A. E. 1929. "The Woman Administrator in the Modern Home." Annals of the American Academy of Political and Social Science 142:21-32.
Root, W. and deRochemont, R. 1976. Eating in America: A History. New York: William Morris and Co.
Rozin, P. 1976. "The Selection of Foods by Rats, Humans and Other Animals." In Advances in the Study of Behavior, eds. J. Rosenblatt, et al. New York: Academic Press.
Stamler, J. 1982. "Current Topics in Biostatistics and Epidemiology." Biometrics Supplement, (March): 95-114.
Steiner, J. E. 1977. "Facial Expressions of the Neonate Infant Indicating the Hedonics of Food-related Chemical Stimuli." In Taste and Development: The Genesis of the Sweet Preference, ed. James M. Weiffenbach. National Institute of Dental Research. DHEW Pub. No. (NIH) 7-1068. Bethesda: DHEW (NIH).
USDA. 1982. Food Consumption: Households in the United States, Spring 1977. Nationwide food consumption survey 1977-78, Report No. H-1. Washington, D.C., Superintendent of Documents, September.
USDHEW/PHS. 1980. Bureau of Foods Research Plan. 1980. Food and Drug Administration, Bureau of Foods.
Walker, W. J. 1977. "Changing United States Life-style and Declining Vascular Mortality: Cause or Coincidence?" New England Journal of Medicine 297(3):163-65.
White, J. G., ed. 1982 The Consumer's Right to Know. Glenview, Illinois: Kraft, Inc., February.
Williams, R. 1956. Biochemical Individuality. New York: Wiley.
Wolff, R. J. 1973. "Who Eats For Health?" American Journal of Clinical Nutrition 26:438-45.

Yew, M. S. 1975. "Biological Variation in Ascorbic Acid Needs." Annals of the New York Academy of Sciences 258:451-57.

Yudkin, J. 1978. "Introduction." In Diet of Man: Needs and Wants, ed. J. Yudkin. London: Applied Science Publishers.

______. 1981. "Objectives and Methods in Nutrition Education--Let's Start Again." Journal of Human Nutrition 35:205-13.

11
Safety Aspects of Processed Foods

Dietrich Knorr and Katherine L. Clancy

INTRODUCTION

There is considerable debate about the excessive use of chemicals in highly processed foods and a concomitant interest in foods processed with fewer chemicals (Binger, 1981; Hall, 1976; Knorr, 1982). A recent study in the Federal Republic of Germany (IFOAM Bulletin, 1982) showed that chemically-contaminated food is the main fear of 47 percent of the German population, while car accidents rank second (45 percent) and war ranks third (34 percent).

Several North American studies have noted the discrepancy among various groups with regard to the actual and perceived hazards in the food supply (Hall, 1979; Bates, 1981). Although microbiological contamination should probably be regarded as the most serious short-term hazard, the concern about food additives and the lack of predictability for their long-term effects poses a need for increased technical, educational, and political emphases on both issues.

A large number of different kinds of compounds that are present in food act to increase nutritional requirements and have been referred to as nutritional stress factors (Lepkovsky, 1953). These nutritional stress factors interfere with the transfer of nutrients from the environment to the cell of the human or animal organism. They are specific and exert their effects by: (1) decreasing food intake, (2) interfering with the digestion of food, (3) decreasing the absorption of nutrients from the gastrointestinal tract, and (4) decreasing the utilization of or increasing the destruction of absorbed nutrients.

Naturally occurring stress factors (Table 11.1) can either be destroyed by heat during processsing or can sometimes be counteracted by restoring the essential nutrients to the diet (e.g., calcium makes oxalic acid insoluble). There are, however, stress factors which are still unknown where the avoidance of excessive intake of such products is the only counteraction known so far (Lepkovsky, 1953).

TABLE 11.1
Some Nutritional Stress Factors Normally Occurring in Foods

Source	Inhibitor	Action
Beans, lima and soybeans	Antitrypsin	Prevents protein digestion
Cabbage, kale, rutabagas	Goitrogens	Prevents synthesis of thyroxine
Cereals	Unknown	Binds niacin
	Phytin	Makes calcium unavailable
Corn, Millet	Unknown	Decreases effectiveness of niacin
Cottonseed oil	Sterculic acid	Interferes with reproduction
Egg white	Ovamucoid	Prevents protein digestion
	Avidin	Binds biotin
	Conalbumin	Binds iron
Fish, Clams	Thiaminase	Destroys thiamine
Milk, Yogurt	Lactose	Diarrhea and loss of nutrients
Onions	Unknown	Produces anemia
Peas	Nitrile	Interference with sulfur metabolism (collagen formation)
Potatoes (immature or sprouting)	Solanine	Vomiting and diarrhea--loss of nutrients
Spinach, Rhubarb	Oxalic acid	Binds calcium

Source: Knorr, 1980a. Reprinted with permission from AB Academic Publishers.

Stress factors which do not normally occur as food constituents can originate (1) in processing (e.g., food additives), (2) in the spoilage of food (see also Table 11.2), or (3) in the environment (soil, water, air).

The effect of spoilage of foods has been reviewed by Lechowich (1979) and by Morisetti (1971). Foster (1982) reported that in 1978 there were 154 incidents of foodborne disease in the U.S. involving 5,000 cases in which the cause of illness could be established, and another 327 incidents where 5,700 cases were recognized as foodborne, but the causal agents could not be identified. Foster (1982) continues:

Not shown... because we have no figures for illness, but potentially one of the most serious biological agents of all, are the fungal poisons called mycotoxins. The best known and probably the most important of these is aflatoxin, which was discovered a little over 20 years ago in moldy peanut meal. Aflatoxin is a potential liver carcinogen in a broad range of animal species. Epidemiological studies in several tropical areas support the assumption that aflatoxin can cause liver cancer in man.

Environmental contaminants have been organized by Pond et al. (1980) as follows:

(1) industrial chemicals accidentally contaminating animal feeds, such as the PBBs (polybrominated biphenyls) and PCBS (polychlorinated biphenyls), which have very long biological half-lives in living systems;
(2) industrial chemical pollutants, toxic trace metals, etc., which may enter ground water supplies and ultimately become incorporated into animal-derived products;
(3) trace chemical contaminants in animal products resulting from ingestion of feeds containing residues of agricultural chemicals, such as pesticides, fungicides, and herbicides, as well as degradation products of these chemicals caused by light; and
(4) classes of chemicals such as plasticizers and residual monomers which may enter food products by diffusion transfer from polymeric package materials used for wrapping, storage, and display of these products.

A very recent example of mainly still unknown pollutants is the occurrence of acid rain and its consequences (Pearce, 1982).

The influence of processing and the environment on nutritional stress factors has been summarized by many authors and will not be discussed here (cf. Clydesdale, 1979; Harris and Karmas, 1975; Synge, 1976; Tannenbaum, 1979).

FOOD CHEMICALS

Estimates for the number of chemicals used in foods at one time or another range from 1,900 to 2,500 (Bernarde, 1979; Kermode, 1972). Reasons for the need of food additives and a classification of their uses (Table 11.2) have been given by Walker (1976), and a classification of food additives is presented in Table 11.3.

TABLE 11.2
The Need For Food Additives

Need	Objective	Examples
Economic	Prevent spoilage Prolong shelf-life facilitate distribution	Preservatives Antioxidants
Technological	Facilitate processing Assist in the formulation of new and convenience foods	Flour improvers, emulsifiers, stabilizers, wetting agents
Nutritional	Replace nutrients lost during processing Supplement diets	Vitamins Minerals
Cosmetic	Render food attractive to the consumer	Colors, flavors

Source: Walker, 1976. Reprinted with permission from The Journal of Biosocial Science.

In addition to the use of food additives during processing a wide range of agricultural chemicals is used in the production of foods (Table 11.4). For example, it has been estimated that approximately 100,000 formulations of about 900 pesticides are being produced in amounts totalling 900,000 metric tons annually as of 1972 (Sissons and Telling, 1979). Extensive reviews on agricultural chemicals have recently been presented by Kadam et al. (1981) and Sissons and Telling (1979).

Food additives, unlike the agricultural chemicals such as insecticides, fungicides, and herbicides, are not designed to be toxic and most of them would have to be ingested in large single doses to produce acute toxicity. It is therefore difficult to determine their possible hazards to man, and as Kermode (1972) explains:

> there will always be an area of doubt concerning the possible effects of ingesting small amounts of additives over the course of a lifetime. One cannot be fully sure of the safety of an additive until it has been consumed by people of all ages in specific amounts over a long period of time and has been shown conclusively by careful toxicological examination to have harmful effects.

TABLE 11.3
Classification of Food Additives

1. Anticaking agents; free flowing agents
2. Antimicrobial agents
3. Antioxidants; antioxidant synergists and stabilizers; oxygen displacers
4. Colors; coloring adjuncts; color fixatives, stabilizers, preservatives; decoloring agents; color retention agents
5. Curing agents; pickling agents and adjuncts
6. Dough conditioners; dough strengtheners
7. Dehydrating agents; dehydrating aids; drying agents
8. Emulsifiers; emulsifier salts
9. Enzymes; immobilized enzymes; entrapped enzymes
10. Firming agents
11. Flavor enhancers; flavor masking agents
12. Flavoring agents, adjuncts and adjuvants
13. Flow treating agents; aging, bleaching, and maturing agents
14. Formulation aids; binders, carriers, fillers, plasticizers, moisture barriers
15. Fumigants
16. Humectants; moisture-retention agents; rehydration aids
17. Leavening agents
18. Lubricants; anti-sticking agents; release agents
19. Non-nutritive sweeteners
20. Nutrient supplements; dietary food supplements; vitamins
21. Nutritive sweeteners
22. Oxidizing and reducing agents
23. pH control agents
24. Processing aids; filters; flocculents
25. Propellants; aerating agents; gases
26. Sequestrants
27. Solvents; vehicles, extracting agents
28. Stabilizers; thickeners; bulking agents; foam enhancers, foam stabilizers; gelling agents
29. Surface-active agents; foaming and defoaming agents; conditioning agents; detergents; solubilizing and wetting agents
30. Surface-finishing agents; coating and sealing agents
31. Synergists
32. Texturizers; texture retention agents
33. Washing, peeling and cleaning aids
34. Freezing agents; crystallization agents
35. Ion-exchange agents; immobilizing agents
36. Fermentation aids

Source: Reprinted with permission from Furia, Thomas, ed. CRC Handbook of Food Additives, 2nd ed., vol. 2., copyright CRC Press.

TABLE 11.4
Main Types of Chemicals Used as Pesticides, Drugs, and Feed Additives For Food Production

Function	Chemical Class	Examples
Insecticides, acaridicides, rodenticides	Organochlorine compounds	DDT, dieldrin, lindane
	Organophosphorus compounds	Malathion, parathion
	Carbamates	Carbaryl, aldicarb
	Nitro compounds	Binapacryl
	Rotenoids	Rotenone
	Pyrethroids	Pyrethrine
	Coumarones	Warfarin
Herbicides	Carbamate, ureas	Linuron, CIPC
	Inorganics	Sodium chlorate
	Nitrophenols	Dinoseb
	Chlorophenoxy	2,4-D, 2,4,5-T, MCPA
	Chloroaliphatic acids	TCA, dalapon
	Triazines	Simazine, prometryne
Fungicides	Inorganic	Sulfur, copper salts
	Dithiocarbamates	Zineb, thiram
	Trichloromercapto compounds	Captan, folpet
	Quinones	Chloranil, dichlone
	Nitrophenols	Dinocap
	Systemic compounds	Benlate, thiophanate
	Substituted diphenyls	Orthophenylphenol
	Antibiotics	Griseofulvin

Fumigants	Organophosphorus compounds	Sulfotep
	Aromatic compounds	Azobenzene
	Miscellaneous	Dichloropropane, dichloropropene, ethylene dibromide, hydrogen cyanide, methyl bromide, carbon disulfide
Plant-growth regulators	Auxin type	3-Indolebutyric acid
	Gibberellins	Gibberellic acid
	Miscellaneous	Maleic hydrazide, naphthalene acetic acid
Growth stimulants	Hormone type	Hexoestrol, zeranol, diethylstilbesterol (DES)
	Antibiotics	Chloramphenicol, bacitracin penicillins, tetracyclines
	Arsenicals	Arsanilic acid
	Goitrogens	Thiouracil
Prophylactics	Coccidiostats	Amprolium, ethopabate, decoquinate
	Antihistominals	Demitradizole, acinitrazole
Therapeutics	Corticosteroids	Cortisones, prednisone, tetracyclines
	Antibiotics	Penicillins
	Halogenated phenyls	Nitroxinil, bromophenophos
Tranquilizers		Azaperon

Source: Reprinted from Nutritional and Safety Aspects of Food Processing, by courtesy of Marcel Dekker, Inc.

Food additives have become a public issue because of recurrent episodes that brought into question the safety of some additives that have been used for some time (Kermode, 1972). Examples of such additives are nitrites, cyclamates, estradiol, and food colorings such as Red No. 2 and Red No. 4. Refined and new analytical procedures as well as advances in toxicological research have led to vexing new problems in hazard assessment of intentionally used chemicals (Pond et al., 1980). Pond and coworkers also indicate that methods to assess the hazards associated with incremental small increases in the normal background burden of potent biologically active substances are currently lacking. Another problem which has been little addressed is the interaction between the various food chemicals.

The wide range of reprocessing pesticides, feed additives, and drugs used in agricultural production, together with substances added to foods for flavoring, coloring, preservation, and other purposes (Kermode, 1972), as well as unintentionally added products and contaminants (e.g., heavy metals or acid rain) provide a large number of potential interactions and synergistic effects among the constituents. The following is a simple example of such possible interaction (Knorr, 1975):

> Different tin compounds can originate from reactions between filling material and packing material containing tin during the production and storage of cans; for instance, tin acetate and/or tin tartrate or organotin compounds causing possible clinical reactions. The toxicity of tin tartrate, for instance, was repeatedly described. Investigations based on an actual case of a clinical reaction with a number of persons (citrus juices with tin content of 200 to 400 mg.kg. and traces of parathion) indicated the possibility of a mutual effect between tin and residues of pesticides as well as the eventuality of the forming of organotin compounds. That the combination of tin tartrate with parathion or paraoxon has an increasing effect on the lethality of white mice could be demonstrated by a biological experiment following up on the influence of small doses of parathion and its metabolite paraxon on the toxicity of tin in orange juice, present as tin tartrate. Reports on the effects of pesticide residues in canned vegetables on the can corrosion--and thus on the dissolution of tin from tin cans--should be taken into consideration.

A classic example of such an interaction between a food additive and food constituents is that between nitrogen trichloride, formerly used in flour treatment, and methionine (a common amino acid) to form methionine sulphoximine. A more recent example involves the use of the preservative diethyl pyrocarbonate which has been found to react with ammonia to form trace amounts of urethane in food. Gangolli (1972) indicated that while levels of urethane so formed are possibly toxicologically insignificant, a reassessment of the entire food safety situation is clearly indicated in light of the cost of regulations and the resources needed for testing such a wide range of still unknown substances.

Many reviews have been published on the use and safety of food additives; for more detailed information see, for example: Birch and Parker, 1980; Carpanini and Crampton, 1976; Day, 1976; Foster, 1982; Francis, 1981; Gangolli, 1972; Hall, 1976; Hall, 1979; Knorr, 1980b; Pond et al., 1980; Tannenbaum, 1979; U.S. Senate Committee, 1979; and Walker, 1976.

FEDERAL FOOD SAFETY REGULATING SYSTEM

The current federal safety regulation system has been synopsized by the Food Safety Council (1982) as follows:

> The laws and regulations governing the safety of the food we eat are complex. Four different laws--the Food, Drug, and Cosmetic Act, the Meat Inspection Act, the Poultry Products Inspection Act, and the Federal Insecticide, Fungicide, and Rodenticide Act (FIFRA)--and three principal agencies--the Food and Drug Administration (FDA), the Food Safety and Inspection Service (FSIS) of the U.S. Department of Agriculture, and the Environmental Protection Agency (EPA)--comprise the bulk of the federal government's food safety system.
>
> The food laws and their corresponding regulations work, for the most part, in conjunction with the Food, Drug, and Cosmetic Act. Under the act and its various amendments, Congress has created seven categories of food and food substances: food additives, color additives, animal food and drug additives, pesticide residues, substances generally recognized as safe (GRAS), prior sanctioned substances, and naturally occurring substances. In addition, FDA administratively created an eighth category (environmental contaminants).
>
> Each category is subject to distinct regulatory requirements, and each is treated differently from the others, particularly with regards to what constitutes an acceptable level of risk or a "safe" use for a substance. Ultimately, however, substances are either permitted, banned, or subject

to specific restrictions on use in foods.

The historic roots of food safety law have been outlined in several recent comprehensive documents (Hutt, 1981; Senate Committee, 1979; Food Safety Council, 1980). Very briefly, the precedent for the present U.S. system is a set of laws and amendments (going back to the Assize of Bread in 1203 in England) passed in 1906--The Pure Food Act; 1938--The Food and Drug and Cosmetic Act; 1958--amendments to the F & DC Act including the Delaney clause; and 1960 and 1962--color additive amendments. At the time of the decision to remove cyclamate from the food supply in 1969 another decision was made to start a review of all the foods which had been placed in GRAS status by the 1958 amendments. In 1972 a select committee on GRAS substances was formed to initiate the review. The report of that committee is recommended for its interesting juxtaposition of scientific and philosophical matters and for its discussion of the concept of "generally recognized as safe" (GRAS Review Committee, 1977). In 1977 the FDA attempted to ban the use of saccharin and since that time an amazing number of formal proposals have been put forth to change the food safety laws, including the following: the IFT paper on the risk/benefit concept as applied to food (IFT, 1978); the National Academy of Sciences report on food safety (NAS/NRC 1979); Principles and Processes for Making Food Safety Decisions prepared by the Food Safety Council (1980); Hatch Food Safety Amendments of 1981, which were written by a consortium of industry representatives (S. 1442, 1981; Mentzer and Harris, 1982); GAO report called "Regulation of Cancer-Causing Food Additives--Time for a Change?" (GAO, 1981); the "Gore bill" to improve food safety laws (H.R. 5491, 1982); the draft Hatch-Kennedy food safety legislation (Anon., 1982b); the Working Group first proposal (sub-Cabinet Administration group) (Bisogni, 1982); and the Working Group second proposal (Anon., 1982a). There have also been innumerable symposia, workshops, and speeches.

We will leave to historians a comprehensive analysis of the verbiage and very briefly describe the six major issues being addressed.

(1) Risk assessment. Reformers argue that consumers make risk/benefit decisions all the time, that there is no way to attain a zero-risk food supply, and that new methods have been developed to perform benefit/risk analyses. Miller (1978), the head of the FDA's Bureau of Foods, has often cautioned against undertaking risk assessment and sums up his concern in the following passage:

> A rigorous risk/benefit ratio is intrinsically impossible with our present knowledge of food

additives. The term calls for dividing a very uncertain probability of risk in an ill-defined series of controversially adverse physiological responses in undefined units, by a heterogeneous collection of so-called benefits, such as economic gains expressed in dollars, convenience expressed in unknown units, palatability expressed in yet-to-be agreed-upon measures, aesthetic appeal not expressible in numbers of any kind and so on."

(2) Definition of Safe. The term "safety" was not used in food legislation until 1958 (Hutt, 1981). Until then the concern was with adulteration. The FD&C Act at this time does not define the term although it requires all additives to be "safe" prior to sale to the public. The FDA working definition is "a reasonable certainty in the minds of competent scientists that (a chemical additive) will not be harmful to the public under its proposed conditions of uses." Finding this unacceptable, most of the proposals have redefined safe; for example, in the Hatch bill it means "the absence of significant risk under the intended conditions of use" and the Working Group suggests it be "a reasonable certainty of no significant risk based on adequate scientific data, under the intended conditions of use." Obviously the key word here is significant, and concern is being expressed about its ambiguity and the implied burden of judgment being placed on the regulatory agencies to decide what is significant.

(3) Delaney clause. The FDA has fairly broad discretionary powers to determine whether an additive is safe. The one exception is additives which are human or animal carcinogens, which the Delaney clause prohibits the FDA from approving. The Hatch amendments (S. 1442) would repeal the Delaney clause by saying that it would not apply if the new additive "does not present a significant risk to human health." The Working Group (Bisogni, 1982) adopts a similar position and raises several major issues, such as the requirement that any new proposal provide for scientific consideration of whether a substance that causes cancer in animals actually poses a risk to humans, and the need to consider health benefits in decision-making. Health benefits would include factors such as the use of a substance for nutritional purposes, dietary management, and "the offering of an adequate food supply." As mentioned earlier the ability to perform benefit/risk analyses is doubted by many people, as is the possibility of identifying animal carcinogens which are safe to humans.

(4) Separate treatment of food categories. The fact that there are seven different food additive categories came under attack in the NAS report. The majority report made the recommendation that the

differences in statutory standards be abolished and a single standard be applicable to all food substances. None of the other proposals has made such a stark and radical recommendation, although redefinition of some food categories in other proposals accomplishes some of the same purposes (see below). Note that if such a proposal were adopted, substances in what are referred to as "natural foods" (potatoes, oranges) would for the first time be regulated.

(5) Administrative procedures. A large portion of the Hatch amendments (S. 1442) is aimed at changing various procedures that the FDA follows in approving new food additives. For example, a phase-out authority would be instituted with no limitations on time; the ability to sanction new uses of food additives not proven safe on an interim basis would be instituted; and the law would mandate that FDA rule on new food additive applications within 180 days (currently this is not a requirement). These provisions have been described collectively as "easy on, slow off" by consumer groups opposed to the legislation (Schultz, 1981). The other procedures which were most extensively addressed by the Food Safety Council are those followed by the agencies in assessing safety (see Figure 11.1). The Council writes (Food Safety Council, 1982):

> At present, FDA generally requires a fairly broad array of toxicological tests for new substances.... The agencies seem to keep reasonably current in the application of new or improved test procedures, but they have not developed any published guidelines for the sequence of testing. The Council proposed a systematic sequence of testing, in the form of a decision tree, to be used for any food substance.... Because the science of toxicology is still evolving, the Council feels it would not be appropriate to require, either by legislation or regulation, this or any other rigid sequence and type of testing.

(6) Designation of exempt foods. Most of the proposals have included a provision to exempt certain categories of food from the definition of food additive, and hence from the requirement that the additive be proven safe (or be GRAS) prior to sale, as well as from the Delaney clause. The language that has been adopted to describe this category is "basic and traditional" foods, which includes common foods with a history of significant food use in the United States (such as potatoes), processed raw agricultural commodities, as long as the processing is "by a method that is generally recognized...as not significantly changing the properties of such commodities," and herbs and spices. The Working Group (Anon., 1982a) also includes natural flavors in its exemption. Even the defenders of the exemption recognize the difficulty of specifying

Figure 11.1

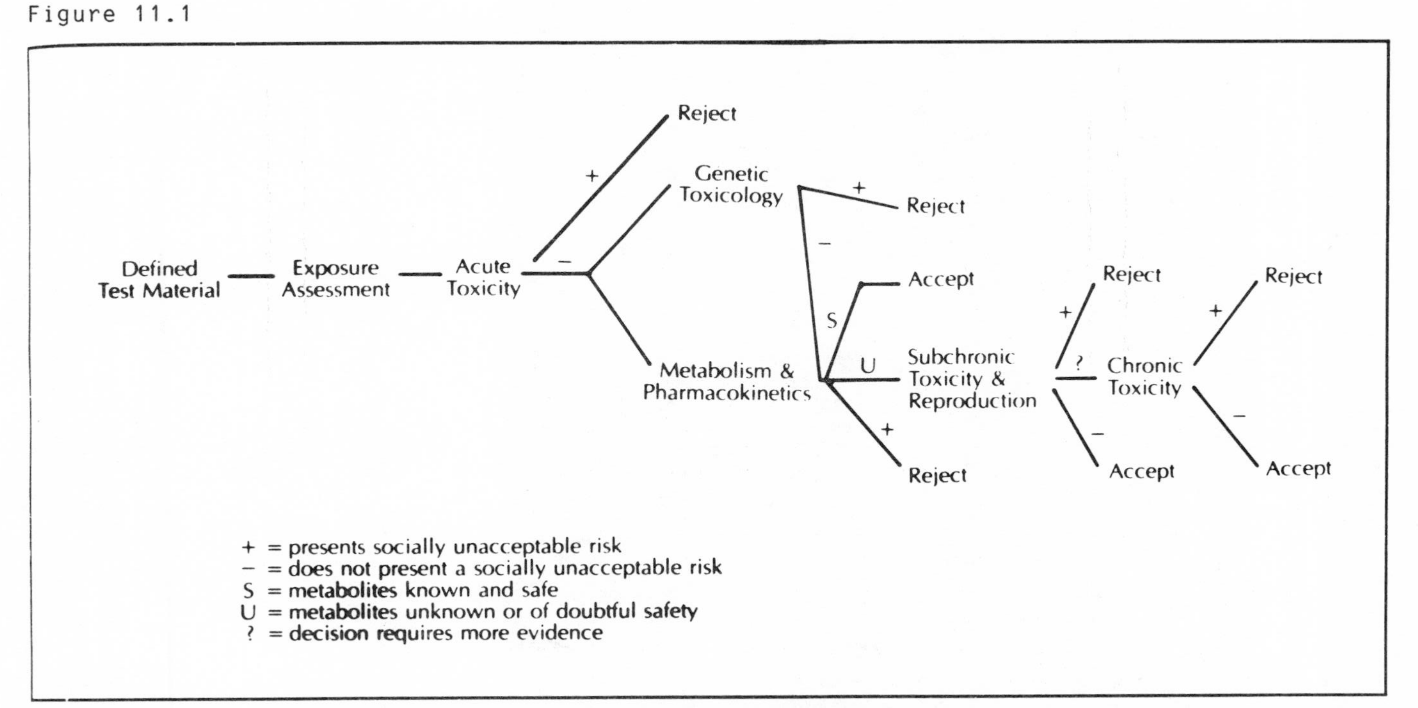

SAFETY DECISION TREE

Reprinted from Food Technology 36(8):169, 1982. Copyright by Institute of Food Technologists.

statutory criteria that will delineate whether something is or is not a traditional food, and suggest that it would be done by administrative and judicial decisions. This strikes defenders of the current definitions as extremely burdensome and unnecessary.

We hope it is clear from this brief analysis that the issues being debated about food safety are serious and quite controversial. At the same time it may be useful to keep in mind E. M. Foster's recent declamation (Foster, 1982) that there is not a "food safety crisis" in the U.S. He stated, "What we do have is a giant controversy with conflict between various factions of society over what we as a nation ought to do about food safety." Interestingly, it is the food industry which is keeping the idea of crisis alive, although its members are not talking about the same problem which led consumers to say there was a crisis earlier. In response to the industry's contention that there is an urgent need to change the law so that necessary new additives and food products can be introduced into the market, the consumer advocate groups have adopted the position that the U.S. food supply is, despite its problems, one of the safest in the world. They are concerned that any changes are likely to be in the direction of decreasing safety rather than increasing it (Schultz, 1981; Moyer, 1981). Furthermore, public opinion appears to have done an about-face on the Delaney clause. In 1977 only 27 percent of those queried on the Roper poll were in favor of retaining it; in 1982 only 32 percent favored changing it (Burros, 1982). Consumers do appear to have more concern about short-term hazards (e.g., microbial contamination). There is also a lot of confusion and disagreement among producers, scientists, legislators, and consumers about most food safety issues. Herman E. Talmadge, as the chair of the U.S. Senate Committee on Agriculture, Nutrition and Forestry (U.S. Senate, 1979), addressed that problem in his introduction to the Committee report on food safety:

> Clearly, there is a need to examine the various charges and concerns being expressed about the safety of our food supply. Major obstacles to this approach are the intensity and emotion of debate and the prevalence of "facts" and "experts" on either side of every question. Even when equally qualified scientists agree on the substance of the facts, they still often disagree on their conclusions. The following exchange between Senator Leahy and two scientists at the nitrite hearings before our subcommittee on Agricultural Research and general legislation last year is a perfect example of the latter point:
>
> <u>Senator Leahy</u>: Do we run a greater risk from the

addition of nitrites to food than from what occurs naturally, generally in the United States, in vegetables and other substances?

Dr. Lijinsky: Yes.

Dr. Weisburger: There I would say categorically, "No."

Senator Leahy: O.K., I understand the answer. I am no better off than before I asked the question but I understand the answer.

Legislators will keep all of this in mind when debate on various food safety proposals finally begins in Congress, and it may be that only minor changes are made after a great deal of discussion.

A NEW APPROACH TO FOOD SAFETY

In the midst of the assessment that is currently going on we would like to suggest that a rational approach would be to attempt to reduce some of the problems rather than changing the existing food safety regulations to get around these problems.

A survey of the recent literature shows the common use of combined terms such as food safety and toxicology, food safety and food hazards, food safety and food contamination, food safety and health risks, etc. (Birch and Parker, 1980; Tannenbaum, 1979; U.S. Senate, 1979; Food Safety Council, 1980, 1982). What does not exist is information on the subject of food quality in combination with food safety. If we assume that there are at least three levels of food safety--first, foods free from immediate (acute) food hazards; second, safe foods; and third, a combination of food safety and food quality--then it might be more appropriate at the given level of advanced technology to deal with problems of food hazards and contamination through the standard of food quality.

The question has been raised whether U.S. food production is overmechanized, and whether our food is overprocessed (Hall, 1976; Merrill, 1976; Taper, 1980). It is suggested that "safe" technologies (e.g., selected unit operations using physical and mechanical modes of preservation; minimal processing) can replace some of the currently applied "convenient" technologies (e.g., use of chemical preservatives). Binger (1981) recently proposed that food product manufacturers begin to shift, possibly on an industry-wide basis, to the production of "minimally processed foods." The author states that his proposal is not to supplement one type of processed food with another that is processed in a different manner. Rather, regardless of the processing technique used, the product, instead of having been "maximally processed," will have been subjected to the minimum of adverse treatment (temperature, pressure, additives, etc.)

without compromising its safety for consumption. Binger points out that some attempts have been made already to produce and market products that make such claims, and the example of Roberts (1979) sets an interesting precedent:

> Although brown rice is more nutritious than milled white rice the outer seed coat contributes a different flavor and texture which is disliked by many consumers. Removal of only three percent of rice bran (instead of the commonly used ten percent removal) resulted in significantly higher yields as well as protein, vitamin and mineral levels as compared to the ten percent milled white rice. Taste tests indicated high acceptance of all samples.

CONCLUSION

It has been argued that years ago when food was scarce no one was concerned about food safety and only now, when food is so abundant and most consumers do not have to be concerned about their daily food supply, is there worry over its safety. There are several objections to this argument: (1) food is not as abundant as is commonly assumed, which can be shown in the declining World Food Reserves as days of world consumption (Table 11.5); (2) since the lifespan of people and the proportion of processed foods in the food supply have both increased, more attention must be given to long-term food safety and quality problems of processed foods; and (3) despite all the advances in food production and in food processing, legitimate concern still exists about food hazards, food toxicology, and food contamination resulting in acute and long-term toxicity of some products.

Taking a new look at existing food processing operations to reduce this severity and exploring means to develop understanding of the real needs for chemical additives in foods are major challenges. A proposed food production/processing matrix (Table 11.6) outlines the complexity of the problem by attempting to organize the terminology that may need to be used for the differently produced and differently processed food products.

Food safety is, of course, not the only area where the need to take a more "systematic," long-term view is evident. Agriculture, nutrition, and food science all need to be seen as parts of the same whole. This chapter has shown that they are components of a system, and that fortunately some scientists have begun to act on the notion. However, many more people will need to employ this type of thinking and action to ensure the long-term safety and security of the population.

TABLE 11.5
Index of World Food Security, 1960-80

Year	Reserve Stocks of Grain (a)	Grain Equiv. of Idled U.S. Cropland (a)	Total Reserves (a)	Reserves as Days of World Consumption
1960	198	36	234	102
1965	143	70	213	80
1970	165	71	236	77
1971	183	46	229	73
1972	142	78	220	66
1973	147	25	172	51
1974	132	4	136	40
1975	138	3	141	40
1976	192	3	195	55
1977	191	1	192	51
1978	228	21	249	62
1979	191	15	206	51
1980	151	0	151	40

Source: Brown, 1981. Reprinted with permission from The World Watch Institute.

(a) Million metric tons

TABLE 11.6
Agricultural Production/Food Processing Matrix

Food Processing	Agricultural Production Organically Grown(a)	Conventionally Grown	Environmentally Controlled
			Proposed terms
Minimally processed(b)	natural food	conventionally grown/ minimally processed food	environmentally controlled/ minimally processed food
Conventionally (highly, complex) processed	organically grown/ conventionally processed food	conventional food	environmentally controlled/ conventionally processed food
Fabricated	organically grown/ fabricated food	fabricated food	environmentally controlled/ fabricated food

(a) A general definition of organic farming was recently presented by USDA (1981): "Organic farming is a production system which avoids or largely excludes the use of synthetically compounded fertilizers, pesticides, growth regulations and livestock food additives. To the maximum extent feasible, organic farming systems rely upon crop rotations, crop residues, animal manures, legumes, green manures, off-farm organic wastes, mechanical cultivation, mineral-bearing rocks and aspects of biological pest control to maintain soil productivity and tilth, to supply plant nutrients and to control insects, weeds and other pests."

(b) According to a proposed definition of natural foods by the Federal Trade Commission (FTC, 1978) minimal processing would include:

(1) the application of physical processes such as the washing, peeling, and cutting of fruits and vegetables; the grinding of nuts; the separation of whole grain flour from whole grains; the expression of juices or oils; the homogenization of milk; and the purification of liquids with inert materials, or (2) certain processes used to make food edible or to preserve it or to make it safe for human consumption, such as canning, bottling, freezing, baking, pasteurization, and drying.

REFERENCES

Andres, C. 1982. "Foods of Tomorrow." Food Processing 43(4):19.

Anon. 1974. "Organic Foods." Food Technology 28:71.

Anon. 1982a. "Coalition of Sixteen Groups Attacks Working Group Food Safety Suggestions." Food Chemical News (June 14):15.

Anon. 1982b. "Sub-cabinet Council Adopts Hatch-Kennedy Food Safety Language." Food Chemical News (September 27): 7.

Bates, R. P. 1981. "The Uneasy Interface Between Food Technology and Natural Philosophies." Food Technology 35(12):50.

Bernarde, M. A. 1979. "Food Safety and Toxicology." In Food Science and Nutrition, ed. F. Clydesdale. Englewood Cliffs: Prentice-Hall.

Binger, H. P. 1981. "Food Research: What Next?" Lebensmittel-Wissenschaft Technologie 14:346.

Birch, G. G., and Parker, K. J. 1980. Food and Health: Science and Technology. London: Applied Sciences Publishers.

Bisogni, 1982 "Food Safety Laws Eyed in Washington." Human Ecology Forum 13:18.

Brown, L. 1981. Building A Sustainable Society. New York: W. W. Norton.

Burros, M. 1982. "Benefits of Carcinogens." New York Times (September 8).

Carpanini, F. M. B., and Crampton, R. F. 1976. "The Testing of Food Additives for Safety in Use." In Health and Food, eds. G. G. Birch, L. F. Green, and L. G. Plaskett. London: Applied Science Publishers.

Clydesdale, F. M. 1979. Food Science and Nutrition: Current Issues and Answers. Englewood Cliffs: Prentice-Hall.

Day, H. G. 1976. "Food Safety--Then and Now." American Dietetic Association Journal 69:229.

Food Technology 36(8):94. 1982. "Food Expo in Print." IFT 82.

Food Safety Council. 1980. "Principles and Processes for Making Food Safety Decisions." Report of the Social and Economic Committee of the Food Safety Council. Food Technology 34(3):82.

______. 1982. "A Proposed Food Safety Evaluation Process." Food Technology 36(8):164.

Foster, E. M. 1982. "Is There a Food Safety Crisis?" Food Technology 36(8):82.

Francis, F. J. 1981. "Public Information: Science and Ethics." Food Technology 35(3):11.

FTC. 1978. "Proposed Trade Regulation Rule on Food Advertising." Staff report and recommendations. Federal Trade Commission. Washington, D.C.

Furia, T. E. 1980. CRC Handbook of Food Additives. 2nd ed., vol. 2. Boca Raton: CRC Press.

Gangolli, S. D. 1972. "Some Aspects of the Toxicology of Foods." Journal of the Association of Public Analysis 10:74.

General Accounting Office. 1981. Regulation of Cancer-causing Food Additives--Time for a Change? GAO, HRD-82-3. Washington, D.C.

GRAS Review Committee. 1977. "Evaluation of Health Aspects of GRAS Food Ingredients." Federal Proceedings 36:2527.

Hall, R. H. 1976. Food for Nought. The Decline in Nutrition. New York: Vintage Books.

Hall, R. L. 1979. "Food Ingredients and Additives." In Food Science and Nutrition: Current Issues and Answers, ed. F. Clydesdale. Englewood Cliffs: Prentice-Hall.

Harris, R. S., and Karmas, E. 1975. Nutrition Evaluation of Food Processing. Westport, CT: AVI Publishing.

H. R. 5491. 1982. Food Safety Amendments of 1982 February. U.S. Congress. Washington, D.C.

Hutt, P. B. 1981. "The Challenge of Risk Management: An Historical Perspective." Proceedings of CNI/FMI National Food Policy Conference: Focus on Food Safety. CNI. Washington, D.C.

IFOAM Bulletin. 1982. "Furcht vor Ruckstaden in Lebensmitteln Wird Immer Groser." IFOAM Bulletin (German Edition) 41:8.

IFT. 1978. "The Risk/Benefit Concept as Applied to Food." Institute of Food Technologists Expert Panel on Food Safety and Nutrition. Food Technology 32(3):51.

Kadam, S. S.; Nerkar, Y. S.; and Salunkhe, D. K. 1981. "Subtoxic Effects of Certain Chemicals on Food Crops." CRC (Chemical Rubber Company) Critical Reviews in Food Science and Nutrition 14:49.

Kermode, G. O. 1972. "Food Additives." Scientific American 226(3):15.

Knorr, D. 1975. "Tin-resorption, Peroral Toxicity and Maximum Admissable Concentration in Foods." Lebensmittel-Wissenschaft Technologie 8:51.

______. 1980a. "The Influence of Food Processing on the Nutritional Quality of Food." In Soil, Food and Health in a Changing World, eds. K. Barlow and P. Bunyard. Berkhemsted, Herts: A. B. Academic Publishers.
______. 1980b. "Lebensmittel und Ernahrung als Krankmachende Faktoren." In Gesundheit im Gesellschaftlichen Konflikt, ed. W. Schonback. Munich, Vienna, Baltimore: Urban and Schwarzenberg.
______. 1982. "Natural and Organic Foods Definitions, Quality and Problems." Cereal Foods World 27(4):163.
Lechowich, R. V. 1979. "Food Microbiology." In Food Science & Nutrition, ed. F. Clydesdale. Englewood Cliffs: Prentice-Hall.
Lepkovsky, S. 1953. "Nutritional Stress Factors and Food Processing." Advances in Food Research 4:105.
Mentzer, R., and Harris, E. 1982. "1982 Legislation--the Food Safety Amendments of 1981." National Food Review - USDA-ERS, Spring, NFR-18. Washington, D.C.
Merrill, R. 1976. Radical Agriculture. New York: Harper and Row.
Miller, S. 1978. "Risk/benefit, No-effect Levels, and Delaney: Is the Message Getting Through." Food Technology 32(2):93.
Morisetti, M. D. 1971. "Public Health Aspects of Food Processing." Process Biochemistry 6(6):21.
Moyer, G. 1981. "The Delaney Clause Draws Fire." Nutrition Action (July 5).
NAS/NRC. 1979. Report on Food Safety. National Academy of Science. Washington, D.C.
Pearce, F. 1982. "The Menace of Acid Rain." New Scientist 95(1318):419.
Pond, W. G.; Merketl, R. A.; McGilliard, L. D.; and Rhodes, V. J. 1980. "Animal Agriculture: Research to Meet Human Needs of the 21st Century." Food Safety. Boulder: Westview Press.
Roberts, R. L. 1979. "Composition and Taste Evaluation of Rice Milled to Different Degrees." Journal of Food Science 44:127.
S.1442. 1981. The Food Safety Amendments of 1981. June. U.S. Congress. Washington, D.C.
Schultz, W. 1981. Comments on S.1442. Mimeograph. Public Citizen Litigation Group. Washington, D.C.
Semling, H. V. 1982. "1982 Food Industry Outlook." Food Processing 43(2):60.
Sissons, D. J., and Telling, G. M. 1979. "Agricultural Chemicals." In Nutritional and Safety Aspects of Food Processing, ed. S. R. Tannenbaum. New York: Marcel Dekker.
Synge, R. L. M. 1976. "Damage to Nutritional Value of Plant Proteins by Chemical Reactions During Storage and Processing." Qualitas Plantarum/Plant Foods for Human Nutrition 26:9.
Tannenbaum, S. 1979. Nutritional and Safety Aspects of Food Processing. New York: Marcel Dekker.

Taper, B. 1980. "The Bittersweet Harvest." Science 80, 1(7):78.
Thijssen, H. A. C., and Bruin, S. 1981. "Scientific and Technical Developments in the Food Industry of the Eighties." Lebensmittel-Wissenschaft Technologie 14:218.
USDA. 1980. Report and Recommendations on Organic Farming. U.S. Department of Agriculture. Washington, D.C.
U.S. Senate Committee. 1979. "Food Safety: Where Are We?" Committee on Agriculture, Nutrition, and Forestry. U.S. Government Printing Office. Washington, D.C.
Walker, R. 1976. "Food Additives - The Benefits and the Risks." Journal of Biosocial Science 8:211.

Transportation and Marketing

12
Domestic Food Security: Transportation and Marketing Issues

C. Phillip Baumel and Marvin Hayenga

INTRODUCTION

Concern over food security is an age old issue. The Bible tells us of Jacob's son, Joseph, building granaries to store grain in times of plenty to hedge against bad years. In medieval times, armies could move only in the direction of available food and fodder. As late as the 1920s, when all distribution broke down in Russia following the Bolshevik Revolution, masses of people roamed the countryside in frantic searches for food. History is full of stories of people migrating to where food was more adequate.

In today's world, and particularly in the United States, we do things the other way around. We move food to the people. Transportation and marketing are the necessary conditions for contemporary specialization and urbanization.

In our predominantly free market system, most commodities are produced in areas where they can be produced more advantageously than other commodities. As a result, corn is grown in the Corn Belt, citrus in California and Florida, broilers in the South and East, while cereal grains are heavily concentrated in the Great Plains and Pacific Northwest. Most of the red meat is located in the Midwest where the feed grains are grown. Milk and egg production are spread widely over the United States; however, egg production is heavily concentrated in the Southeast, and in Arkansas, Missouri and California; milk production is largely located around the Great Lake states. Vegetable and fruit production is heavily concentrated in the Great Lakes states, California, the Pacific Northwest and Florida.

Most food processing and storage activities are located in the production areas because in most cases it is cheaper to process or store the raw products in the production areas and transport the processed products to the consumption areas. In a few cases, it is cheaper to process the food in the consuming areas. Examples are potato chips and bakery products, which are usually manufactured near the consuming market to avoid the high cost of transporting bulky lightweight commodities.

Food consumption typically takes place in a person's home or community. Figure 12.1 shows the location of the population and hence the major consuming areas. Figures 12.2, 12.3, and 12.4 illustrate the regional specialization of the production of beef, potatoes, and vegetables. A comparison of population centers and the food production areas shows that most of the food is produced in regions located hundreds or thousands of miles from the major population centers. Moving and transforming the raw commodity to the food on the consumer's table is an extensive process subject to disruptions from a variety of sources. The task of transporting food from producers to consumers typically consists of moving the raw food products relatively short distances from farms to processing or storage facilities and then moving the processed products from the processing and storage facilities to storage or retailing facilities in consuming regions. Therefore, almost all food products must be transported at least twice and, in some cases, more often. Few consumers carry more than a few days' inventory of many perishable products, so interruptions of food shipments could have a quick impact.

Long distance transportation requires time, which creates the possibility of food deterioration, shrinkage, and loss during transport. Transporting food long distance requires that fresh fruits and vegetables be harvested early to ensure high quality at the market. Food scientists assure us that harvesting fresh fruits and vegetables early (or green) preserves the nutritional quality better than harvesting at maturity. Once the product reaches maturity, it begins to deteriorate rapidly. Nutritional and textural quality could be maximized if the product were harvested in the consumer's backyard. However, climatic conditions in many areas would require major investments in greenhouses, which would substantially increase food costs.

CONDITION OF THE FOOD TRANSPORTATION SYSTEM

The four modes of transportation which haul food products in the United States are railroads, trucks, barges, and airplanes. Transportation of food products by air is limited to small quantities of high value perishable products such as fresh fruits and produce from Hawaii and California. The major food products moving by barges are raw grains and processed grain products destined for export. Most of our domestically consumed foods are transported by railroads and trucks. Railroads haul large quantities of grain and grain

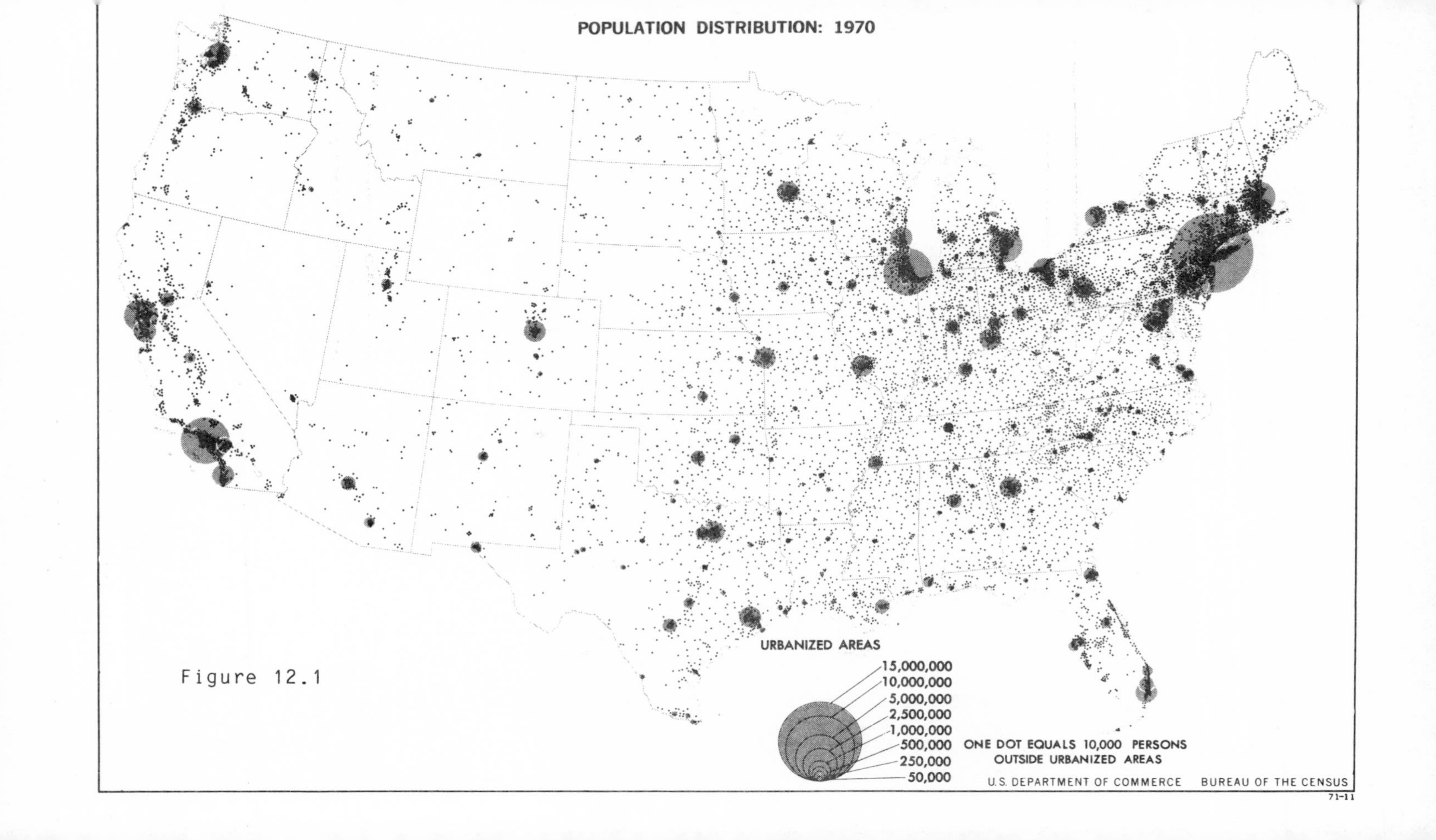
POPULATION DISTRIBUTION: 1970
URBANIZED AREAS
15,000,000
10,000,000
5,000,000
2,500,000
1,000,000
500,000
250,000
50,000
ONE DOT EQUALS 10,000 PERSONS
OUTSIDE URBANIZED AREAS
U.S. DEPARTMENT OF COMMERCE
BUREAU OF THE CENSUS
71-11

Figure 12.1

Figure 12.2

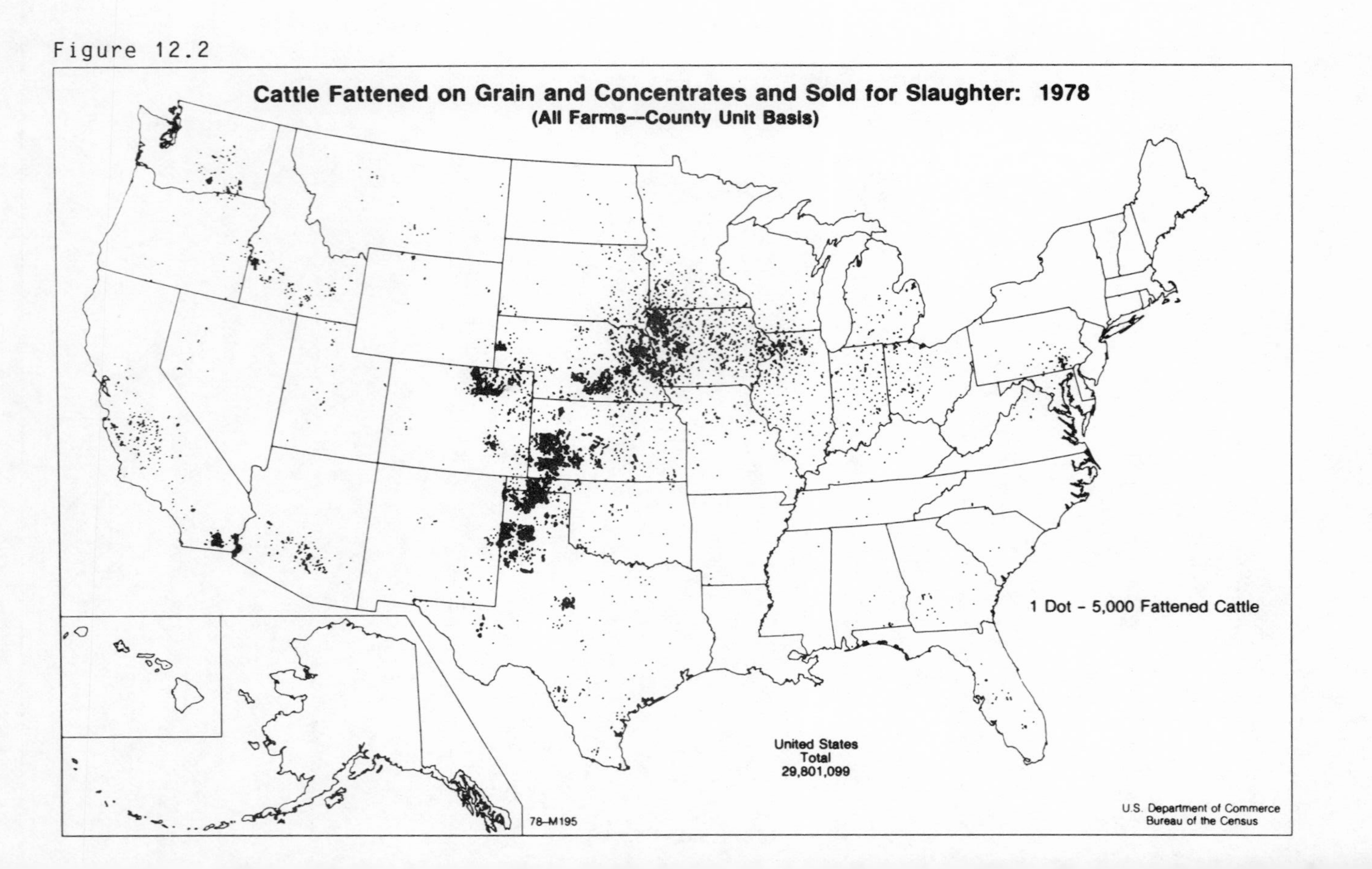

Cattle Fattened on Grain and Concentrates and Sold for Slaughter: 1978
(All Farms--County Unit Basis)
1 Dot - 5,000 Fattened Cattle
United States
Total
29,801,099
78-M195
U.S. Department of Commerce
Bureau of the Census

Figure 12.3

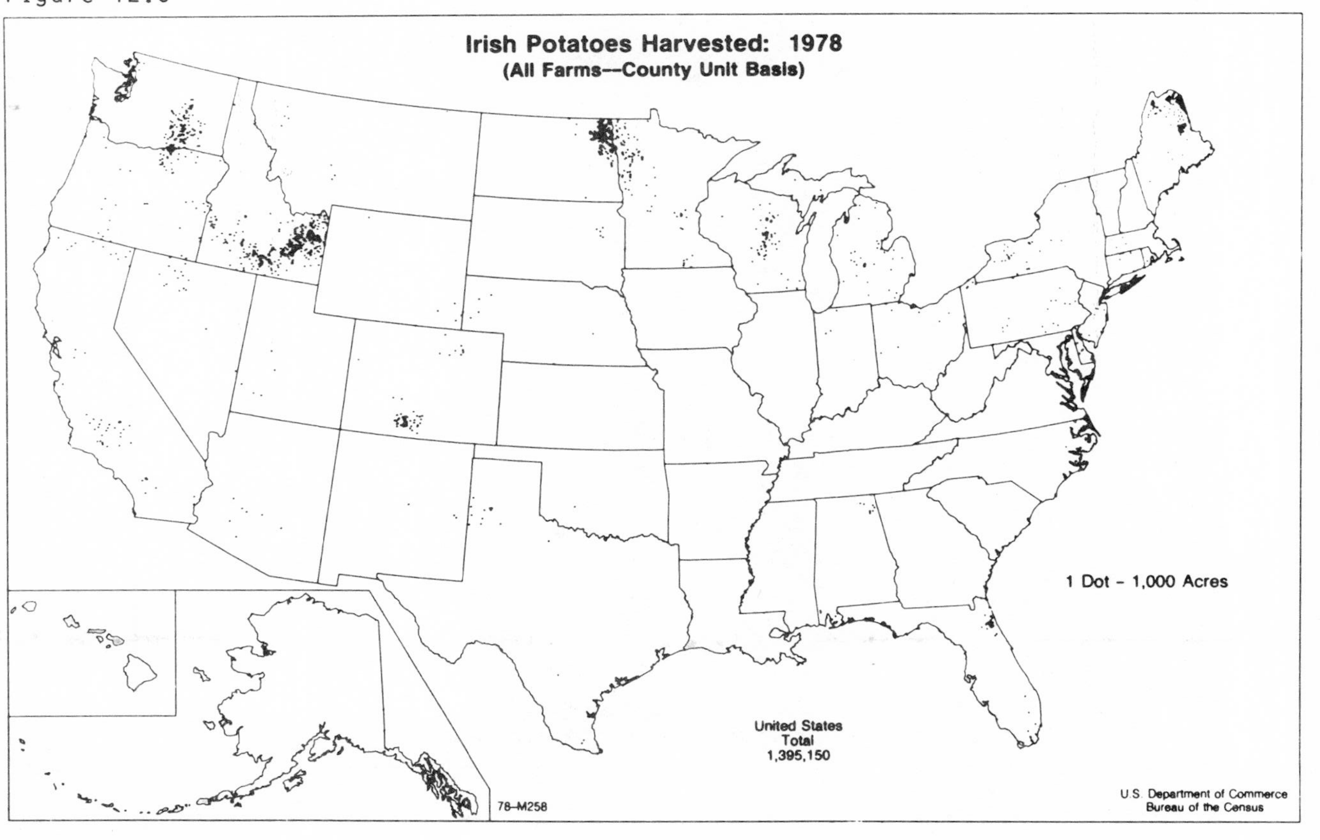
Irish Potatoes Harvested: 1978
(All Farms--County Unit Basis)
1 Dot - 1,000 Acres
United States
Total
1,395,150
78-M258
U.S. Department of Commerce
Bureau of the Census

Figure 12.4

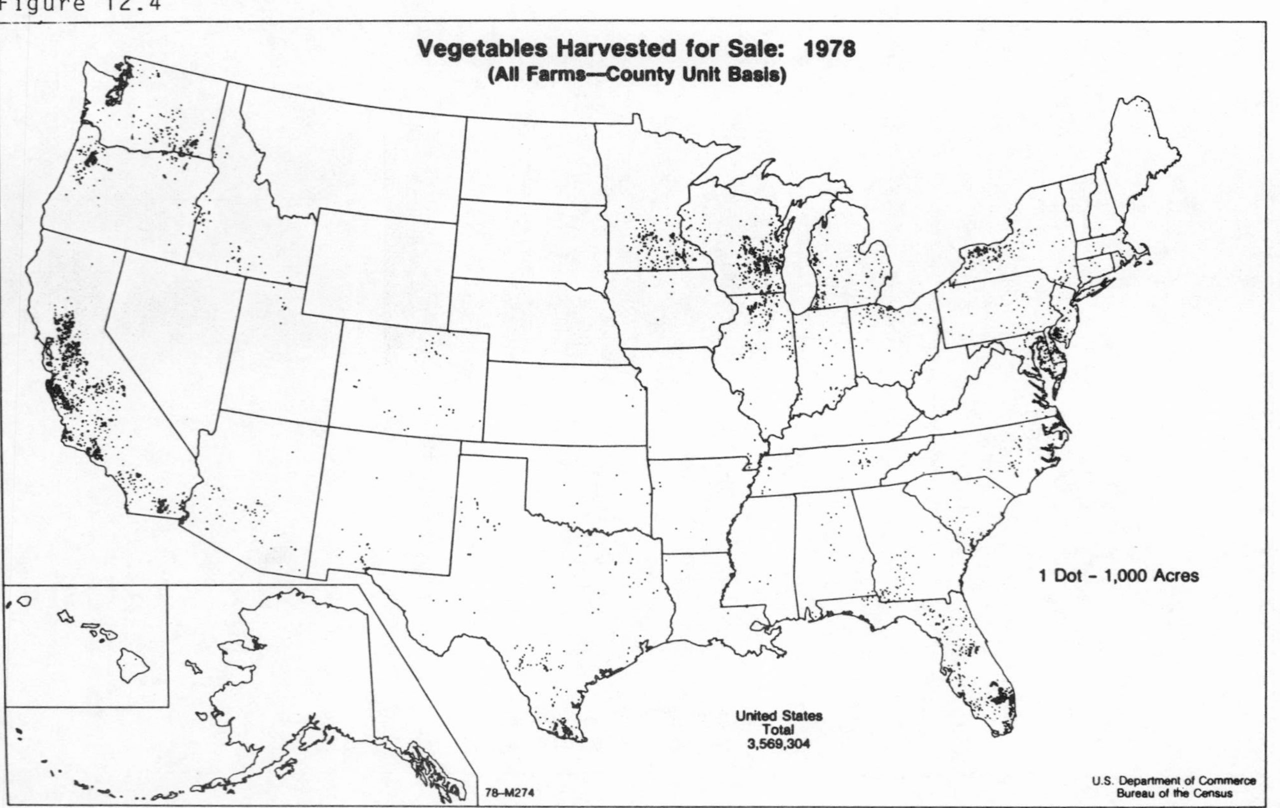
Vegetables Harvested for Sale: 1978
(All Farms—County Unit Basis)
1 Dot - 1,000 Acres
United States
Total
3,569,304
78-M274
U.S. Department of Commerce
Bureau of the Census

products, edible oils, some fresh fruits and vegetables, and large amounts of dry and canned food products. Trucks haul almost all of the livestock and fresh meats, large quantities of fresh fruits and vegetables, and almost all milk products. Therefore, our analysis of the condition of the transportation system will be limited to railroads and trucks.

Railroads

The general public's conception of the condition of the railroad system is that it is deteriorating rapidly and is in a near state of collapse. This view is reinforced by the numerous railroad bankruptcies, massive rail line abandonments, and the poor earnings performance of the railroad industry during the 1970s. The railroad industry has abandoned 73,000 miles of track since 1929--an average of 1,400 miles of track per year. Although the rate of abandonment increased during the 1970s, abandonment has been a consistent characteristic of the industry over the last half century, primarily to reduce the size of the system to be consistent with the demand for rail transportation and to adjust to unit train operations and economies of concentration in railroading (Association of American Railroads, 1982).

Although the size of the rail system is shrinking, the railroad industry is investing increasing amounts of capital into rail line rehabilitation. After reaching a low point in 1961, annual railroad capital expenditures in current dollars increased almost six times over a 20-year period. Annual installations of new rail track increased 3.6 times and annual installations of crossties doubled between 1961 and 1979. The railroad industry is investing more money in fewer miles of track which, in turn, has resulted in a much improved railroad physical plant. The net result is that between 1971 and 1980, the railroad industry moved almost 25 percent more ton-miles of freight on about 12 percent fewer miles of rail. At the same time earnings, as measured by return on investment, increased 2.5 times. The increased freight hauling capacity of the railroad system combined with massive increases in rail car purchases and the current recession has resulted in a surplus of about 120,000 rail cars (Association of American Railroads, 1982). A large share of these surplus rail cars have food hauling capacity. Thus, the major threats to the ability of the railroad system to transport domestic food supplies should not come from the deterioration of the railroad physical plant; rather, the primary threats might be civil or military destruction of parts of the physical plant, or short-term disruptions caused by fuel shortages or labor strikes.

Truck Transportation

Trucks have become the dominant mode of transport of most perishable food products such as fresh meat, fruits and vegetables. Even where railroads have cost advantages in long haul movements of bulky products such as grain, grain products, processed food products, and fertilizer, trucks are almost always involved in hauling these products to and from the rail line or from the warehouse to the supermarket or restaurant. Thus, there is some truth in the American Trucking Association motto, "If you got it, a truck brought it." The implication of this motto is that trucks were involved in transporting almost all commodities at some stage in the distribution process.

The total number of trucks in the United States has increased rapidly. The privately owned truck fleet increased from 26.6 million vehicles in 1976 to 30.4 million vehicles in 1978--an increase of almost 2 million vehicles per year. Except for personal ownership for personal transportation--which accounts for about half of the number of trucks--hauling food to the nation's dining tables tops the list of uses for trucks. In 1977, about 4.3 million trucks were used in some phase of agriculture. The next largest category of use--2 million trucks--was in wholesale and retail trade. The major advantages of trucks are their speed, reliability, and flexibility in transporting commodities to almost any point desired (Motor Vehicle Manufacturers Association of the U.S., 1980).

Trucks must, of course, travel on the highways. And the highway system in the United States is deteriorating rapidly. The major reason for the rapid decline in both the federal-aid system (basically the interstate and state highways) and the off federal-aid system (basically the rural county roads and city streets) is the concomitant decline in the real dollars available to rebuild the highway system. The decline in real dollars is a result of: (1) rapid rate of inflation in recent years; (2) relative stability in the rate of highway use taxation; (3) increased fuel efficiency of cars and trucks; and (4) increased weight limits on trucks.

The rural road and bridge system (off federal-aid system) is deteriorating even more rapidly than the federal-aid system. Because the off federal-aid system basically consists of county roads serving small towns and rural areas, a large portion of all agricultural commodities must move over the off federal-aid system. About 61 percent of all off-system bridges are deficient. The states with the highest number of deficient bridges are Arkansas, California, Illinois, Indiana, Iowa, Kansas, Mississippi, Missouri, Nebraska, New York, North Carolina, Oklahoma and Texas. Most of these states are major agricultural producing states (U.S. Dept. of Transportation, 1982).

Although the highway system is still tolerable--i.e., only a few trucks are falling through bridges--there are indications that the highway system is following the same deterioration pattern that the railroad industry experienced in the 1950s and 1960s. If this is the case, then continued highway deterioration will first result in highway officials being accused of doing a poor job. Then as trucks and farm equipment fall through bridges, and transport costs increase, the problem of inadequate revenues to maintain the system will be recognized and alternative solutions will be discussed.

Given the general condition of the transportation system, what are the transportation and marketing implications for domestic food security? The general public probably considers food security in terms of (1) the adequacy of food supply to the consumer when and where it is needed; and (2) the price and affordability of food. Even if food were available but not affordable for some consumers, the public would consider that situation to be inadequate "food security." When food security is viewed in these terms, and the logistical problems of taking raw agricultural commodities through the marketing and distribution process are considered, then the possibility that disruptions in the marketing system could affect the food security seems to deserve a more careful analysis.

POTENTIAL DISRUPTIONS OF THE TRANSPORTATION AND MARKETING SYSTEM

Some sources of potential disruptions from the transportation and marketing system that may deserve consideration include: (1) excessive export demand brought on by production failures abroad; (2) elimination of U.S. food or fertilizer imports due to production failures, political trade restrictions, or disruption of shipping channels due to war, strikes, or energy shortages; and (3) disruption of food shipments within the United States due to strikes, civil disturbances, war, severe weather, or energy shortages.

Excess Export Demand

Several studies prepared in conjunction with the 1974 World Food Conference pointed to the growth in demand for food products in the less developed countries with high population growth rates (see United Nations, 1974; and the University of California, Food Task Force, 1974). Concern was expressed regarding the inability of food production to keep pace with increased consumption requirements, and the corresponding need for sharply increased imports from the developed countries. The

United States has faced a few situations when weather adversities here and abroad led to concerns about possible shortages of some commodities. The response in each case was sharply higher prices to ration available supplies among foreign and domestic consumers on the basis of their ability to pay, and government intervention through export controls to ensure that domestic supplies did not run out. With the United States frequently serving as the "supplier of last resort" for many major commodities, production shortages abroad can have a significant impact on domestic prices and consumer perceptions of food security--especially among low income consumers--in our "free market" system. However, high export demand has prompted temporary government intervention to "protect" the domestic consumer. But this protection has been at the rather significant long-term cost of alienating important trading partners and reducing future export demand. This raises the issue of "food security at what price level for consumers, for our trading partners, and for U.S. agricultural producers?" Clearly, the tradeoffs have not always been well thought out when short-term "emergencies" have occurred; however, some persons may argue that "good economics is not always good politics."

Import Interruptions

A wartime import disruption clearly could have a significant impact on defense and other industries dependent on unique metals and minerals not available on the North American continent. Although the United States is a net exporter of agricultural products, we also import a large part of our food supply. And in many cases, we are unable to produce the imported products in the United States, especially some of the tropical products such as coffee, cocoa, and bananas. Thus, a disruption of imports for whatever reason could leave an unfilled vacuum in our food supply. However, very few food products are sufficiently unique that they do not have effective substitutes, and the shortage of one product with several substitutes clearly is of less concern than a product with no substitutes. Even then, few products are so essential that "going without" would be nutritionally or psychologically harmful to many consumers. An examination of the list of imported products suggests that a complete cut-off of food imports could have some psychological impact on consumers, but little nutritional impact because of the ready availability of other protein, fat, or carbohydrate sources. The more pronounced impact would be on the processing companies dependent upon particular products which might not have effective substitutes in the short term.

Although consumers might have caffeine withdrawal symptoms, or have to shift to domestically produced corn sweeteners as processing capacity is built up, the trauma would generally be manageable. For example, soybean or other vegetable oils could serve as substitutes for coconut or palm oil, synthetic chocolate is available, and other fruits could replace bananas in a fruit salad. Eating patterns might be disrupted temporarily, but the ingenuity of consumers and food processors together would quickly lead to second-best eating patterns and product specifications that would suffice, perhaps at high cost, while more effective substitutes were developed over time.

Import disruptions could also lead to problems in fertilizer supplies. About 88 percent of all potash consumed in the U.S. is imported, mainly from Canada. Only small amounts of phosphate fertilizer are imported. Nitrogen imports represent almost one-fourth of total nitrogen consumption. And the amount of nitrogen imports is growing. Total nitrogen imports tripled from about 900,000 tons in 1970 to 2.7 million tons in 1980; and imports are projected to increase to 5.0 million tons in 1985 and to 8.2 million tons in 1990. The main sources of nitrogen imports are Canada, Mexico, and the U.S.S.R. The major reason for the growth in imports is the increasing cost of natural gas in the United States. Since July 1, 1976, 36 nitrogen plants with annual capacities of 5.3 million tons have closed in the United States; and additional plant closings are under consideration (Lyon, 1982). Since nitrogen is a major input into food crop production, cheap natural gas supplies in foreign countries are creating the potential for disruption of U.S. fertilizer and food supplies.

Energy Disruptions

Energy is extremely important in agricultural production and home food consumption. In addition, interruptions in energy supplies would severely curtail fertilizer production. Thus, food security is closely tied to energy security.

The food processing and marketing system is also a highly energy intensive system. But it uses only one-fourth of the food and fiber system energy consumption. Energy consumption in the marketing function is only about 10 percent of the total food and fiber fuel consumption and less than 1 percent of total U.S. energy consumption (Van Dyne, 1979). Nevertheless, a disruption in energy supplies could result in the spoilage of large quantities of available food supplies. Almost half of the per capita food consumption, especially meat and dairy products, require some kind of refrigeration. A disruption of electricity could result in the spoilage of up to half of our per capita food consumption.

Transportation consumed about 19 percent of the total petroleum used in the food and fiber system and less than one percent of total energy consumption (Van Dyne, 1979). Although the food transportation system uses only a small share of total petroleum consumption, it is almost totally dependent on petroleum as a source of energy. And the system is shifting rapidly toward an essentially all-diesel system.

Destruction of Transportation Facilities

Natural disasters such as earthquakes and floods, civil insurrection, or military action are the most likely potential causes of destruction of parts of the transportation system. The key railroad and highway systems needed for defense purposes have been identified by the Military Traffic Management Command. The key railroad system, called the Strategic Railroad Corridor Network--STRACNET--consists of 32,500 miles of mainline track. In addition, 5,000 miles of connector lines were identified to connect STRACNET with 216 military bases, arsenals, and other military installations. A September 7, 1982 article in the Chicago Sun-Times described STRACNET as in "astounding" good shape and one of the Defense Department's better kept secrets. The Department of Defense reports that 95 percent of the 32,500 miles of STRACNET can handle train speeds of 40 mph or better. The STRACNET system includes most but not all of the nation's well-maintained mainline rail system. And additional mainline track outside the STRACNET system is being upgraded. The system is well dispersed and redundancy in the nation's rail system permits alternative routings (U.S. Department of Defense, 1981).

The Military Traffic Management Command identified a Strategic Highway Corridor Network. Called STRAHNET, this highway system of 54,500 miles consists mainly of the interstate system and 12,000 miles of additional highways. The military analysis of the highway system identified several problems, including:
(1) deteriorating and substandard bridges;
(2) inadequate maintenance, which is becoming a serious threat to the highway system; (3) inadequate vertical clearance on public highways for the oversized equipment in the defense inventory; and (4) the need for careful planning and control of special defense use of public highways because of oversize and weight restrictions of individual states (Military Traffic Management Command, 1981). The special attention that the military is giving to the STRACNET and STRAHNET systems raises the question of the availability of these systems for food transport purposes in the event of a national emergency.

In addition, the major weaknesses of both systems are the railroad and highway bridges over the Missouri and Mississippi Rivers. These bridges would undoubtedly become military targets. Since a large portion of our domestic food supplies flow west to east, destruction of these bridges could disrupt our food supply lines.

Developing a System to Coordinate Food Movements During Emergencies

Railroads are particularly effective in hauling large quantities for long distances. Thus, during various types of short-term emergencies, one effective means of getting food quickly to consumers would be to use railraod unit trains for the long distance movements and to use trucks to gather a product at the train loading stations or to distribute the food around the consuming areas. This type of emergency effort would require a coordinated effort between the railroad and trucking industries.

The United States railroad industry consists of fewer than 40 Class I railroads that operate about 94 percent of the railroad mileage and haul about 98 percent of the rail freight (Association of American Railroads, 1982). These railroad companies coordinate their interline freight movements and operate over each others' rail lines through agreements arranged through the Association of American Railroads and through the regulatory power of the Interstate Commerce Commission. The Association of American Railroads operates a car control system that is relatively effective in providing current information on the number, type, location, operating status, and ownership of all railroad cars. Thus, the essential machinery needed to coordinate the railroad movement of food in an emergency is in place.

The trucking industry, particularly the segment which hauls most of the food products, is relatively unorganized. There is little information on exactly what types of trucks exist, who owns them, where they are and the operating status of each truck. Thus, utilizing the trucking industry in a food emergency situation would create major problems in determining the locations of the various types of needed trucks.

CONCLUSIONS

The primary characteristics of the food transportation and marketing system which are relevant to domestic food security follow.

- Most of the food production and processing industries are located long distances from the major population centers.

- Food production is generally located where climatic conditions favor production and where that production enterprise has an economic comparative advantage.
- Most of the domestic food supplies are transported by railroads and trucks.
- The railroad industry has made major improvements in the physical and operating conditions of the rail system. However, the total miles of the railroad system has declined and will continue to decline.
- Trucks are becoming increasingly important in moving food products. The condition of the highway system in the United States is deteriorating rapidly.
- The food transportation and marketing system is highly dependent on a steady supply of energy in the form of petroleum for transportation and natural gas and electricity for marketing and fertilizer inputs.
- The railroad industry is relatively well organized and could be coordinated for emergencies. The food trucking industry is relatively unorganized. A major effort would be needed to effectively coordinate the trucking industry during a food emergency.
- The destruction of railroad and highway bridges over the Mississippi and Missouri Rivers could create major problems in transporting food to consumers in the event of civil insurrections or military action.
- A reduction of food imports would create some disruption in our food system and diets. The foods with the greatest potential for import disruptions are sweeteners, coffee, tea, and cocoa. A reduction in nitrogen imports could also disrupt our domestic food supply, and result in higher prices or lower food exports.

POLICY OPTIONS

The following are several domestic food security policy issues related to transportation marketing:

- Should the U.S. government establish policies encouraging regional self-sufficiency in most food products? This would involve producing many food products at locations where there is a comparative disadvantage in production and where processing facilities are not now located. This would result in higher production costs and consumer prices. Government restrictions or subsidies would be required to implement these policies. We doubt that the likelihood of disruptions is sufficiently great to warrant such costs or be accepted by consumers or politicians.

- Should U.S. government policy place limitations on the economic power of transportation and longshoremen labor unions to reduce the risk of paralysis in the food transport system brought on by nationwide strikes? Nationwide truck strikes could be extremely disruptive to perishable food supplies. Long-term rail strikes would cause a major disruption in coal supplies for electricity generation and could impact the quality of perishable food supplies. Longshoremen strikes would have the most impact on imports. Placing regional limits on the jurisdiction of truck and rail unions would help reduce the risk to food security, but achieving such limits clearly might be politically difficult.
- Should the government continue the policy of market interventions and export interruptions in times of extremely high prices to protect domestic consumers? The disruptions to trade relationships have been costly to agriculture in the past. And such disruptions could trigger reciprocal interruptions of some of our fertilizer and energy imports. We believe that such policies should be the absolutely last resort, triggered only in a clear, "immediate danger" national emergency.
- Should a domestic food reserve be established near population centers? If so, what foods should it contain, and who should bear the cost? Observing the civil defense bomb shelter and emergency food supply system, we feel the costs of a domestic food reserve near population centers would be quite high.
- How should the massive investments needed to improve the road and bridge system be financed?
- Is there a need to ensure that the Strategic Railroad Corridor Network and the Strategic Highway Corridor Network will be open to food transportation during military emergencies?
- Should an emergency transport coordination system be developed? A coordinated transport system would require an improved statistical system for the trucking industry.
- How can the public ensure that the agricultural transportation and marketing system will receive adequate energy supplies during emergencies? Should public policy encourage railroads to electrify mainline tracks or shift to other alternative fuels?

REFERENCES

Association of American Railroads. 1982. Yearbook of Railroad Facts. Washington, D.C.
Cartmill, R. 1982. "The Grain Trading Industry Today." Paper presented at the 1982 Spring Conference of Robert Morris and Associates, Missouri Valley Chapter, April 22-23, 1982, Lincoln, Nebraska.
Ingersoll, B. 1982. "U.S. Railroads Pass Pentagon Test." Chicago Sun-Times (September 7).
Lyon, F. D. 1982. "Forecast Nitrogen Supply and Demand Balances." Unpublished report, CF Industries, Inc. Chicago.
Military Traffic Management Command. 1981. An Analysis of the Highways for National Defense Program. May. Transportation Engineering Agency, Newport News, Virginia.
Motor Vehicle Manufacturers Association of the United States. 1980. Facts and Figures for 1980. Washington, D.C.
United Nations World Food Conference. 1974. "Assessment of the World Food Situation, Present and Future." November 5-16. Rome: United Nations.
University of California Food Task Force. 1974. A Hungry World: The Challenge to Agriculture. Summary report. Berkeley: Division of Agricultural Sciences, University of California.
USDA. 1981. Food Consumption, Prices, and Expenditures: Statistical Bulletin No. 256. February. Economics and Statistical Services. Washington, D.C.
U.S. Department of Commerce, Bureau of the Census. 1972. "Part 1: Graphic Summary." 1978 Census of Agriculture, vol. 5. Washington, D.C.: U.S. Government Printing Office.
U.S. Department of Commerce, Social and Economic Statistics Administration, Bureau of the Census. 1973. "Characteristics of the Population." 1970 Census of Population, vol. 1. United States Summary. Washington, D.C.: U.S. Government Printing Office.
U.S. Department of Defense. 1981. A Study of Rail Lines Important to National Defense for the Armed Services

Committee of the Congress. June. The Military Traffic Management Command, United States Army for the Department of Defense. Washington, D.C.

U.S. Department of Transportation. 1982. Third Annual Report to Congress. March. Washington, D.C.: Highway Bridge Replacement and Rehabilitation Program, Federal Highway Administration.

Van Dyne, D. L., et al. 1979. Energy Use and Energy Policy in Structural Issues of American Agriculture. U.S. Department of Agriculture, Economics, Statistics and Cooperatives Service, Agricultural Economics Report No. 438, November. Washington, D.C.

13 What Happens When the Grocery Stores Don't Open? Stockpiles, Subsidy Programs, and Access Issues

James A. Christenson, Paul D. Warner, and Sally J. Lawrence

What if a truckers' strike or natural disaster occurred and grocery stores did not open? Store reserves might last three days. How many days of food supplies do people keep at home? What segments of the population would be most affected through an interruption of the food delivery system? Are food stockpiles available in times of crisis? Are there contingency plans for meeting shortfalls of food?

Americans seldom worry about the availability of food. Although rationing occurred during the world wars and citizens were urged to keep a two-week supply of food during the cold war period of the 1960s (USDA, 1964), food access never became a serious issue. Cost was considered the prime factor in determining whether citizens would have adequate food supplies. Food cost was controlled somewhat through (1) agricultural commodity programs aimed at stabilizing food prices and (2) government programs for the poor to help them purchase food staples. It seems unthinkable to question the possibility of having no food available. This issue of access is of central concern in this chapter.

We will discuss how the processing and marketing of food have created a food delivery system acclaimed for variety and low costs but fragile because of consolidation and specialization. Modern technology, processing, and delivery of food have lulled Americans into new patterns of buying and storage that may cause major problems if the food delivery chain is interrupted. We will look at (1) changing consumer buying and storage habits, (2) the food delivery system, (3) existing public and private distribution plans for food in times of emergency, (4) citizen feelings about domestic and foreign food policies, (5) how much food people keep in their homes, and (6) policy implications of current practices and government programs.

FROM ROOT CELLAR TO MINIFREEZERS

Americans traditionally have prided themselves on their self-sufficiency. A hundred years ago, people produced and stored much of their own food supplies more by necessity than by choice. The media often depict the self-sufficient farms of yore with root cellars piled high with apples, carrots, potatoes, beets, hams, and so forth. A problem often overlooked is the difficulty that farmers had getting their produce to market. Movies do depict the long cattle drives and the riverboat transportation systems. For a developing industrial society, delivery and storage of food were major problems. Railroads, highways, and air travel gradually helped to resolve delivery problems. The transcontinental railroad opened the way for inter- and intra-national delivery of food. Rutted roads have been replaced by nearly 4 million miles of modern highways. And airplanes rush perishables to consumers in a matter of hours. This infrastructure has facilitated today's efficient and inexpensive food delivery system. People now can store food by choice and not by necessity.

Canning and freezing of food in conjunction with the standardization and mass production of containers and centralized processing systems were major factors in the development of modern food storage and distribution systems. These factors, combined with the economies of scale in the purchases of large quantities of a product, precipitated the demise of the small family grocery store and the rise of the supermarket. A wide variety of food at quite reasonable prices became available to the buyer. People could feel secure in having a few days of food supplies rather than enough for a winter season, because food is conveniently available.

Although storage practices have changed, the public has not abandoned the practice of maintaining food reserves. There has been a substantial increase in the amount of food stored through freezing. The number of households with freezers increased from 23 percent in 1960 to 31 percent in 1970 to 45 percent in 1979 (Statistical Abstract of the U.S., 1981:766). We would expect those individuals with ample freezer space to keep substantial food supplies in their homes. However, given the American preference for fresh or frozen food, a sizeable portion of the population without freezers may have little food on hand. For the more than half of the households without freezers, the only food stored is in the minifreezer in their refrigerator and in the cupboard.

Although food gardening has decreased slightly in the past few years (49 percent of households had gardens in 1975 and 43 percent had gardens in 1980), a sizeable segment of the U.S. population grows some of its own food. About 65 percent of rural residents are engaged in gardening while only 31 percent of central city residents and 34 percent of suburban residents have gardens (U.S. Bureau of the Census, 1981:232). There has been little change in the proportion of households (about 35 percent) canning for home use over the last 15 years; however, an internal shift in the types of households canning has occurred. Home canning has shifted from a predominantly lower income phenomenon to an upper income practice (Hatfield, 1981:24). During this same time period, freezing of food for home use has more than doubled, increasing from 24 to 55 percent (Hatfield, 1981). Likewise, home freezing has also increased with household income levels. Kaitz (1979) suggests that the lack of storage space for apartment dwellers is the reason that such individuals seldom store extra food items. Kaitz also observes that volume buying generally is more prevalent in households of adults 35 to 65 years of age who have children.

This information can be viewed in two ways. Close to half of the population have gardens, freezers, or both. Such households likely would have a supply of food to draw upon if delivery of food supplies to supermarkets were cut off. Of course, if electricity were interrupted for any period of time because of some man-made or natural disaster, frozen food would quickly spoil. The other half of the population have no gardens or freezers and may have very limited food reserves. Apartment dwellers, those of lower incomes, the young and the old, and particular combinations of these groups (e.g., the elderly lower income apartment dwellers) would be expected to be particularly vulnerable to an interruption in food distribution.

THE SUPERMARKETS

When we run short on food, we go to the store. Over the years, independent family grocery stores have given way to large supermarkets. Although the privately run family markets are still important in some neighborhoods and communities, these stores handle a very small percentage of food staples. It should be noted that many of these small family-oriented stores do include and encourage the sale of locally grown food. This aspect of small private stores has implications for subsequent discussion of governmental policies concerning regional food security.

Regional specialization in food production, an elaborate interstate transportation system, and large-scale buying and selling have stimulated a rather rigid structure in today's food system. Four major food firms accounted for 67 percent of all supermarket sales in metropolitan areas in 1972 (Grinnell, 1978). Overall, supermarkets handle more than 60 percent of the food consumed in the U.S. and thus "provide the critical link between consumers and the rest of the food delivery system" (Van Dress and Grinnell, 1980:10).

The current marketing system discourages the purchase of locally grown food. Food chains occasionally advertize fresh, locally grown foods such as corn or strawberries in season, but this is the exception rather than the rule; it constitutes a very small proportion of the food handled by a store. Food firms maintain contracts with suppliers who, in turn, have contracts with farmers and cooperatives. Food items are standardized in form and contracted in bulk on a year-around basis. A locally grown tomato has little chance of ending up on a nearby kitchen table unless it is bought at a local farmers' market or roadside stand. Very little (less than 20 percent for the Northeast) of what we eat is grown and processed within the state in which we reside (New York Times, 1981). Most food staples travel by truck, train, or plane thousands of miles from where they were produced to where they are consumed.

Supermarkets generally restock food staples on the average of two or three times a week. And few modern supermarkets have space other than on the shelves in which to store substantial quantities of food. So essentially what you see on supermarket shelves is only enough food to maintain a supply of about three days based on normal customer buying habits. The limited amount of food available is apparent during the winter when the weather forecasters predict a major storm approaching an area. Customers rush to the stores to stock up on all sorts of staples. Milk, eggs, bread, meat, cereal, and many other basic foods are emptied from stores within hours. The same can be said of a single product. If the evening news announces that coffee or some other item is going to be in short supply or that the price of a certain product is rising, the next-day rush to the store can exhaust the supply of that item from the shelves. Grocery stores are but the end of a chain of delivery points, and they are not structured to withstand fluctuations in customer demand precipitated by such external conditions as changes in the weather, costs, and availability of an item.

Another trend has been the relocation of supermarkets to the suburbs. Most large grocery stores have moved out of inner city areas and have been replaced with convenience stores and small speciality shops that keep a smaller variety of foodstuffs at higher prices. This trend particularly affects the elderly and the low-income persons living in the inner city who then must pay more for less.

A third trend is the growth of chain convenience stores. Supermarkets are feeling some pressure from convenience stores in that today, there are more convenience stores than supermarkets (Stafford, 1979). In a manner similar to the major supermarkets, convenience stores dominate the small store aspect of the food delivery system. For example, 93 percent of convenience stores are owned by companies with 11 or more stores (Stafford, 1979). Such stores have incorporated the economies of scale of purchasing and of fixed contracts much like the supermarkets, and in like manner seldom include any locally produced food.

THE DELIVERY SYSTEM

Baumel and Hayenga (Ch. 12) note major problems with the infrastructure of American transportation systems. While much of the road system of America is deteriorating at a rapid rate and the railroad system has been reduced drastically, both systems, even with minimal repairs, generally do not inhibit the delivery of food. Potential disruption of the food system could come from the interruption of food shipments within the U.S. due to strikes, civil disturbances, severe weather (e.g., major earthquake), energy shortages, or war. Experience has shown that actions such as excessive export demand or trade restrictions can influence the price of food, making it difficult for low-income people to afford; but such events do not greatly affect availability. Although food might cost more and variety of selection could be curtailed, basic food items would remain available.

The trucking industry is a good example of where a major interruption in food availability could occur. Trucks are the dominant mode of transportation for most food commodities. And even when aircraft, trains, or barges are used for long-distance delivery, trucks are still used to transport the products to the long-distance carriers and then to local stores. The trucker is the ubiquitous middleman carrying food from the producer to the processor, to the distributor, and to the retailer. If the truckers were to have an effective national strike, food delivery would be drastically curtailed. The U.S. food delivery system is extremely vulnerable to such an interruption.

FEDERAL POLICY ON FOOD

The USDA commodity distribution program is a result of legislation that began in the mid-1930s to aid farmers who were not able to market all their produce; it thereby served to stabilize farm prices. The government purchased the excess products and established programs for food assistance to needy persons (Longen and Allen, 1981). However, the government has found it administratively and economically problematic to directly give food to people. Particularly in domestic programs, the direct provision of food competes with that offered in the private marketplace and the established distribution system. To ameliorate the fears and objections of food producers, processors, and distributors, the U.S. government has provided money, or some form of money such as stamps or coupons, through which eligible recipients could purchase food. A recent exception is the distribution of dairy products such as cheese.

Even with an abundant and relatively cheap food supply, hunger and starvation exist within our society. When this issue has been brought to the public's attention by the media and concerned leaders, government response has been to develop food programs to help those in poverty. However, the focus is not on the availability of food, but rather on the ability to purchase it. Although the proportion of income the average family spends on food has declined from 20 percent in 1960 to 17 percent in 1980 (Gallo, 1982), low-income families still spend 30 to 45 percent of their income on food (Leathers and Zellner, 1982).

Food stockpiles were generated during World War I and World War II as a hedge against any major shortage and to ensure adequate supplies for the military. However, since that time, stockpiles have resulted from periods of overproduction and governmental commmodity programs to stabilize prices. This stockpiled food has been sold or given to foreign countries in the form of aid or has gone into such domestic programs as the school lunch program. For example, in the fiscal year 1980, schools received 87.5 percent of food purchased by the U.S. Department of Agriculture (Longen and Allen, 1981).

America's stockpiles of grain were almost depleted in the Russian grain deal of the late 1970s. However, in 1983, the government owned approximately 270 million bushels of corn and 180 million bushels of wheat. This government reserve plus the commodities stored by farmers contributes to a sizeable stockpile of food. However, most of these reserves have not been processed nor are they readily available for distribution if a major interruption in the food supply system should occur.

PUBLIC AND PRIVATE RESPONSE CAPACITY

What happens if roads and bridges are washed out by a flood and all food delivery is interrupted to a community? In the public sector, the USDA's Department of Food Distribution maintains food storage warehouses in most states for food assistance programs to the poor and school lunch programs. These food supplies include staples such as flour, sugar, dried milk, salt, cereals, dry beans, spaghetti, rice, and canned fruits, vegetables, and meats. Although local and state offices of civil defense no longer store food products, they do provide educational material and communication mechanisms to facilitate the distribution of food supplies in times of crisis. In addition, the National Guard is available to facilitate food delivery. Thus, in the public sector, the Office of Civil Defense, the National Guard, and the storage of food by the USDA could facilitate the provision and distribution of food and could ease a shortage of food to an isolated community or area.

In the private sector, non-profit organizations such as the Red Cross have historically supplied food in times of natural or man-made disasters. Most food comes from donations from individuals and organizations away from the affected area, food purchased from nearby communities (funds being provided by private donations and groups such as the United Way), and from commodity products of USDA. In addition, emergency assistance is usually forthcoming from church and civic organizations, businesses, and individual citizens. In general, organizations and agencies are available both in the private and public sector to help a community in times of disaster. However, this type of public and private aid is usually short-term and limited to specific areas. A national strike by truckers or a major winter storm over the Midwest or Northeast that lasted for a period of time likely would exceed the response capacity of the aforementioned private and public organizations. In short, food reserves are available for limited geographical locations and for short durations. A major interruption such as a strike or a major long-term natural disaster that would affect a large segment of the public could not be adequately handled by existing methods. Long-term survival and relief would depend on the private sector, particularly the reserves of individual citizens.

THE EMPTY CUPBOARD

What are the food reserves of individual citizens? Do Americans keep much food in their homes to guard against crises? The convenience of supermarkets has resulted in most Americans keeping fewer food products in the home. Most staple foods such as dairy, meat, and bread products are perishable, and little attempt is made by the majority of consumers to keep dried or preserved products for long periods of time. People desire fresh or frozen meat, fruits, and vegetables. Because of these consumption patterns, many households are likely to have limited reserves of food. And, the supplies people do keep depend upon their social, economic, and locational situations.

We would expect that people living on a farm or in a rural area would maintain larger supplies of food than those who reside in large cities. First, supermarkets are less accessible to rural residents and thus they make fewer trips to replenish food supplies. Second, as noted earlier, people in rural areas are more likely to have gardens and to can or freeze their produce. And third, people in rural areas are more aware of and experienced in the disruption of food supplies. From time to time, rural residents have had their food supplies interrupted because of such things as bad weather. Thus, rural people, as a matter of course, keep more food on hand.

Characteristics of household members affect the size of food reserves. Married people with children keep more food on hand than married people without children (Kaitz, 1979). Of course, they need a larger quantity to feed more people. Single people, whether young or old, tend to eat out more than married people. When both husbands and wives have jobs outside the home, they tend to eat out more often than households where only one person is working outside the home. Although eating out taps into another aspect of the food distribution system (restaurants), those who eat out often tend to maintain fewer food supplies at home.

In considering the characteristics of household members in their relationship to food reserves, we are more concerned about who gets hurt, if and when disruption occurs in the food delivery system, than who can claim the largest number of days of food supplies. Since awareness of food availability and inter/intranational food distribution policy may influence personal reserve patterns, we also examine the attitudes of citizens about food policy issues.

NATIONAL SURVEY

To gather information on home food reserves and attitudes on related food policies, a national (random digit dial) survey was conducted in 1982 by the Survey Research Center at the University of Kentucky. The framework for the national telephone sample was provided by the University of Michigan's Institute for Survey Research. Procedures for sampling the United States' adult population were based on procedures developed by Groves (1978). Information was gathered from 1,048 respondents for a response rate of 70 percent. The sample as a whole should provide a ±3 percent accuracy for the total U.S. adult population.

Most respondents lived in SMSA counties (76 percent), were married (60 percent), owned homes (64 percent), were white (84 percent), and were females (58 percent). Twenty-two percent of the respondents had total family incomes (before taxes) of less than $10,000, 26 percent had incomes of $10,000 to $19,999, 25 percent had incomes of $20,000 to $29,999, and 27 percent had incomes in excess of $30,000. These survey data correspond closely to Bureau of the Census data. For example, census data indicate that 75 percent of the population live in SMSA counties, 64 percent own homes, 83 percent are white, 52 percent are females (19 years or older), and 21 percent have incomes under $10,000(1).

FALSE SENSE OF FOOD SECURITY

Four out of five citizens feel that there is enough food available to feed all Americans. When asked, "In your opinion, is there enough food available in the U.S. to feed all Americans well, or is there a shortage of food?" eighty percent felt that there is enough food available. However, in a follow-up question only 52 percent felt that there will be enough food 10 years from now (44 percent felt that there will be a shortage and 4 percent qualified their answer). When the same questions were asked nine years earlier in a national telephone survey conducted by PACER, the findings were about the same (PACER, 1974:75). In 1973, 77 percent felt that there was enough food available in the U.S. to feed all Americans well. However, in 1973 only 40 percent felt that enough food would be available ten years from then. One conclusion from this comparison is that Americans are less fearful today of a future food shortage than they were in 1973. The public attitude toward this issue of food availability/shortage was quite uniform; no one segment of the population was significantly more or less concerned about the issue. For example, households with incomes of less than $10,000 were not significantly more concerned than households with incomes exceeding $30,000.

	1973	1982
Enough food available	77%	80%
Enough food will be available 10 years from now	40%	52%

A false security exists among the public in their belief about food storage in stores. When asked, "If the delivery of food supplies to local stores was disrupted because of natural disasters, strikes, or fuel shortages, how many days of food supplies do you think that local stores would have?" the common response was one week. Actually, perishables are often delivered daily and hard goods once or twice a week. On the average, supplies could be expected to last about one-half week. And, in a crisis situation, most grocery store food staples would be depleted in several hours.

PUBLIC ATTITUDES TOWARD GOVERNMENT FOOD POLICIES

The public was asked several questions concerning the issues of domestic and international food programs. Looking at domestic programs, there was strong agreement (77 percent) that "government should provide food subsidy programs such as food stamps and the school lunch program to low-income families." There was also support for maintaining food reserves. Seventy-one percent thought that the "government should have a policy of keeping a one-month supply of food staples in reserve in case of a national crisis."

Note that on the issue of government reserves variation did exist between the four regions of the U.S. Although 79 percent of the respondents from the Northcentral and West wanted a month's food supply, 72 percent in the South and only 66 percent in the Northeast favored a month's food supply. It is rather ironic that the region of the country (Northeast) most vulnerable to an interruption is also the least concerned with keeping a supply of food staples.

There was little consensus (29 percent agreeing, 71 percent disagreeing) that "the United States has a responsibility to supply food for the rest of the world." Neither was there agreement (37 percent) that "the U.S. government should use food as a political weapon when dealing with other countries." In short, the public expressed strong support for domestic issues such as food subsidy programs for the poor and adequate food stockpiles, but felt that food issues should be kept out of international politics.

HOUSEHOLD FOOD SUPPLIES

"If the delivery of food supplies to local stores were disrupted because of natural disasters, strikes, or fuel shortages, how many days of food supplies would you have in your home?" When asked this question, 49 percent indicated they had one week (7 days) or less of food on hand; 23 percent had one to two weeks of food; and 28 percent had at least a two-week supply. In this latter group, seven respondents reported a one-year (365 days) supply. In other words, half of the households in the U.S. have a supply that would be expected to last no more than one week, a fourth have from a one to two week supply, and another quarter have enough to last at least two weeks.

The median number of home food reserves is 8.2 days for the nation as a whole. The major difference in household food supplies is according to whether the person lives in a metropolitan (SMSA county) or in a nonmetropolitan area. Nonmetro residents maintain twice as much food in their homes as do metro residents (14 days supply versus 7 days). Respondents also differ significantly (median test, $p \leq .01$) in the amount of food kept in the home according to whether they rent or own their home, their age, and whether they are married or not (Figure 13.1). Homeowners, older persons, and married couples keep more food in their homes. Education, income, sex, and race in and of themselves do not have a significant impact on home food supplies. Food policy issues traditionally have focused on metro or nonmetro location, level of income, marital status, and age. Thus, these variables will be examined more closely.

Within metro and nonmetro counties, lower income households maintain slightly less food supplies than upper income households. Homeowners generally maintain more food supplies than renters. One exception is that low-income renters in nonmetro counties maintain twice as large a food supply as homeowners in the same income group. In metro and nonmetro counties, the pattern of food supplies by age grouping is unclear. Only those more than 60 years of age, of upper income who rent in metro areas manifest a very limited food supply (three days). Finally, although Figure 13.1 reveals some variation in median days of food supplies for different components of the U.S. population, it only begins to suggest those who would be most hurt in a food access crisis.

PROFILE OF THOSE WITH EMPTY CUPBOARDS

To exemplify those who could be most severely affected by a food crisis, we examined the characteristics of persons reporting three days or less of food supplies. These persons represented 13 percent of the respondents, and when projected to the U.S. population, account for more than 10 million households.

Young, single renters, living in metropolitan areas, typify respondents with small food reserves. Four out of five persons with three days or less of food provisions live in metro counties, 39 percent have incomes of less than $10,000, 64 percent are renters, 67 percent are between the ages of 18 and 39, and 60 percent are single (never married, divorced, or separated).

SUMMARY

Several summary points can be drawn from this presentation. First, Americans are not overly concerned about food availability, but second, they are concerned about domestic food programs. Third, the public is maintaining only limited food reserves in their homes. Fourth, the private food delivery system maintains very limited reserves. Fifth, the response capacity of

Figure 13.1: Median Number of Days of Household Food Supply

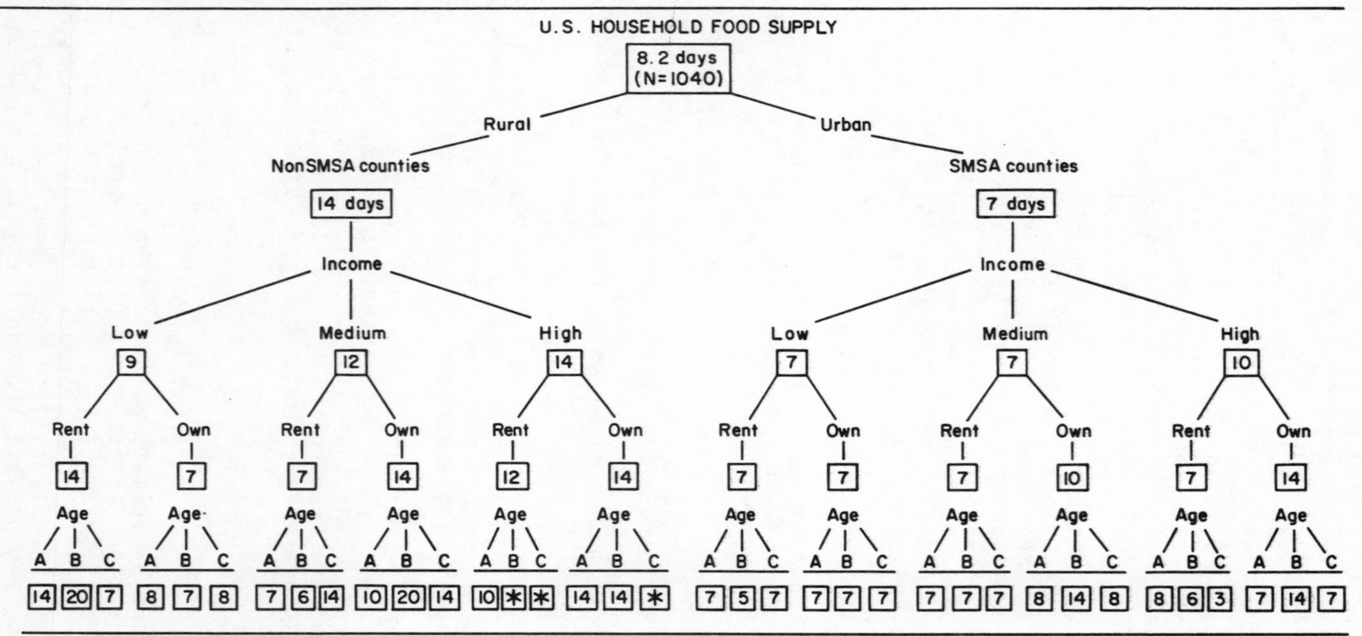

* NA - Less than 3 respondents.

Income: Low = <$10,000, Medium = $10,000 - $29,999, High = $30,000+

Age: A = <40, B = 40-59, C = 60+

public and private agencies to crisis situations is confined to those of short duration and of limited scope. Sixth, the majority of households would be greatly affected by an interruption in the food delivery system that lasted more than ten days.

Americans are fairly optimistic about food availability today and are more positive about future availability of food than they were a decade ago. The public is quite concerned about the government's role in domestic food programs but feels that government should not use food as an instrument of international policy. For example, about four out of five citizens favor subsidy programs for the poor and favor a national stockpile reserve in case of a crisis, while only one of three favors using food as a weapon in international policy.

Americans, overall, keep a little more than a week's supply of food in their homes. Metro residents have a one-week supply, whereas nonmetro people have about two weeks of food reserves. Renters in metro areas who are young and single are more likely to be affected by an interruption in the food delivery system than other segments of the public.

IMPLICATIONS

Most Americans can handle a ten-day interruption in the food distribution system with their home reserves coupled with store reserves. However, one out of every seven households has fewer than six days of food supply available to them (three days home - three days store). This assumes that the three-day store reserves would be available to these people but, in fact, people who keep small reserves may not have ready access to the three-day store reserves. Even a shut down of a few days could cause real problems for renters in urban areas, particularly those who are single.

An interruption of the food delivery system can not be substantially alleviated through inclusion of locally generated food because of the (1) seasonal nature of production, (2) absence of local a processing and distribution system, and (3) lack of an existing marketing system that incorporates locally grown foods. Even if areas could fall back on locally produced food, most regions simply do not produce sufficient volume nor variety of food products to support the population. In short, small quantities of local reserves would be available if the food delivery system is interrupted but would have a small impact in times of crisis.

Federal food policy has focused on commodity programs for the purpose of stabilizing food prices and providing subsidies for low-income people to enable them to purchase food at a reduced cost. The government has given very little attention to supplying and distributing food in times of crisis. It has been assumed that the private sector can care for these needs; but, in reality, private suppliers have very limited quantities available for such contingencies. If more foods were produced, processed, and consumed either regionally or locally, or if food reserves were stockpiled in areas of population concentration throughout the U.S., then there would be less need to be concerned about an interruption in the delivery system. Never in U.S. history have so many people had so little food in reserve; nor does the public seem aware of the fragile nature of the food delivery system.

The issue of government food programs for the poor relates to the problem of interruption in the food delivery system only to the extent that lower income households would be more quickly and drastically affected by an interruption. A major curtailment in the food delivery system would affect most households to some extent; the rich, the poor, the young, and the old. However, those who live in inner cities who have a low income, are either young or old, are single and rent would be most vulnerable because of their limited reserves and because few supermarkets are located in those areas. An interruption in the food delivery system would be most acutely felt in the large metropolitan areas of the Northeast region of the U.S.

Food security problems could be alleviated, at least to some extent, through (1) educational programs on the need to maintain home food reserves, (2) a government policy specifying a minimum level of reserves, (3) plans for the dissemination of stockpiles in time of crisis, and (4) government policies that would encourage food firms to incorporate local production in their delivery system.

Why do people keep so little food at home? Why do they seem so unconcerned about its availability? Most Americans have never experienced a major interruption or inconvenience in the food delivery system. Food is cheap and convenient. Why worry? Americans consider the possibility of a major interruption in the food delivery system unthinkable.

The food delivery system in the U.S. is so fragile that it would not take a major crisis to cause people to worry about survival. Although it is hoped that major interruptions in the food distribution system will never occur in American society, in actuality the potential for such an occurrence is quite high. There exists no need to hoard, no need to panic, but a need to be aware of the situation and to have contingency plans and programs to deal with a possible interruption in the food delivery system.

NOTES

* This research was partially supported by the Kentucky Agricultural Experiment Station and the Kentucky Cooperative Extension Service. In addition, support for the development of this chapter was provided by the Ford Foundation.

1. Comparisons with Census data include: income: U.S. Bureau of the Census, Population Characteristics, Series P-20, No. 374, p. 52, Table 9-3, Population Profile of the U.S.: 1981. U.S. Department of Commerce, Washington, DC; sex: U.S. Bureau of the Census, Current Population Reports, Series P-25, No. 917, Table 2, Preliminary Estimates of the Population of the United States, by Age, Sex, and Race: 1970 to 1981. U.S. Government Printing Office, Washington, DC; race: U.S. Bureau of the Census, PC80-S1-3, Table 2, Race of the Population by States: 1980. U.S. Government Printing Office, Washington, DC; tenure status - U.S. Bureau of the Census, Data User News, Vol. 17, No. 2, p. 1. February, 1982. Department of Commerce, Washington, DC; SMSA: U.S. Bureau of the Census, PC80-S1-5, Standard Metropolitan Statistical Areas and Standard Consolidated Statistical Areas: 1980, Table A. U.S. Department of Commerce, Washington, DC.

REFERENCES

Gallo, A. 1982. "Food Expenditures and Income." National Food Review 14(Spring):2-9.

Gallo, A., and Boehm, W. T. 1978. "Food Purchasing Patterns of Senior Citizens." National Food Review (September):42-45.

Grinnell, G. 1978. "Trends in Grocery Retailing." National Food Review (January):17-21.

Groves, R. M. 1978. "An Empirical Comparison of Two Telephone Sample Designs." Journal of Marketing Research 15(November):622-31.

Hatfield, K. H. 1981. "Changing Home Food Production and Preservation Patterns." National Food Review 13(Winter):22-25.

Kaitz, E. 1979. "Getting the Most From Your Food Dollar." National Food Review (Winter):26-29.

Leathers, H., and Zellner, J. 1982. "The Beneficiaries of Food Stamp Expenditures." National Food Review 14(Spring):23-24.

Longen, K., and Allen, J. 1981. "Domestic Food Programs." National Food Review 15(Summer):2-4.

New York Times. 1981. "New York State Food Imports." Sunday, July 26. 1981.

PACER, Inc. 1974. What the Public Says About Food Farmers and Agriculture. Monograph. Princeton: Response Analysis Corporation.

Stafford, T. H. 1979. "The Convenience Store Industry." National Food Review (Fall):19-21.

United States Bureau of the Census. 1981. Statistical Abstract of the U.S. 102nd ed. Washington, D.C.

USDA. 1964. Home and Garden Bulletin #77. Food Supply for Homes. Washington, D.C.

Van Dress, M., and Grinnell, G. 1980. "Grocery Stores and Eating Places: Industries in Transition." National Food Review (Spring):9-10.

Research and Extension

14
The Role of Agricultural Research for U.S. Food Security

William B. Lacy and Lawrence Busch

From its modest beginnings in the nineteenth century, agricultural research has emerged as a major component of the United States' food system and a key force in determining future U.S. food security. Although farmers have tilled the soil, domesticated animals, invented various mechanical devices, selected high yielding plant varieties, built irrigation networks, and developed methods of pest control for thousands of years, agriculture has been a subject of scientific investigation for little more than a hundred years. In that relatively short period of time, U.S. agricultural research has proven itself a productive enterprise, responsible for the technological advances underlying the past successes of our food system.

In this chapter we briefly trace the historical role of agricultural science within the U.S. food system. Next we examine the context for agricultural research in terms of (1) the current structure of agriculture and its implications for food security and (2) the changing assumptions for science in society. Then we suggest research policies and priorities which are likely to contribute to a sustainable, nutritious, adequate, and equitable food supply in the future. The core of the paper consists of an examination of the capacity of the contemporary research system to address these priorities and to contribute to U.S. food security. The issues explored are (1) public and private roles, (2) federal and state roles, (3) funding, (4) institutional insularity, (5) the aging scientific community, (6) basic science capacity, (7) interdisciplinary capacity, (8) long and short term research, and (9) policy making and assessment capacity. Following the discussion of each issue we conclude with some observations and suggestions to improve the research system's ability to meet the needs for a secure food system.

HISTORICAL CONTEXT

The entry of science into agriculture was at first a gradual process. By the late 1830s stationary steam engines had been developed by the private sector and were in use for threshing, ginning, sawing wood, and grinding sugar cane. At about the same time, the patent office received a budget of $1000 and began to collect agricultural statistics and distribute new varieties of seed. The first university agricultural research in the United States began at Harvard and Yale in the 1840s, stimulated by the pioneering work of von Liebig in soil chemistry and fertilizer development.

The research institutions that would later become an integral part of the agricultural system began to emerge during the middle of the nineteenth century. In 1855, the first land-grant college, Michigan State University, was founded. Seven years later, a national system of agricultural colleges was created with the passage of the Morrill Act. In the same year the U.S. Department of Agriculture (USDA) was established with a primary focus on developing a research capability. In 1875 the first U.S. agricultural experiment station was established as an independent institution in Connecticut. The passage of the Hatch Act in 1887 completed this fundamental change in the structure of U.S. agricultural research. The Hatch Act, which gave each state $15,000 a year for the study of problems in agriculture, established a decentralized system of experiment stations at the land-grant colleges across the country.

Until this institutionalization of research, science in the Western world had been primarily the domain of intellectually curious, isolated individuals. The scientist's role in this Cartesian style of science was much like that of the atomistic, laissez-faire entrepreneur or a self-employed master craftsman, pursuing what he/she felt to be important, and confident that the findings would be building blocks of knowledge. With the establishment of USDA, the land-grant colleges, and the experiment stations, the U.S. Congress institutionalized science along the lines of Bacon's corporate model. In this model science became industrialized and scientists became employees of the state and industry. Moreover, this shift from Cartesian to Baconian science (Busch and Lacy, 1981; Haberer, 1969) changed the nature of scientific knowledge: "The dominant mode of production of scientific knowledge has become that of knowledge-as-commodity, as a marketable good with a cash value" (Rose and Rose, 1976:12).

The motivations for the establishment of the experiment stations were diverse and sometimes contradictory, but included a utilitarian concept of agricultural science. Knowledge was to be pursued for its usefulness in improving the material conditions of the population. Research was to be conducted by large numbers of persons who had learned the proper methods, and empirical results were to be emphasized over theory.

With this act as a foundation for the contemporary structure of public sector agricultural science, the complex set of institutions supporting agricultural productivity and growth evolved and flourished over the next century within the context of an ever-changing social, economic, and political environment (Rosenberg, 1976). During the latter part of the nineteenth century the agricultural sciences developed as part of the expansion of colonial empires and the shift from subsistence to capitalist farming.

Like their contemporaries in other sciences and the general public, agricultural scientists in the early years of this century tended to see the products of science as undiluted good. While society might be guided by science, it seemed that science itself was autonomous--a type of knowing that, through the use of certain special methods, ensured the emergence of truth and social progress (Jordon, 1907). To this end, the leaders of the agricultural colleges felt compelled to embark upon a program for remaking rural society through the application of scientific knowledge. The increasingly specialized, commodity-specific nature of scientific inquiry encouraged the development of a commodity orientation among clients (Rossiter, 1979). Among the unanticipated consequences of this development were the underfunding or abandonment of problems not directly related to commodities. Gradually, over the course of a century, this relationship began to appear to many as "natural" (McCalla, 1977).

The claim of scientific autonomy made science the pawn of vested interests because it served to divert scientific attention away from questions of ends--what is food security--to questions of means--how can productivity be increased? Both the belief in the autonomy of science and the organization of influential clientele along commodity lines made it possible for productivity and efficiency to be treated as ends and for scientific work to be gauged more and more by the degree to which it contributed to these "ends." The broad range of goals provided in the laws relating to public agricultural research (e.g., Hatch Act of 1887, Adams Act of 1906, Bankhead-Jones Act of 1935, and the Agricultural Marketing Act of 1946) became merely a backdrop for the increasing emphasis on productivity from the mid-1930s through the 1970s.

CHANGING CONTEXT FOR RESEARCH

Structure of Agriculture

Agricultural research became one of the unquestioned

success stories of public investment in science. Research and the new technology it helped to generate were major factors in the transformation of U.S. agriculture to a high technology, mechanized, science-based industry. The new technology utilized relatively inexpensive and plentiful chemicals, petroleum, and capital in contrast to the shrinking agricultural labor force (Hayami and Ruttan, 1971). This new agricultural system became the most productive in the world. While one farm worker's production fed only six other people in 1900, this had increased to 10 others by 1940. By 1980, with extensive use of water, energy, fertilizers, pesticides, improved varieties, and labor saving mechanization, one farm worker fed 60 other people (Wittwer, 1982). In that same year gross farm income was $100 billion; agricultural exports were valued at $40 billion, accounting for 19 percent of all U.S. exports; and the food and agricultural sector as a whole was the largest of all U.S. industries (Rockefeller Foundation, 1982).

However, as the chapter authors of this volume clearly note, there is trouble amidst this bounty. In the 1970s U.S. agriculture entered what many people regard as an era of limits and critical choices, requiring significant adjustments in the use of resources to ensure long-term sustainability of our food system. The output per person of virtually every major commodity produced by forests, fisheries, and grasslands appears to be declining worldwide (Brown, 1981). Productivity of the major food crops has plateaued with yields of wheat, maize, sorghum, soybeans, and potatoes in the U.S. unchanged since 1970 (Wittwer, 1978). Scientists have even suggested that plant breeders in the advanced industrial countries have already raised yields as far as they can (Jensen, 1978). Furthermore, pests and diseases still destroy a third of the country's total potential harvest despite the use of chemicals by most farmers (Budiansky, 1984).

Related to the changing U.S. production capacity and efficiency is the growing recognition of the constraints on nonrenewable natural resources as production inputs and the necessity to maintain minimum levels of environmental quality. First, there are the growing and competing demands for other uses of prime farmland, water, forests, and other natural resources (e.g., conversion of prime cropland to nonfarm use has been occurring at the rate of one million acres per year since the 1960s [Brown, 1981]). Second, as remaining reserves of readily available, relatively fertile land are depleted, the expansion of agriculture will mean the elimination of previously fallow acreage and movement onto less productive and more fragile soils. This will likely entail greater production inputs, such as irrigation water and fertilizers, and increased risks of erosion and other environmental damage. One third of all U.S. cropland is being eroded faster than it can be replaced by natural processes and a significant percentage of soils has the potential for serious losses in production due to erosion (Larson, et al., 1983).

Dwindling water supplies in some regions, increasing salinity and waterlogging in irrigated lands, water pollution, potential environmental and health hazards related to pesticides and herbicides, and other health hazards from past and current production practices have become increasingly problematic. Several writers, such as Berry (1977), Brown (1981), Friedland, et al. (1981), and Hightower (1973), have argued that current agricultural practices have contributed to environmental degradation and may be inappropriate for ecological sustainability.

Farmers continue to adopt technological advances and new practices based on intensive use of energy and petroleum based chemicals, to develop land and to increase land holdings despite having reached or surpassed the size needed to attain most economies of scale (Miller, et al., 1981). As a consequence farms today are fewer, larger, more specialized, more capital intensive, and generally more highly in debt. The number of farms in 1980 was 2.31 million, down from nearly 7 million in the 1930s. In addition, just 7 percent of the farms generated 56 percent of the total value of all food and fiber. However, spiraling farm costs and widely fluctuating world markets have made these farmers increasingly economically unstable. Furthermore, regional concentration and specialization have increased crop vulnerability to adverse and fluctuating climate conditions; and since 1970, economic and environmental pressures, which are, in part, the result of the focus of agricultural research, have eliminated an average of 45,000 farms per year (Budiansky, 1984).

The growth in the specialization, capital intensity, and concentration of the farm sector is matched by a growing concentration throughout the entire food industry. Agricultural input suppliers, the agricultural marketing system, food processing, distributing, and retailing are all becoming increasingly concentrated. The processing, distribution and marketing sectors, which now account for a substantially larger portion of the consumer food dollar than the farm sector, make many basic decisions concerning the kinds and volume of food products to be produced. This economic concentration in the nonfarm sectors of the food system, facilitated by new technology, raises additional concerns about the adverse effects of monopolistic power on both consumers and producers.

The agricultural sector and the research community are also being confronted with increasing consumer interests and concerns about the products of this food system. Changing farm technology, new hybrids bred for crop storage and shipment, and widespread use of pesticides, as well as increased processing of raw foods and numerous food additives, have led to concerns about the nutritional quality and safety of food. Finally, the public has begun to raise broad questions about the fundamental goals for the nation's food and agricultural research system, including such issues as equity, efficiency, sustainability, flexibility, conservation, and consistency with other objectives of U.S. society.

Science, Technology, and Society

Paralleling these dramatic changes in the structure of U.S. agriculture, radical changes have also occurred in the broader context of science itself. Until recently, science and technology were viewed as the chief architects of human progress. As a consequence, during the last century, huge sums of money were devoted to scientific research, development, and training throughout the world in the confident belief that science was an indisputable good that creates knowledge free from the constraints of human circumstance.

Recently, however, the assumptions, as well as the products, of science have come under attack. First, the scientific claims to objective knowledge have been challenged. Several social scientists and philosophers (e.g., Busch and Lacy, 1981; Ravetz, 1977; Rose and Rose, 1976) have asserted that science shares the ideologies prevailing in a given society at a given time. These writers have proposed that the neutrality of science is a myth, and with a host of suggestive examples have shown how both scientific choice and scientific concepts are ideologically and politically influenced. They have argued that values and ideologies, as well as economic factors, affect: (1) the organization of the sciences into broad areas and specific disciplines; (2) scientific research styles or orientations; (3) the degree of permissible deviation from the established scientific orthodoxy; (4) the way in which research results are reported; and (5) even the choice of research problems.

These challenges to an autonomous, objective science have been bolstered by two related changes: the decline of the belief in the unity of science and the changing role of science in society. Kuhn (1970) challenged the notions of the unity of science, scientific rationality, and progress. It could now be argued that knowledge of certain aspects of the world must be developed at the expense of knowledge of others. Consequently many separate sciences could develop out of many cultural systems but all would still be bound by the phenomena in that world.

The status of science also has changed from that of

a peripheral to that of a core institution with a corresponding change in the role of science in the larger society. With the development of the corporate form, science became industrialized. This "big" science (Price, 1963) began to claim resources and budgets so large as to be politically significant and consequently constrained by general social priorities. In the 1950s and '60s growth rates were often in excess of that of the gross national product. Governments also became interested in science policy, since solutions to politically perceived problems, such as food and energy, were viewed as directly technological. Therefore, governments developed science policies to encourage the development of scientific and technical research and to exploit the results of this research for general political objectives (Salomon, 1977). Both the state and large corporations have become increasingly reliant upon scientific knowledge to bolster and maintain their legitimacy. This increase in scale, change in structure, and increasing centrality of the relations between science and the state have reduced the autonomy of science.

By the late 1960s and early 1970s science was also faced with several practical considerations. The idea that progress in the basic sciences was bound to guarantee successful future innovations ceased to be an article of faith. While feasible, some innovations might not be totally desirable, either politically or socially. There was disappointment in science, for it often promised more than it delivered. Some concluded that science was not the sole or even the supreme tool for improving the world. Furthermore, there was a growing concern in society about the adverse consequences of our use of science and technology, from environmental pollution and fears of genetic engineering to the awesome techniques of warfare.

From this rising concern with science came a realization of the need to reexamine the role of science and technology in society to include broader social objectives. The central issue in the field of science, technology, and public policy shifted from unequivocal support to concern for monitoring, directing, and controlling the development of science. The emphasis became one of how science and technology could be channeled for social purposes. Procedures for technology assessment included an examination of technology's social implications and a delineation of the advantages and disadvantages of alternative solutions. Further, these assessment procedures sought to anticipate possible undesired effects and to analyze the real costs of new technologies in relation to obvious or disregarded social needs (Nelkin, 1977).

The effort to assess science and technology has raised new difficulties for research programs. Technology assessment requires a closer link among the natural sciences as well as between the natural and social sciences. For a variety of reasons this is often difficult to achieve. In addition, when one moves from strictly technological objectives to include social objectives, the means for achieving them are less clear because these objectives are diffuse, fluid, and given to partial and progressive solutions. Further difficulties in the identification of appropriate technical solutions to social objectives are created by divergent interest groups and economic and political constraints. Consequently research which engenders social objectives is compelled "to develop in successive stages of adaptation to various conditions determined not only by the progress of skills and techniques but also by the understanding of problems and changes which intervened in the balance of political, social and economic forces" (Salomon, 1977:60).

RESEARCH PRIORITIES FOR FOOD SECURITY

Given the historical context for U.S. agricultural research, the changing assumptions about science and society, and the threat to U.S. food security posed by the current structure of agriculture, what should be the future role of agricultural research in ensuring a long-term, adequate, and secure food supply in the United States? Several researchers, research administrators, and science policy analysts view the role of agricultural science and technology as central in the transition to a secure, sustainable, and equitable food system (Bergland and Sechler, 1981; Brown, 1981; Ruttan, 1983; Wittwer, 1982). A recent workshop of agricultural research leaders concluded, "The critical importance of agriculture to the vitality and strength of this country, and the increasing diversity, complexity and intractability of problems facing American agriculture make it imperative that the agricultural research system be able to sustain its level of past performance" (Rockefeller Foundation, 1982).

Overall, the science and technology requirements for a secure food system will be diverse. There can be no easy technological fixes or simple breakthroughs. At the center of this long-term, national scientific effort will be research and technologies that conserve energy and nonrenewable resources, result in stable food and fiber production at high levels, are nonpolluting, add to rather than diminish the earth's resources, are capital efficient, and are beneficial to farms of all size. Research must also attend to issues of quality, safety, and nutrition in food products and consider the consumer as well as the producer in addressing these issues. Finally, scientists should incorporate a broader sense of society's social objectives and goals into their research agendas. This will require a major involvement with the issues of equity, quality of life, and human development.

The redirection of priorities will likely entail movement away from research on food production technologies that are based on a high degree of mechanization with extensive use of land, water, and energy resources. Instead, emphasis should be placed on research which contributes to technologies dependent on mission-oriented basic research relating to the biological and biochemical processes that control and limit crop and livestock productivity (Brady, 1982).

According to the National Academy of Sciences' National Research Council (1977) and the Joint Council on Food and Agricultural Sciences (1983), research to ensure greater food security should address the problems of enabling plants and animals to more effectively utilize present environmental resources through (1) improved biological nitrogen fixation; (2) more resistance to competing biological systems and environmental and climatic stresses; (3) greater photosynthetic efficiency; (4) new techniques for genetic improvement such as cell and protoplast culture, somatic hybridization, embryo transfer, and such recombinant DNA approaches as gene identification, characterization, splicing, replication, regulation, and transfer; and (5) more efficient nutrient and water uptake.

Since crop production is limited more often by water than by any other production variable, a key factor in long-term food security is an adequate and safe water supply. Therefore research is called for to provide more efficient management of surface and groundwaters, including irrigation, drainage, and quality as well as regional modeling, water control, and allocation.

In addition, research should assess the economic impact of erosion on the productive capacity of U.S. agriculture, examine the relationship between erosion and reduction in soil productivity, and develop ways to maximize plant growth without jeopardizing conservation of the soil resource. Further, socioeconomic evaluation research that reviews alternative incentive measures to encourage adoption of soil conserving farming practices is a priority.

In the area of human nutrition, research is needed to determine nutritional requirements, the relationship of nutrition to performance and good health at all ages, and the relationship between dietary habits and chronic diseases. Food security also requires improved technology to detect and remove pathogens and toxins from food products, to evaluate the quality and acceptability of foodstuffs to consumers, and to determine the effects of processing, transportation, and storage on nutrient levels in food.

Furthermore, technologies and systems should be

developed that can reduce costs and improve efficiency in the marketing, processing, and distribution industries, while contributing to the maintenance of fuller competition in the nonfarm, agribusiness sector. This may entail research and development focused on alternative marketing and distributions systems, such as food co-ops and farmers markets (Schaaf, 1983), and reduction of food wastes through recycling nutrients from organic waste (Knorr, 1983).

To ensure greater economic stability and survival for U.S. farmers increasingly involved in world agricultural and forest product markets, agricultural economics research is needed to develop longer-term anticipatory policies. Issues to be examined include projections of supply and use of crops, factors affecting import demand in foreign markets such as trade barriers, macroeconomic conditions, and changing export marketing practices.

Many smaller technological advances and adaptations of existing technologies will be important for a sustainable food system. Scientists and engineers will need to reassess the full consequences and potential payoffs from relatively greater attention to the needs of smaller farms. Furthermore, researchers should explore the long-run potential for nonconventional production practices (e.g., perennial crops, organic farming). Successful research and development of these nonconventional practices and small-scale technologies may also have an indirect positive effect on U.S. food security by providing technologies appropriate to the resources and needs of those not now self-sufficient in food production. As Wittwer (1982) points out, the U.S. cannot and should not plan as a long-term policy to be the breadbasket of the world. This, he argues, would require an exploitation of land, water, mineral, and energy resources that neither we nor the rest of the world can afford for long and would jeopardize our own food security. Therefore, research and development which assists developing countries to improve their food systems indirectly contributes to long-term sustainability of our system.

Another major priority for agricultural research is to develop appropriate mechanisms for planning to meet long-term food security needs and to allocate resources sufficient to support those efforts while retaining the flexibility to respond to urgent short-term problems. This will necessitate more careful assessment of long-term food security priorities, anticipation of short-term problems through effective crop forecasting and early warning systems, and better coordinated determination of overall priorities at the national, regional and local levels. This will also entail an improved understanding by the public sector planners of private sector science plans and investments to ensure the most efficient and effective use of all science funds and research capacity.

The final priority is to develop technology assessment capability in the research system. As noted earlier, technological change is seldom neutral. It frequently provides an advantage to those who seek or can readily adapt to change and a disadvantage to those who did not or could not readily adapt to change. Therefore, mechanisms need to be developed within the research process to seek and to more consciously and deliberately ensure the fulfillment of environmentally and socially desirable goals (Friedland and Kappel, 1979). Although formal social impact assessment of changing technologies in U.S. agriculture is still in its infancy, researchers utilizing this approach have provided insights on such topics as the socioeconomic consequences of automated vegetable harvesting (Friedland and Barton, 1975; Friedland, et al., 1981), tobacco harvesting, center pivot irrigation, and organic and no-till cultivation (Berardi and Geisler, 1984). This type of research requires a substantial monitoring effort and the development of the projective capacity of scientists and engineers to understand the consequences that derive from research. Finally these impact assessments should be integrated into the initial research problem development, and continue to be a part of the research process as the work progresses.

CAPACITY OF RESEARCH SYSTEM

To assess the role of agricultural research for U.S. food security, it is necessary to briefly review its current organization, resources, scientific personnel, and capacity to address the broad range of research priorities identified in the previous section. Today the U.S. agricultural research system consists of a complex and diverse set of both research institutions and supporting public policy, and a wide range of potential clients. The research is conducted primarily by the land-grant agricultural colleges, the U.S. Department of Agriculture (USDA), and the privately supported research sector. Some additional agricultural research is also performed by other government agencies (e.g., U.S. Agency for International Development) and other public and private universities. Recent estimates of U.S. agricultural research and development expenditures suggest that nearly $3 billion is allocated yearly. In 1979 support for public research performed by the USDA and the state agricultural experiment stations amounted to approximately $1.2 billion, while research and development expenditures by private firms in the agricultural input, food processing, and distribution industries were about $1.6 billion. Ruttan (1982b) has argued that in 1979 a complete accounting of private sector research and development would probably reveal expenditures in excess of $2 billion.

Public and Private Roles

Several general criteria have been used to gauge the appropriate roles of the public and private sectors. In general the private sector focuses upon those aspects of research that are likely to involve short-term commitments of capital and to lead to proprietary products, such as machines and processed food products. Consequently, most private sector research is focused on product development. In contrast, the public sector's investment in agricultural research is in areas where the social rate of return exceeds the private rate of return. A large share of the gains from research are captured by firms and by consumers in the form of (1) nonpatentable improvements, such as new agronomic practices, improved seed varieties, and better management and marketing; (2) long-term, high risk or basic research products in such areas as soil physics and chemistry, plant physiology, and insect ecology; and (3) broader issues of the well being of the public, such as environmental protection, nutrition, and community development. Furthermore, the public sector maintains a research component that not only meets broader research goals but also trains and educates people to do research. Thus, in the agricultural sciences, a synergistic interaction exists between research and education (Committee for Agricultural Research Policy, 1981).

The appropriate roles of public- and private-sector agricultural research have generated widespread discussion and controversy in recent years, and the appropriate balance between the two sectors continues to be subject to intensive scrutiny. The relative emphases placed on various fields of agricultural science and technology by the private and public sectors differ markedly. Approximately two-thirds of the private sector research and development is concentrated in the physical sciences and engineering, although this may change with heavy investments in biotechnology. In contrast, public agricultural research is much more heavily concentrated in the biological sciences and technology, with three-fourths of the state experiment stations' research in these areas (Ruttan, 1982b). Since there is often substantial complementarity between research in the two sectors, a freer flow of communication in the future may be important to avoid unnecessary overlap and waste.

Although private-sector research in support of agricultural production and the processing and marketing of agricultural products has been expanding, its capacity is neither documented nor understood. In addition, given the general mandates for the two sectors, much of the challenge of addressing issues of U.S. food security among agricultural scientists will fall primarily to the public sector. The remainder of this section, therefore, examines the will and capacity of U.S. public-sector agricultural research to address the full range of issues related to U.S. food security.

Despite the accomplishments and high regard worldwide for the U.S. public sector research system, a number of questions have arisen regarding its future research capacity and directions. Since the early 1970s a series of internal and external reviews have criticized the adequacy of funding and the quality and focus of research being conducted by the USDA and the state agricultural experiment stations (Buttel, et al., 1983; Hightower, 1973; National Research Council, 1972; Rockefeller Foundation, 1982; Office of Technology Assessment, 1981). Surplus capacity in American agriculture during the 1950s and '60s generated complacency in the federal government and USDA concerning the returns from increased research investment. The U.S. scientific and technological leadership for maintaining food production and contributing to a long-term, sustainable and secure food system can no longer be taken for granted. As Congressman George Brown notes, "Agricultural research is going through a mid-life crisis. We've emphasized applying existing knowledge and failed to replenish our intellectual capital" (Budiansky, 1984:66).

Funding

Instead of expanding to meet the new challenges for a secure food system, the funding for agricultural research has remained woefully inadequate. Several researchers (Bonnen, in press; Brown, 1981; Ruttan, 1983; Wittwer, 1982) have pointed out that annual rates of return on research expenditures in agriculture range from 35 to 50 percent, well above the returns to other public investments. A reasonable long-term equilibrium rate of return is estimated at about 10 percent. Yet appropriations for agricultural research are well below the level consistent with this payoff and its value to society.

This inadequate funding is contributing to depletion of the scientific resource base and undermining the system's capacity to sustain the productivity growth of the present system and to address the new research priorities. With limited resources, an ever-increasing portion of the budget is devoted to maintenance and adaptive research. Maintenance research is work that protects previous increases in plant and animal productivity against forces that would otherwise result in productivity loss (e.g., disease, weeds, pathogens, and various adverse environmental conditions). Araji, et al., (1978), estimated from a survey of experiment station researchers and extension specialists in the West, that 10 to 35 percent of their research is maintenance research. These authors contend that if no such research were done for a decade or more, productivity would decline by 10 to 40 percent depending on the commodity and time period. Bonnen (in press) observed that maintenance research and adaptive research (that which adapts science and technology to specific local ecosystems) probably constitute between 35 and 80 percent of the typical experiment station's effort. Ruttan (1983) reported that maintenance research is likely to grow concomitantly with the level of productivity. Therefore, during periods of increasing productivity and a relatively static research budget, maintenance research will constitute a larger share of the total budget each year. While this research is essential, additional funds must be allocated now for research which addresses the increasingly urgent issues of long-term food security. White and Havlicek note that "inadequate funding of agricultural research and extension activities in one period will be very difficult to overcome later" (1982:55).

This underfunding of agricultural research is particularly acute at the federal government level. Real federal support for agricultural research remained essentially unchanged between 1965 and 1980. In fact, since the early 1970s, the purchasing power of federal monies for agricultural research declined at the rate of 2 percent per year. In contrast, federal funding for other types of research and development, such as military, space, health, and energy, grew rapidly and in 1980 averaged a 7 percent increase (Wittwer, 1982). In that same year agricultural research, which constituted a major share of the federal funding for research and development prior to World War II, was only 1.25 percent of the total federal research budget of $30 billion (Rockefeller Foundation, 1982). Within USDA the relative importance of research has also declined precipitously. In 1916 agricultural research activities accounted for one quarter of the USDA budget. By 1980 a far larger USDA research program accounted for about 2 percent of the total budget. Research has become a secondary mission within USDA as programs which provide direct economic benefits to a specific set of farm and business interests have become the dominant activity (e.g., price supports, Payment in Kind).

Despite this lack of growth in federal funds, state funding of agricultural research has continued to grow in both current and real dollar terms. Of the total public appropriations for agricultural research in 1979, 42 percent ($480 million) were provided by the states, and approximately 58 percent ($600 million) were appropriated by the U.S. Congress. Recent federal budget cuts have brought the proportions closer to 50 percent for each. In fact, because a portion of the federal budget is channeled directly to the states, in 1979 about two dollars ($708 million) were expended for agricultural research at the state level for each one dollar ($380 million) at the federal level (Office of Technology Assessment, 1981). However, even this growth at the state level has been inadequate to meet the research needs of the state experiment stations.

Federal Institutions

As a consequence of declining federal funding for agricultural research, the state experiment stations and colleges of agriculture have assumed expanded financial responsibility as well as scientific leadership and research priority setting for agricultural and food research. Unfortunately, each experiment station must often emphasize research on relatively short-term problems of high priority to state agricultural interests. Thus, insufficient attention is given to regional, national, and international issues regarding food security and inadequate attention is often given to basic research to replenish the knowledge base. Furthermore, according to recent critiques, the federal agricultural research system, which includes the Agricultural Research Service, the Economic Research Service, and the Cooperative State Research Service, currently lacks the scientific capacity or the bureaucratic structure to exercise national leadership (Ruttan, 1983; Rockefeller Foundation, 1982).

Agricultural Research Service (ARS). For years the Agricultural Research Service performed much of the basic science research related to agriculture, set standards for research quality, and provided scientific leadership for agricultural research. The ARS has been weakened in part due to reduced resources and manpower, deemphasis of basic food research, and increased emphasis on short-term applied research. As Wittwer (1982) noted, 40 percent of USDA scientists became eligible for retirement between 1977 and 1982, but due to personnel hiring ceilings, many were not replaced. The scientific force was reduced from 3,300 to 2,850 during that period.

The Agricultural Research Service has also regionally decentralized its administration and fragmented its research efforts by establishing a large number of field sites more as a consequence of "pork barrel" politics in Congress than research need. A General Accounting Office report published in 1983 noted that the ARS operates 148 research facilities, many of which are used well below capacity. Several have fewer than 10 scientists, but when ARS discusses consolidating and closing facilities, area Congressmen apply pressure to maintain these facilities (Norman, 1983).

Cooperative State Research Service (CSRS). Along with the weakening of the ARS, the partnership between USDA and the state agricultural experiment stations has also eroded primarily out of neglect and attrition within the CSRS of the federal system. CSRS was established to represent the federal government in its partnership with the state agricultural experiment stations. However, due to past personnel and budget constraints and a lack of substantial support from the states, CSRS's activities have been largely limited to administration and oversight of formula funds, instead of dialogue on substantive research issues.

Suggestions. The leadership and strong federal support for agricultural research should be reestablished and sustained. In addition, the ARS research capacity needs to be strengthened and its basic science, long-term research, and national missions reemphasized. The existing ARS facilities need to be systematically evaluated in terms of the research missions and retained only when they have a feasible possibility of contributing to the research priorities. Simultaneously the CSRS's leadership capacity for coordinating and stimulating regional and national research among states and federal research agencies needs to be strengthened.

State Institutions

While the state agricultural experiment stations have suffered neither the serious manpower losses of the federal system nor the large decrease of funds, they too have experienced a loss in their research capacity for growth and innovation. In the experiment stations the scientist years in support of agricultural research have not changed since 1966. While enrollment in colleges of agriculture in the land-grant universities tripled during the late sixties and seventies, there was little increase in faculty (Wittwer, 1978). Often research has been neglected in order to meet teaching needs. In addition, research capacity has also declined in terms of new equipment and facilities. Federal support in the form of formula funding to the states has been static since the late 1960s. Traditionally much of this money has been allocated to maintain and replace research equipment and renovate and improve facilities. Wittwer (1982) maintains that research equipment and facilities are now woefully inadequate and outmoded. Furthermore, he contends that indirect and overhead charges in competitive grants are not sufficient to cover ever-rising maintenance and replacement costs for these research needs nor are they used by universities to meet these costs. This failure of the federal government to support any growth in state experiment station systems, coupled with the real growth of state funding, may have resulted in research priorities that place too great an emphasis on short-term, site-specific applied research. As Bonnen (in press:18) notes, the clear lesson "from successful agricultural development the world over is the necessity for a centralized national investment in agricultural research complemented by and coordinated with a decentralized capacity for adapting agricultural research to the highly varied local ecospheres within which agriculture is practiced."

Suggestions. Consequently, federal formula funding for the states as well as state appropriations for agricultural research should be increased significantly. Ruttan (1983) has suggested that, given recent funding trends, the incentive effect of formula funds from the federal government might be greater if the federal government matched state expenditures rather than the current, reverse practice.

Institutional Insularity

The research capacity of the agricultural research system may have been further eroded by its isolation from the larger scientific community. About a decade ago, Andre and Jean Mayer (1974) characterized the system as an "island empire" and observed that the closed, politically self-sufficient nature of the agricultural research institutions, while historically a strength, has led to the isolation of agricultural research from the rest of science. Other research universities and industry lie outside traditional communication channels of the agricultural sciences. However, science and its communication seem to be inextricably bound together, so that the production and dissemination of the results of research go hand in hand. Formal and informal communication appear to shape the very way in which scientists pursue their research and choose research agendas. These forms of communication have become highly specialized among agricultural scientists as they have throughout the scientific community. Formal communications (various written sources and professional publications) have become increasingly dominated by domestic agricultural disciplinary journals to the general exclusion of other sources of information (Lacy and Busch, 1982).

Informal communication (e.g., conversations, telephone exchanges, personal correspondence) has become a strong competitor for the limited amount of time that any scientist can devote to receiving information. However, informal communication about research among agricultural scientists is relatively infrequent (usually less often than once a week) and limited primarily to contact with scientists in one's own department. Informal contact with nonagricultural scientists occurs less often than once a month.

Further, agricultural researchers report that these nonagricultural scientists are relatively noninfluential in any of their research decisions, from the type of problem and the research methods to the key concepts and theoretical orientation (Lacy and Busch, 1983).

Further evidence of agricultural scientists' isolation is reflected in their educational and career patterns. The total educational and career experience of most agricultural scientists tends to be restricted to the land-grant institutions. Indeed, 79 percent of the scientists employed at these institutions received all their education at land-grant schools. In addition, nearly one-third (32 percent) of the scientists currently at these institutions received at least one degree from the institution at which they are employed and 18 percent received at least two degrees from that same institution. Only a small subset of these institutions provided training for the overwhelming majority of practicing agricultural scientists. A dozen land-grant colleges of agriculture have provided the doctoral training for nearly three of every five agricultural scientists (Lacy, 1982). In contrast, the top twelve Ph.D. granting institutions for the natural sciences provided 33 percent of the doctorates (Harmon, 1978).

This insularity in educational and career patterns results in minimal mobility among research institutions, which is posited to be an important determinant and stimulant of a scientific career (Harmon, 1978). For science organizations, mobility may be an important source of innovation, since the movement of a knowledgeable individual from one organization to another is perhaps the most efficient way of transferring knowledge. The insularity and minimal mobility is likely to limit scientists' horizons by restricting their exposure to ideas originating outside the system.

Moreover, many scientists outside agriculture are both ignorant and critical of the system and some seem even to resent its existence. Lewontin (1980) contends that this has resulted in two cultures, with agricultural research relegated almost exclusively to the state experiment stations and the federal agricultural research facilities. Nevertheless, meeting the future agricultural research needs will require the very best scientific expertise and a free exchange of ideas and information. A number of scientists in private universities, state universities without agricultural colleges, and land-grant universities outside agricultural colleges have overlapping research interests and could carry out agricultural research if there were available funds.

Suggestions. Innovative steps should be taken (1) to increase the substantive interaction between the land-grant agricultural scientists and those from other research settings through seminars, workshops, fellowships, and exchanges; (2) to ensure that significant and relevant research results of nonagricultural scientists are incorporated into the research activities of the experiment station and the federal system; (3) to encourage experiment station and federal government scientists to broaden their research training through such mechanisms as sabbaticals, post-doctoral leave, workshops, and fellowships; and (4) to attract nonagricultural scientists to the fields of food and agricultural research through collaborative efforts, exchanges, and grants.

Aging Scientific Community

Another problem facing all levels of the public sector research community is the increasing age of the scientific staff reflecting a failure to recruit new scientists. According to the U.S. Department of Commerce Bureau of the Census (1978), the highest median ages for persons in science and engineering in 1978 were for agricultural scientists (45.3), engineers (45.8), and earth scientists (46.0). A representative national sample of principal investigators in the U.S. public sector agricultural research community revealed a median age of 48 (Lacy, 1982). This pattern is even more pronounced in the federal system. The Office of Technology Assessment (1981) reported that in 1976 only 2 percent of USDA scientists were aged 30 or less, compared with 25 percent at the National Institutes of Health (NIH). In addition, 39 percent of scientists at USDA were over 50, while only 15 percent were in this category at NIH. In the coming years the average age is likely to increase as the agricultural sciences continue to experience a period of limited growth and fiscal restraint. Wittwer (1982) reports that the average age of career scientists in the federal agricultural research system is now 49 and increasing by a third of a year per year.

Mulkay (1977) has noted that new scientists help in the exploration of new lines of inquiry. Their absence is likely to diminish levels of competition among scientists, reducing the likelihood of risk-taking and the professional incentive for mature researchers to move into new and, therefore, unpredictable fields. Furthermore, younger scientists appear more likely to address emerging agricultural research issues for food security and may more readily accept new ideas (Hull et al., 1978; Lacy, 1982).

Suggestions. Substantial recruitment of new scientists will be important in the coming decades if the agricultural sciences are to retain their vitality and increase the influx of new ideas. Unfortunately, as Ruttan (1983) points out, for American students, graduate enrollment in the agricultural sciences, and particularly in the basic sciences, declined in the late 1970s and 1980s. Additionally, private industry is now successfully competing for these same scientists, particularly those in certain high technology fields, such as biotechnology. This is in part the result of the uncompetitive salary structure in the public sector. Consequently, the ARS and the land-grant institutions will be limited in their ability to address new research agendas and to attract and train new entrants due to a lack of trained faculty (see Walsh, 1981).

Basic Science Capacity

The ability to address the research agenda for food security not only depends on adequate funding and strong research institutions at the federal, state, and private levels, but also relies heavily on the nature and substance of the research itself. Concern over the adequacy of basic science in the experiment stations and USDA facilities to meet the new research agendas has been voiced by a number of critics both within and outside the system. They describe the research structure as incapable of keeping pace with modern science and one that has systematically excluded most of the country's best research institutions. By focusing more narrowly on solving practical problems important to relatively few farmers, researchers have often neglected the basic research on which all farmers will ultimately depend. W. David Hopper, vice president of the World Bank, speaking at a symposium on the genetic engineering of plants, said, "In the twenty-to-thirty year perspective, we will be unable to support food demand from the potential of traditional research and infrastructure improvement. We must get back to the biological materials" (Budiansky, 1984:66). However, this may be a difficult task to accomplish. Lowell Lewis, director of the California Agricultural Experiment Station, recently observed, "The same system that by and large has excelled in sustaining effective relationships with the agricultural community has over the last few decades contributed to the declining association of agricultural research with basic science" (Budiansky, 1984:66).

From a representative national survey of more than 1,400 agricultural scientists in the public sector (Busch and Lacy, 1983), the influence of the basic sciences on the other agricultural sciences appeared minimal. Although agricultural basic scientists perceived other disciplines as the most important beneficiaries of their research, other agricultural scientists reported infrequently reading basic science journals, rarely speaking to these scientists about their research, and perceiving basic scientists as uniformly noninfluential for nearly all research decisions. As a consequence Richard Cauldecott, dean of biological sciences at the University of Minnesota, contends that the National Institutes of Health "has brought about advances in animal science by paying attention to basic biology. USDA hasn't done that in the plant sciences and they are seriously lagging behind" (Norman, 1982:1228). Lowell Lewis estimates that perhaps as many as forty state experiment stations could not be competitive in obtaining funding for basic research on the basis of scientific merit (Budiansky, 1984). Since an important portion of the cutting edge research proposed to address long-term food security issues is basic molecular biology and biochemistry, the capacity of the current system to address these issues may be severely limited.

Suggestions. The basic science research capacity of the public sector agricultural research system needs to be strengthened. This may entail active recruitment of personnel and increased investments in facilities and equipment. In addition, since many of the best biological basic science researchers are outside the system, mechanisms need to be developed to stimulate their pursuit of agriculturally related issues and to establish personal and institutional linkages among the public system, the private universities, and the industrial sector. These mechanisms may include fellowships, sabbaticals, workshops, and competitive grants. However, the mere development of a basic science capacity is not sufficient. Basic science must be interactively linked to applied research and technology development within the context of the broader goals of food security.

Disciplinary and Interdisciplinary Research

The research agenda to meet long-term U.S. food security calls for a more integrated approach to research both in terms of substance and goals. However, agricultural research is increasingly fragmented along disciplinary lines. Scientists typically receive all or most of their education within the same discipline with little exposure to fields not closely allied to their own (Lacy and Busch, 1982). As one frustrated industry spokesman complained, "There has been no real attempt on the part of graduate schools to teach students how to work in a multidisciplinary team. The language is different, the method of approach to problems is considerably different, and for all practical purposes, it is very difficult for members of a team to be able to communicate among themselves" (Bauman, 1979:30). In fact, the disciplinary societies that accredit agricultural curricula often make so many demands for disciplinary courses that even basic education in the humanities and social sciences is neglected (Kellogg and Knapp, 1966).

In addition, agricultural scientists rarely subscribe or publish outside disciplinary boundaries. For example, fewer than 6 percent of crop or animal scientists subscribe to or publish in each other's journals, while only 8 percent of agricultural engineers subscribe to at least one crop science journal, and an even lower percentage subscribe to an animal science journal. Finally, social scientists and agricultural economists are isolated from other agricultural scientists and appear to be largely unaware of work in other disciplines. Moreover, longitudinal data on publications suggest that disciplinary standards have replaced organizational standards in assessing the quality of scientists' work. In nearly all fields, the primary emphasis for promotion and tenure given to the publication of disciplinary journal articles reinforces the role of the disciplines in certifying research results (Busch and Lacy, 1983).

This insularity among the agricultural sciences may have important scientific and social significance. First, disciplinary problems are likely to receive more support than those that cross disciplinary lines (Ruttan, 1971). Such problems are more easily defined, easier to assess in terms of (disciplinary) significance, and are more likely to contribute to (disciplinary) knowledge. However, this also implies that the knowledge in each of the disciplines may become divorced from that of other disciplines.

Second, by focusing on only those aspects of the world that are deemed relevant by a particular discipline, scientists may ignore problems that lie outside their competence. As Berry (1977) suggested, disciplinary insularity may give scientists the illusion that total control over agricultural phenomena is possible. "The specialist puts himself in charge of one possibility. By leaving out all other possibilities, he enfranchises his little fiction of total control. Leaving out all the 'nonfunctional' or otherwise undesirable possibilities, he makes a rigid exclusive boundary within which absolute control becomes, if not possible, at least conceivable" (1977:70). Once having divided the world for study, it may be impossible to reintegrate it.

Third, the narrow disciplinary focus may limit a scientist's perspective and goals. In our national study of public sector agricultural scientists (Busch and Lacy, 1983), researchers were asked how important they considered each of eleven comprehensive goals currently identified by USDA. Scientists overwhelmingly emphasized the creation of disciplinary knowledge and the increase of agricultural productivity as the most important goals for agricultural research. Many other goals such as human nutrition, improving rural levels of living, and improving communities were seen as neither the subject of their own inquiry nor as intrinsically important. Furthermore, these other goals often tended to be relegated to one or two disciplines. For example, improving nutrition and protecting consumer health was seen as an important goal only by food scientists and nutritionists. Research on the improvement of the level of living in rural America and the promotion of community improvement appeared to be overwhelmingly the province of scientists in the social sciences. The improvement of marketing efficiency seemed to be the sole province of agricultural economists. In general, scientists appeared to undervalue or be unaware of the importance of the research of other agricultural disciplines.

This raises the further issue of potential conflict between goals. The various official statements of research goals tend to assume that all may be equally pursued and that little or no conflict exists among them. Official statements and scientific pronouncements notwithstanding, in a world with limited resources and time, some research goals must necessarily conflict. Consequently, any attempts to maximize one disciplinary goal may seriously reduce the likelihood of realizing others or optimizing the full set of goals necessary for food security. The establishment of new goals and priorities, even that of food security, entails tradeoffs and questions of who benefits.

Finally, the increased disciplinary emphasis appears to be partially responsible for increased specialization on the part of farmers. Much like agricultural scientists, American farmers have tended increasingly to specialize in the production of a single crop or commodity, while they pay scant attention to soil erosion, farm runoff, and other long-term problems. In fact, scientific agriculture has resulted in the adoption of scientifically validated processes and products by farmers as well as the transposition of the social organization of science to farming. Yet, ironically, this organization of science and farming may not be the only way or even the most effective and sustainable way to pursue science and to produce food and fiber for the U.S. or the world.

Suggestions. Increasing the capacity of the current system to engage in interdisciplinary research may require not only changes in the training of scientists, but modifications of (1) research strategies and methodologies as well as (2) organizational structures. Some creative efforts are just beginning to develop. For example, Integrated Pest Management (IPM) is an innovative research strategy which promotes an interdisciplinary effort. IPM has begun to affect the direction of research in economic entomology by shifting major research efforts from chemical control measures to the development of control strategies employing a combination of biological, mechanical, and chemical means. Acceptance of the IPM strategies has made it far more acceptable, but still not imperative, for entomologists to collaborate directly with their colleagues in agronomy and horticulture in an effort to develop better systems of management.

A second strategy is farming systems research (FSR), which has developed over the last decade largely out of the experiences of the International Agricultural Research Centers. These centers offer an alternative approach for organizing research which focuses multidisciplinary research efforts on increasing food production in particular ecological zones. The ideal of farming systems research even in these centers is far from a reality. However, the primary aim of the FSR approach is "to increase the productivity of the farming system in the context of the entire range of private and societal goals, given the constraints and potentials of the existing farming systems" (Gilbert, 1980:2). Effective farming systems research requires not only the creation of multidisciplinary research teams, but also the reconstruction of the very epistemology employed in research. If taken seriously, this should tend to refocus the emphasis in agricultural research away from disciplinary and commodity concerns toward complex interactions among and between people, crops, soil, and livestock.

New institutional structures which promote interdisciplinary research are also emerging. One example is Solutions to Environmental and Economic Problems (STEEP), a multidisciplinary research effort to develop new techniques and strategies to control soil erosion in the croplands of Washington, Oregon, and Idaho (Oldenstadt et al., 1982). With a USDA grant to the three state agricultural experiment stations as well as supplementary state and federal funds, STEEP awards intermediate-term (15 year) grants for research in tillage and plant management, plant design, erosion and runoff prediction, pest management, and socioeconomics of erosion control. Most of these research projects require multidisciplinary collaboration and in some instances multi-state efforts. Participants argue this is an important alternative since "neither the competitive grant nor the formula fund model are effective in stimulating multidisciplinary research across state boundaries with the aim of solving regional or national agricultural problems" (Oldenstadt, et al., 1982:904).

A second example of a new institutional structure can be found in the evolving Collaborative Research Support Programs (CRSP) that are administered by the Board on International Food and Agricultural Development. U.S. social scientists and natural scientists are participating in inter-university programs which attempt to collaborate on both basic and applied research with similar groups from developing countries. The major goal of these programs is to contribute to more stable food systems for all countries by addressing issues of production, storage, distribution, utilization, and sociocultural impacts. Research emphasis is placed on strategies and technologies that are sustainable, ecologically sound, energy efficient, and socioeconomically appropriate. In addition to offering an alternative way of organizing research the program may have significant implications for a secure, sustainable U.S. agriculture.

Long- and Short-term Research

The consideration of long-term versus short-term approaches to research has been an underlying theme throughout this discussion of the capacity of the research system to address emerging research issues. The time frame for research may be as important for the problems facing U.S. food security as the human resources, capital resources, and the organization and substance of the research. Generally, the more fundamental concerns of agricultural science, such as biological nitrogen-fixation research, alternative technology, and germ plasm alterations, require years of work and long-term planning and funding.

U.S. agriculture may require fundamentally different practices and technologies that are likely to be developed only through commitment to long-term research. In our national study (Busch and Lacy, 1983), numerous scientists in a variety of disciplines offered unsolicited observations regarding the time frame for research. An entomologist commented that "attention is focused on 'brush fires' and pressures of the moment with relatively little attention to long-term biological research on population dynamics, behavior, taxonomy" (Busch and Lacy, 1983:224). A soil chemist criticized the fickle nature of research funding:

> The hop on and off bandwagon approach taken by the Congress and administration dissipates energy and funds so that basic understanding is bypassed for collection of data that will be meaningless in five years (Busch and Lacy, 1983:225).

One forester offered a number of explanations for the lack of long-term research:

> Long-term research, such as needed in forestry, is difficult because of the short-term trend in funding projects. [There is also] increasing pressure for short discrete problems and quick publication to satisfy needs for publication records.... Extremely heavy teaching loads and restricted funding have

made the research process more difficult to initiate and carry to completion (Busch and Lacy, 1983:225).

In summary, these scientists note that as a consequence of funding practices, reward structures, immediate needs and demands of clientele, and organizational priorities, current research appears limited to short-term disciplinary projects. Coincidental with the demands raised by emerging long-term issues are the more frequent short-term problems that will continue to face the current specialized agricultural system.

Suggestions. Agricultural research needs to develop appropriate planning and sufficient resources for sustained efforts to meet long-term needs, while retaining the flexibility to respond to urgent short-term problems and coordinating the overall effort.

Policy and Assessment Capacity

Finally, it is important to examine the capacity of the research system to develop research plans and priorities and to assess the effectiveness and impact of research agendas. To address the range and complexity of the emerging research agenda for food security, the development of appropriate planning and assessment will be necessary.

Several members of the research community (Bonnen, in press; Busch and Lacy, 1983; Ruttan, 1982a, 1983) have questioned the ability of the current system to engage in effective planning, particularly at the national level. Universities are fragmented along institutional and disciplinary lines. Ruttan (1982a) noted that by the 1960s "most agricultural experiment station directors had given up any pretension about exercising significant intellectual leadership over the research activities that were funded by the stations.... These functions were left to the heads or chairpersons of the disciplinary departments." At the same time the private sector must respond to the pressures of product development and marketing. Finally, as noted earlier, the federal agricultural research system may lack the scientific capacity or the bureaucratic structure to exercise such leadership.

Nevertheless, the agricultural research planning staffs have been strengthened at the national level in part as a result of legislative acts. For example, the Food and Agricultural Act of 1977 mandated the establishment of the Joint Council on Food and Agricultural Sciences to foster improved planning of federal and state agricultural research. Often, however, the process used by these planning bodies to set research priorities largely serves to aggregate what researchers are already doing. When planning of research resource allocation does occur, it is largely an intuitive process. For example, the new six-year plan of ARS may meet many criticisms that have been directed toward the U.S. agricultural research system. Terry Kenny, the ARS administrator, described it as "the most significant planning activity that ARS has ever completed" (Norman, 1983:1046). It would begin to shift more of the ARS's $420 million budget into basic research and reorder priorities among the agency's major programs. However Ruttan (1983) notes that the ARS utilized few analytical strategies in making judgments about the value of alternative research priorities.

A number of increasingly powerful methodologies are being developed for interpreting scientific, technical and economic information in order to increase the effectiveness of research efforts. (See Ruttan [1982a:262-94] for discussion of these formal methodologies and alternatives.) In addition, methodologies are beginning to emerge which attempt to assess the results of policies, forecast the consequences, and integrate technical, economic, social, aesthetic and moral considerations (Berardi and Geisler, 1983; Friedland and Kappel, 1979).

Suggestions. Resources must be devoted to these new methodologies if they are to move beyond their infancy and contribute to effective research resource allocation and priority planning. Furthermore, efforts must be devoted to preparing scientists to utilize these methodologies and assessment strategies for agricultural research. Finally, research planning will require closer collaboration among natural and social scientists as planners address not only the possibilities of advancing knowledge or technology, if particular resource appropriations are made, but also consider the value to society of the new knowledge or technology.

CONCLUSION

The role of agricultural research in the transition to a long-term, sustainable, safe, nutritious and equitable food system in the United States is a central one. In the past the agricultural research system has demonstrated its capacity to produce the technological advances underlying the success of American agriculture. Today it is facing new, diverse, complex and often intractable problems coupled with institutional structures, research goals and missions, and human and physical resources which will make it extremely difficult to play that central role.

To meet the research agenda for a secure food system will require a major replenishing of our intellectual capital and a strengthening of the institutional bases of the research system at all levels. It will entail change and adaptation that retains the strongest elements of the existing system and responds to the new scientific frontiers, the involvement of institutions and scientists outside the system, and the changing social, political, and economic context for research. It will require a strengthening of the articulation between advances in knowledge and technology development and between new technologies and the adaptive response to their utilization. It will necessitate more effective interdisciplinary efforts and commitment to more high risk, long-term research. Finally, it will require a sustained national policy and effective planning and assessment strategies.

To accomplish this will necessitate institutional change and entirely new relationships and linkages between disciplines, institutions, and potential clientele, as well as concomitant changes in what constitutes knowledge in the agricultural sciences. Finally, it will entail levels of public funding of agricultural research more commensurate with its value to society and with the research needs for a secure food system.

REFERENCES

Araji, A. A.; Sim, R. J.; and Gardner, R. L. 1978. "Returns to Agricultural Research and Extension Programs: An Ex-ante Approach." American Journal of Agricultural Economics 60(5):964-68.

Bauman, H. E. 1979. "Changes in Industrial Research Require Changes in Graduate Education." Food Technology 33(12):30-31.

Berardi, G. M., and Geisler, C. C. 1984. The Social Consequences and Challenges of New Agricultural Technologies. Boulder: Westview Press.

Bergland, R., and Sechler, S. 1981. Time to Choose: Summary Report on the Structure of Agriculture. Washington, D.C.: USDA.

Berry, W. 1977. The Unsettling of America: Culture and Agriculture. Totawa, New Jersey: Sierra Club Books.

Bonnen, J. T. In press. "Technology, Human Capital and Institutions: Three Factors in Search of an Agricultural Research Strategy." In The United States-Mexico Relations: Economic and Social Aspects, eds. C. W. Reynolds and C. Tello. Palo Alto: Stanford Press.

Brady, N. C. 1982. "Chemistry and World Food Supplies." Science 218(4575):847-53.

Brown, L. R. 1981. Building a Sustainable Society. New York: Norton.

Budiansky, S. 1984. "Trouble Amid Plenty." The Atlantic Monthly 253:65-69.

Busch, L., and Lacy, W. B. 1981. "Sources of Influence on Problem Choice in the Agricultural Sciences, the New Atlantis Revisited." In Science and Agricultural Development, ed. Lawrence Busch, pp. 113-28. Totawa, New Jersey: Allanheld, Osmun.

______. 1983. Science, Agriculture and the Politics of Research. Boulder: Westview Press.

Buttel, F. H.; Kenney, M.; Kloppenburg, Jr, J.; and Cowan, J. T. "Problems and Prospects of Agricultural Research: the Winrock Report." The Rural Sociologist 3(2):67-75.

Committee for Agricultural Research Policy. 1981. Agricultural Research in the Private and Public Sectors: Goals and Priorities. Lexington: Kentucky Agricultural Experiment Station.
Friedland, W. H., and Barton, A. 1975. Destalking the Wily Tomato: A Case Study in Social Consequences in California Agricultural Research. Research monograph no. 15. Davis: Univ. of California, Department of Behavioral Sciences.
Friedland, W. H.; Barton, A. E.; and Thomas, R. J. 1981. Manufacturing Green Gold. Cambridge: Cambridge Univ. Press.
Friedland, W. H., and Kappel, T. 1979. Production or Perish: Changing the Inequalities of Agricultural Research Priorities. Santa Cruz: Project on Social Impact Assessment and Values, Univ. of California.
Gilbert, E. H.; Norman, D. N.; and Winch, F. E. 1980. Farming Systems Research: A Critical Appraisal. Rural Development Paper no. 6. East Lansing: Michigan State Univ., Department of Agricultural Economics.
Haberer, J. 1969. Politics and the Community of Science. New York: Reinhold.
Harmon, L. R. 1978. A Century of Doctorates: Data Analysis of Growth and Change. Washington, D.C.: National Academy of Sciences.
Hayami, Y., and Ruttan, V.W. 1971. Agricultural Development: An International Perspective. Baltimore: Johns Hopkins Univ. Press.
Hightower, J. 1973. Hard Tomatoes, Hard Times. Cambridge: Schenckman.
Hull, D. F.; Tessner, P. D.; and Diamond, A. M. 1978. "Planck's Principle: Do Younger Scientists Accept New Scientific Ideas with Greater Alacrity Than Older Scientists?" Science 202:717-23.
Jensen, N. 1978. "Limits to Growth in World Food Production." Science 201(4353):317-20.
Joint Council on Food and Agricultural Sciences. 1983. FY 1985 Priorities for Research, Extension and Higher Education. Washington, D.C.: U.S. Government Printing Office.
Jordon, W. H. 1907. "The Authority of Science." In Proceedings of the 21st Annual Convention of the Association of American Agricultural Colleges and Experiment Stations, pp. 60-66. Lansing, Michigan.
Kellogg, C. E., and Knapp, D. C. 1966. The College of Agriculture: Science in the Public Service. New York: McGraw-Hill.
Knorr, D. 1983. "Recycling of Nutrients from Food Wastes." In Sustainable Food Systems, ed. D. Knorr, pp. 249-78. Westport: Avi Publishing.
Kuhn, T. S. 1970. The Structures of Scientific Revolutions. 2nd ed. Chicago: Univ. of Chicago Press.
Lacy, W. B. 1982. "Profile of U.S. Agricultural

Scientists in the Public Sector. Analysis of Their Origins and Nature of Their Work." The Rural Sociologist 2(2):85-94.
Lacy, W. B., and Busch, L. 1982. "Problem Choice in Agricultural Research: Scientist's Initiatives." In Enabling Interdisciplinary Research: Perspectives from Agriculture, Forestry, and Home Economics, ed. M. G. Russell, pp. 51-66. St. Paul: Univ. of Minnesota, Agricultural Experiment Station.
______. 1983. "Informal Scientific Communication in the Agricultural Sciences." Information Processing and Management 19(4):193-202.
Larson, W. E.; Pierce, F. J.; and Dowdy, R. H. 1983. "The Threat of Soil Erosion to Long-term Crop Production." Science 219(4583):458-65.
Lewontin, R. C. 1980. "Agricultural Research in Non-land-grant Universities." Commissioned paper V for Office of Technology Assessment.
Mayer, A., and Mayer, J. 1974. "Agriculture, the Island Empire." Daedalus 103(3):83-95.
McCalla, A. F. 1977. "Politics of the U.S. Agricultural Research Establishment." Paper prepared for the Agricultural Policy Symposium. Washington, D.C.
Miller, T. A.; Roderwald, G. E.; and McElroy, R. 1981. Economics of Size in Major Field Crop Farming Regions of the United States. Washington, D.C.: USDA, Economics and Statistics Service.
Mulkay, M. J. 1977. "Sociology of the Scientific Research Community." In Science, Technology, and Society, eds. I. Spiegel-Rosing and D. D. S. Price, pp. 93-148. London: Sage.
National Research Council. 1972. Report of the Committee on Research Advisory to the U.S. Department of Agriculture (Pound Report). National Technical Information Service, PB 213 338. Washington, D.C.
National Research Council. 1977. Supporting Papers: World Food and Nutrition Study: Agricultural Research Organization. Washington, D.C.: U.S. Government Printing Office.
Nelkin, D. 1977. "Technology and Public Policy." In Science, Technology, and Society, eds. I. Spiegel-Rosing and D. D. Price, pp. 393-442. London: Sage.
Norman, C. 1982. "A New Pot of Money for Plant Sciences." Science 217:1228-29.
______. 1983. "ARS Floats a Plan." Science 219:1046.
Office of Technology Assessment. 1981. An Assessment of the United States Food and Agricultural Research System. Washington, D.C.: U.S. Government Printing Office.
Oldenstadt, D. L.; Allan, R. E.; Bruehl, G. W.; Dillman, D. A.; Michalson, E. L.; Papendick, R. I.; and Rydrych, D. J. 1982. "Solutions to Environmental and Economic Problems." Science 217:904-9.
Price, D. D. S. 1963. Little Science, Big Science. New York: Columbia Univ. Press.

Ravetz, J. R. 1977. "Criticisms of Science." In Science, Technology and Society, eds. I. Spiegel-Rosing and D. D. S. Price, pp. 71-92. London: Sage.
Rockefeller Foundation. 1982. Science for Agriculture. New York: Rockefeller Foundation.
Rose, H., and Rose, S. 1976. "The Problematic Inheritance: Marx and Engels on the Natural Sciences." In The Political Economy of Science, eds. H. Rose and S. Rose, pp. 1-13. New York: Holmes and Meier.
Rosenberg, C. E. 1976. No Other Gods: On Science and American Thought. Baltimore: Johns Hopkins.
Rossiter, M. W. 1979. "The Organization of the Agricultural Sciences." In The Organization of Knowledge in Modern America, eds. A. Oleson and J. Voss, pp. 211-48. Baltimore: Johns Hopkins.
Ruttan, V. W. 1971. "Research Institutions: Questions of Organization." In Institutions in Agricultural Development, ed. M. G. Blase, pp. 129-38. Ames: Iowa State Univ. Press.
______. 1982a. Agricultural Research Policy. Minneapolis: Univ. of Minnesota Press.
______. 1982b. "Changing Role of Public and Private Sectors in Agricultural Research." Science 216:23-29.
______. 1983. "Agricultural Research Policy Issues." Paper presented at annual meeting of the American Society for Horticultural Science, McAllen, Texas.
Salomon, J. J. 1977. "Science Policy Studies and the Development of Science Policy." In Science, Technology and Society, eds. I. Spiegel-Rosing and D. D. S. Price, pp. 43-70. London: Sage.
Schaaf, M. 1983. "Challenging the Modern U.S. Food System: Notes from the Grassroots." In Sustainable Food Systems, ed. D. Knorr, pp. 279-301. Westport, Connecticut: Avi.
U.S. Department of Commerce, Bureau of the Census. 1978. Selected Characteristics of Persons in Fields of Science or Engineering, 1976. Special Studies P-23, No. 76. Washington, D.C.: U.S. Government Printing Office.
Walsh. 1981. "Biotechnology Boom Reaches Agriculture." Science 213:1339-41.
White, F. C., and Havlicek Jr., J. 1982. "Optimal Expenditures for Agricultural Research and Extension: Implications of Underfunding." American Journal of Agricultural Economics 64(1):47-55.
Wittwer, S. H. 1978. "The Next Generation of Agricultural Research." Science 199(4327):375.
______. 1982. "U.S. Agriculture in the Context of the World Food Situation." In Science, Technology, and the Issues of the Eighties: Policy Outlook, eds. A. H. Teich and R. Thornton, pp. 191-214. Boulder: Westview Press.

15
Food Security and the Extension of Agricultural Science

James M. Meyers

Food security remains a persistent human concern and priority. Despite the impressive gains in both domestic and world food production which have been achieved in the past two generations, not all in this world are yet adequately fed. Further, as the articles in this volume and elsewhere demonstrate, new concerns about food supplies continue to emerge. The dimensions and salient issues of our concern for food security are plastic, changing across conditions, perceptions, and time.

Nearly seven decades ago, this nation's Agricultural Extension system was organized to apply the results of research in agricultural science towards the improvement of the efficiency of human labor, the supply of available food, and the welfare of the nation's people, especially those in rural areas. Until the reemergence in the last decade of a progressive conservation movement, increased farm production has been a little-questioned primary objective of virtually all agricultural science. As a result, significant improvements have repeatedly been made in volume of supply. The issues given preeminence in this volume, however, include: (1) the long-term sustainability of production potential (i.e., across generations); (2) equity of access to the food supply; (3) nutritional quality of available foods; and (4) the social, economic, and health costs of our current food system. In the contexts presented here, these issues represent emergent emphases and suggest that new perspectives must be taken by the agricultural science system, for such concerns have not been as intensely nor productively addressed as have production volume objectives.

It is the purpose of this chapter to explore to what degree the nation's extenders of food and agricultural science are giving priority to these four issues and to suggest how they might be most productively addressed. To do this, a brief inquiry into how the extension process works and how its priorities are influenced is necessary. It is also important to note two fundamental perspectives that will influence this review. First, it is the position of this review that virtually all extenders of food and agricultural science can and should do more regarding the issues addressed in this volume. They are serious and complex, and their implications for our future demand full and responsible attention. But, having said that, the issue of exactly what new work to begin, how much of our constrained resources to devote to it, and how best to go about it quickly becomes clouded by unknowns and complexities of both the issues and of how the innovation process works. It is important to note that none of these issues is totally unaddressed, nor is there reason to believe that extenders of agricultural science lack concern for the security of our food supply or its future. As with most emergent issues, these generate controversy as to how best they should be defined and approached, and that fact may prove to be the key to addressing them most productively.

Second, it is the position of this paper that the extension of science has to do more with the development and use of innovations (in ideas, practices, or technologies) than with the simple provision of general information access or educational opportunity. For the most part modern extenders, as well as students of extension, subscribe largely to the idea that development and diffusion of innovations are separate processes and that diffusion consists essentially of communications (or demonstrations) that inform and persuade users to adopt the innovation proferred by the extender. This model has been well described by Rogers and Shoemaker (1971), who view the diffusion process from the perspective of communication theory.

The view expressed here differs in suggesting that development and diffusion of innovations are not best viewed as separate processes. Rather, it is here presumed that not only will and should innovation development be influenced by what information, effort, and experience is shared between researchers and users but that both development and diffusion will be more efficient and appropriate where direct user-researcher interaction takes place before and during development itself. This is an uncommon perspective on the extension of science. It suggests, in part, that not only the appropriateness of an innovation but research and extension priorities and products are greatly influenced by the identity and interests of those users having direct and interactive access to the development process.

THE EXTENSION FUNCTION

Although much has been written about the extension function, it remains incompletely understood. Traditional extension rhetoric has emphasized dissemination of research ideas and information as if research groups were factories producing ready-to-use innovations needing only to be described or delivered to potential users. Similarly, extension theory, in the form of the adoption/diffusion model, emphasizes the adoption of selected innovations as a function of communication and persuasion, reinforcing the misconception that development of innovations is a separate and somehow autonomous process. From this perspective, one must suppose that innovation ideas emerging from the research establishment are real improvements over current practices and any unintended consequences of this supplanting of current practices are either acceptable or necessary and tolerable. It is from this perspective that most extension groups evolve programs which emphasize one-way communication, demonstration, or persuasion strategies to encourage innovation adoption. Such an orientation leads the extension group to focus largely on a small, selected audience of those most likely to adopt the proffered new technology, and whose adoption will positively influence others. In fact, the adoption/diffusion literature (Rogers and Shoemaker, 1971) characterizes users primarily by their propensity to adopt innovations on the basis of information (e.g. early-adoptor, mid-adoptor, late-adoptor).

Just as there is confusion about education as a process of only information acquisition, there has been a continued tendency to see the extension function primarily as one of information provision. Over the past two decades public Agricultural Extension has tended more and more to define its role as that of primary information provision, not unlike the private sector's view of a sales or marketing unit. Two factors have contributed to this tendency. First, extension mechanisms are expected to "reach" numerous diverse audiences with information and services across ever more numerous topical areas with little increase in staffing. Informing, rather than collaborating on development, thus becomes a way of stretching resources. Second, the prevalence of the adoption/diffusion model, itself derived from communication theory, helped focus Extension thinking on its informational role. The adoption/diffusion model, however, tends to treat innovation diffusion as separate and distinct from research and development. Thus, theory itself has contributed to the tendency for participants in the innovation process to truncate their roles and operate in increasing isolation from one another.

This view of the extension function is as common to the public sector as to sales and marketing groups in the private sector. Contrasted to it is recent work on the process of technological innovation (Sahal, 1981) which suggests that the development and implementation of new technologies and practices is a highly interactive process. Applications and understanding of knowledge are improved with successive experiences, as old and new ideas compete and evolve. In this view of the innovation process, there is no clearly distinct separation between scientific discovery, development of applications, and their useful implementation. A premium is placed on interactive flow of information, from researcher through user and back. Not only are innovations not expected to spring forth from an autonomous research group, but research itself is seen to be dependent on cooperative effort, reciprocal exchange of information, and improvements between competing ideas and technologies. This view emphasizes an active role by the extension unit in adapting research knowledge to solve problems, to take advantage of opportunities, and even to create new opportunities. By interacting directly and cooperatively with researchers and users the extension group makes problem definition and resolution a fully shared and collaborative effort. In the increasingly discrete role specialization of modern science, the extension mechanism is often the primary (if not sole) directly interactive interface between researchers and potential users. If the extension mechanism merely disseminates information from an isolated research unit to distant users, research effectiveness, appropriateness, and priorities must suffer as well as the adoption process.

An obvious implication of this view is that during the process of innovation development and adoption the technical environment will reflect differing and even contending ideas. The job of the scientific extender is much more complex than simple information dissemination. Major roles include: helping knowledge users communicate effectively with and influence research project definitions and priorities, developing and testing applications of research knowledge cooperatively with researchers <u>and</u> users to solve designated problems, and helping researchers influence users' understanding, adaptation, and modification of new knowledge and its applications. Additionally, as problems increase in complexity, the extension unit needs to provide for collaborative representation of increasingly multiple interests and perspectives in the development and adoption process. Likewise, as technical ideas compete, cooperative interaction with proponents of alternate approaches helps to speed the evolution of full understanding and practical application development. Where this interaction flow is slowed, innovation development and resolution of flaws or unintended side-effects will likewise be slowed. Thus, when the development process seems sub-optimum and/or significant issues do not seem fully addressed, one cause to look for is obstruction to full and direct interactive collaboration and communication between the whole range of relevant researchers and users.

EXTENDERS OF AGRICULTURAL SCIENCE

Extension mechanisms are needed in addition to research units because the whole process of developing and using innovations tends to become subdivided into separate and distinct organizational tasks. Increasingly, researchers have come to see their ultimate contribution as the publication of their findings, leaving it to others to develop pragmatic applications and to generate a demand for adoption. The innovation process itself, however, can nonetheless be viewed as a whole--from discovery of knowledge through its application to a given use, to adoption and resultant feedback.

The lines differentiating the extension function from the research end of the innovation process are as nebulous and variable as those drawn between basic and applied research. As a result the differing major types of research performers provide for somewhat differently defined and focused extension mechanisms. The National Science Foundation notes four categorical types of research performing organizations: industry, government, colleges and universities, and other nonprofit organizations. Reviewing each of these in turn demonstrates some of the differences.

Industry has long recognized the development of applications of new knowledge as a distinct function and has historically emphasized applied research and development. Influenced heavily by the profit potential in its research decisions, industry depends largely on marketing and sales units to provide user access to new developments, as well as feedback information to influence future development directions. Advertising and sales contact and reinforcement are common industry extension strategies. Client audiences tend to be selected for their probable ability to afford and adopt products and technologies. Theoretically, any group manifesting sufficient demand for a product will be addressed. In fact, volume and price considerations eliminate some groups and encourage competitive attention to others, especially existing major market groups.

Government is as inherently and inescapably influenced by political issues as industry is influenced by profit considerations. Much government research is applied in nature and conducted to support preferred policies and regulations, or to pursue issues of interest to the government and those influencing it. As a result, much of the audience for government research consists of public agencies or selected constituencies. The government has tended to use information provision, regulation, policy statement, funding, and political influence to disseminate and encourage use of its findings. As in the case of industry, theoretically any group evincing sufficient need and demand will be addressed. However, the potential or real political strength and sophistication (or lack) of some groups can eliminate some interest groups and pit others against one another. Like profit potential, these factors act to encourage competitive attention to existing major political interests.

Colleges and universities perform most of the nation's basic research. Much of this research is performed in agriculture, with nonspecific government funding, which allows much latitude in the choice of subjects for research. Academic research units have tended to focus their efforts on the discovery of basic knowledge, expressing less interest in development of applications. Simultaneously, research has tended to be influenced more by academic interests and structures than by worldly problems or even funding sources. As a result, academic research units generally see other academics as their primary audience and emphasize academic publication and dissemination as access mechanisms. Partly as a consequence, public universities have established organized extension units. These separately administered research and extension units often have some problems in coordination and responsiveness. In fact, the very existence of an organized, separate extension unit may add to a research unit's sense of legitimate isolation from external (or nonacademic) issues and interests.

University agricultural extension units have a long history of one-to-one and one-to-small-group contact in local settings, acting both to facilitate adaptation of developments and to encourage adoption. Extension tradition has been for local agents to work most closely with selected "influence leaders" in developing and sponsoring "demonstrations." Local agents are thus oriented to major accepted local interest groups and issues. In recent decades, Extension has made increasing use of information delivery media and strategies to contact wider audiences. This seems to have been accompanied by some decline in Extension's emphasis on active organization and education of emergent interest groups. These factors, coupled with increasing academic orientation to disciplinary communities, have weakened Extension's influence on research decisions, making Extension a more one-way, outward-flowing information service aimed at a broader total audience, still, however, interactively engaged primarily with its original local constituency vested in traditional outlook issues.

Other nonprofit organizations are largely foundations and study-activist groups which, in general, are minor research performers, usually focusing on or employed for specific interests. Not infrequently they have a directly supporting constituency which exercises substantial influence. Nonprofits utilize popular and scientific publication of their work, political or funding influence, and engage in direct intervention via local activists and/or organizations. Partly because they tend to have narrower audiences or interests, some nonprofit organizations have been active regarding minority viewpoints and emergent issues. However, because of often limited research capacity and support, many nonprofits function more as issue "highlighters" or definers than as research extenders.

These comments represent rather gross generalizations, but they should help illuminate some of the salient differences and similarities in how different research performers seek to provide for development and extension of applications. Perhaps most important for this discussion is the observation that despite differences in functional definition and in audiences exercising access, virtually all performers can be said to utilize some form of extension mechanism. Ideals about scientific values aside, both public and private sponsors assign priorities to investment in science with the express expectation that new knowledge and its applications will result in tangibly enhanced human welfare.

TYPES OF EXTENSION ACCESS

In either view it is clear that those research users or clients with direct, interactive access to research, development, and extension staff will benefit the most, and will influence priorities the most. Thus, the identity and interests of directly interactive users do much to contribute to the system's perspective on needs, its priorities, and where its effectiveness potential as an innovation system is greatest. Such users tend to be those for whom innovations or services were specifically designed or those who have expressed effective demand. However, these are not always the same groups. For example, forest products development work is heavily influenced by consistent and effective user demand, which _pulls_ research and extension attention with both influence and funding. Multiple use forestry and conservation work, by contrast, tends to be _pushed_ at targeted user constituencies by government and interest groups while those users of most interest remain undemanding. The chief factor of interest here is that, regardless of specificity of purpose or audience, only where researchers _and_ users share in problem definition and resolution effort can the relationship be said to be truly directly interactive.

Obviously, directly interactive users will tend to be few in number compared to other users due to the time consuming one-to-one nature of direct interaction.

At least two other types of user access are distinguishable. The first might be referred to as direct receivers. While these users enjoy direct contact, the flow of influence and information tends to be much more one-way than in a truly interactive relationship. For example, many of public Extension's nutrition education programs offer only information and predetermined experiences, rather than interactive collaboration on definition and resolution of nutritional problems. This is similar to private sector sales contacts who may be offered information and the opportunity to adopt, but not cooperation in definition or development of products. Another type of direct receiving user is the multiplier or receiver/transmitter--essentially a kind of technological middle man--who serves as a technological distributor, but who is not, individually, the explicitly intended user. Media operators become direct access users of this sort; so do adult 4-H volunteers. While they may not be directly interested themselves in applying certain information, they are of considered importance to the diffusion of information. These multiplier users (and other direct receivers) do exert influence on research and extension priorities simply by advocating their own interests and demands as part of an information or technology delivery system. Similarly, both public and private R&D tend to treat food or product distribution systems as receiver/transmitter users through whose ostensibly neutral services the bulk of public beneficiaries or audiences are reached. Interestingly enough, government agencies working in overlapping or parallel fields frequently regard each other as one or another type of passive or secondary access user.

A third type of user access may be termed indirect diffusion and includes a variety of methods. In some cases, the users constitute groups whose interests are in some manner taken account of, but are themselves not directly addressed in the development/adoption process. One type of indirect access user would be the adoption/diffusion model's mid- to late-adopters. That is, users expected to adopt an innovation after its use becomes relatively accepted.

Some other indirect users are those whose participation is hoped for or expected via some indirect means. For example, farmers reading an Extension-generated article on conservation methods are expected to undergo an awareness or attitude change. Likewise, it is expected homemakers and their families will benefit from improved nutrition at lower cost from consumer information provided via television or radio. This group has very real importance since any new technology or practice expected to be adopted by very large audiences will of necessity involve some form of indirect contact or communication. This is an area in which the private sector's sales and marketing experience has made them considerably more effective.

A final audience group should be included here.

These are the beneficiaries of innovation adoption decisions. For example, the mass of food consumers benefits from production research through wider food availability at lower cost. Generally, beneficiaries are carefully distinguished from research or technology users and as a result generally have little access to the development process. However, increasingly, final beneficiaries or consumers are expressing active interest in development processes which will impact their welfare and practices. It would seem that the public sector especially has an obligation to provide for some beneficiary representation in development and adoption processes since these institutions hold specific responsibility to act in the public interest and welfare.

Types of research access can doubtless be categorized in alternative ways. The point of this discussion is that all contacts with an extension mechanism are not methodologically the same, and should not therefore be expected to have the same impacts. In general, we should expect that direct, interactive users would gain the most, and influence the pace and priorities of the development process the most, while indirect diffusion users would gain the least and exercise the least influence.

PRIORITIES IN THE EXTENSION OF AGRICULTURAL SCIENCE

Disregarding for the moment the niceties of functional discrimination among types of access or other functions, what can be said regarding current activities of various extension mechanisms on the topics of interest here? In truth, given the decentralized nature of agricultural research and extension, and the official reporting categories which include a nearly infinite diversity of specific topics, it is difficult to pinpoint priorities with empirical certainty. However, a study performed for the National Agricultural Research and Extension Users Advisory Board in 1980 (USDA, 1980) reported estimates of all private and public food and agricultural research and development investments according to the following priorities:

Production research is, to a great extent, work to increase volume of production and to reduce handling costs while reducing waste. The study reports that "management, conservation and economic research received the least attention, while breeding (seeds and animals) and all product development R&D the most" (USDA, 1980, p. 1, citing a report by Agricultural Research Institute, 1977).

Federal and state research and extension constitute the major portion of public agricultural science. Table 15.2 shows proportional resource investments by federal

TABLE 15.1
Public and Private Food and Agricultural Research and Development Activity

Category	% of Private	% of Public	% of Total
Production (on farm)	48.9	46.2	47.0
Production (off farm)	40.5	15.0	29.0
Nutrition	0.2	11.0	5.0
Rural/community development	--	7.2	3.2
Family/people programs	--	0.6	0.3
Health and safety	0.8	3.0	1.8
Natural resource conservation	9.5	16.8	12.5
	100%	100%	100%
Total ($ billion)	$1.8	$1.4	$3.3

and state research and extension performers. This table raises some question as to what extent and how the issues of emerging interest with respect to our food supply are being addressed. Statistics as of 1981 for federal and state Extension work in the agriculture and natural resources area make clear that crop and livestock production work predominate above all other concerns at 68.8% of expended effort. Natural resources and agricultural economics, where priority work on three

TABLE 15.2
Federal and State Research and Extension Priorities

Category	Research	Extension	Total
Production (on farm)	50.7	36.0	45.3
Production (off farm)	13.8	2.1	9.5
Nutrition	12.6	16.3	14.0
Rural/community development	6.5	20.4	11.5
Family/people programs	0.8	20.9	8.1
Health and safety	3.6	1.5	2.8
Natural resource conservation	12.0	2.9	8.7
	100%	100%	100%
Total	$1.2 (billion)	$6.9 (million)	$1.9 (billion)

of the four food security issues explored here would be reported, have declined in proportional funding since 1978 (USDA-ES, 1982). The issue of nutritional quality of the food supply is the only one for which national statistics show a slight increase. (This is due to Extension work in the area of home economics, generally focused on providing low-income and other consumers with advice on food purchasing and diet.)

Given the earlier discussion of the dependence of the innovation process on an interactive relationship between researchers and users, a second way of checking Extension priorities is to consider what types of access users of differing interest appear to enjoy.

Congress recognized the diversity among potential users of food and agricultural research in its 1977 Farm Bill, mandating that membership on the National Agricultural Research and Extension Users Advisory Board include representatives of the following:

- producers of agricultural commodities, forest products, and agricultural products;
- consumer interests;
- farm suppliers and food and fiber processors;
- food marketing interests;
- environmental interests;
- rural development work;
- human nutrition work;
- animal health interests;
- transportation of food and agricultural products;
- labor organizations primarily concerned with the production, processing, distribution, or transportation of food and agricultural products; and
- organizations involved in development programs and issues in developing countries.

Considering the types of access suggested earlier by each of these major user groups, the following observations give insight into the type and extent of access each group is currently afforded and exercises.

(1) Producers of commodities. Represent a high proportion of direct-interactive users for both public and private research and extension. There is some emphasis on increasing farm income through increased volume of production. The most aggressive, commercial producers seem to dominate direct-interactive access. Smaller, less aggressive producers--or those with other than conventional technology agendas--tend to be addressed more through indirect access.

Audience: 30 million farms, 4 million small woodlot owners.

Time Input: 36 percent of all Cooperative Extension time nationally (1980-1981 data) (USDA-ES, 1982).

Contact Numbers: Cooperative Extension reports some form of contact with two-thirds of all producers

(Cooperative Extension estimates it contacts 75 to 80 percent of all producers with over $100,000 and 49 percent of those with under $2,500 annual income).

(2) Consumers. Represent the bulk of indirect access audiences (i.e., beneficiaries) for public and private research and extension. Private sector marketing groups interact directly in limited fashion with selected focus groups. Consumer organizations occasionally act as direct diffusion users, transmitting information in both directions, though not always with a great deal of influence on priorities or collaboration. In summary, significant Extension effort is intended to benefit consumers, but mostly indirectly, with little interactive development contact.

Audience: 226 million consumers.

Time Input: 22 percent of Cooperative Extension time (9 percent nutrition, 13 percent other home economics) (1980-1981 data).

Contact Numbers: The Gallup Poll reported that 10 percent of the adult population reported some Cooperative Extension consumer contact in lifetime; estimated at time of poll as 17 million persons (75 percent white, 17 percent black, 5 percent Hispanic).

Cooperative Extension records report 27 million information contacts (not individuals contacted directly) for Fiscal Year 1978.

(3) Suppliers and processors. Actually two groups, both treated in multiple ways. Suppliers tend to be treated via direct diffusion access strategies, and are largely viewed as transmitters of innovations. Processors are treated similarly, except on those occasions when they engage in direct interaction. With increasing concentration and larger corporate entities in the processing industry, these users frequently have in-house R&D capability. Overall, a small amount of Extension time and interactive contact.

Time Input: Difficult to aggregate, but estimated to be about 1.6 percent (1976 data).

Contact Numbers: Difficult to aggregate, but estimated to be about 1.2 million informational contacts (not individuals) or about 1 percent of all Cooperative Extension contacts (1976 data) (USDA-SEA, 1980).

(4) Food marketing interests. These entities are increasingly divided into larger corporate units with direct-interactive access to varied private sector extension mechanisms--which are treated via diffusion access by public R&D--and smaller units which, to the extent addressed, are generally dealt with via indirect, diffusion strategies. Smaller marketing interests do receive some direct-interactive access through selected special efforts. Overall, food marketing interests represent a very small portion of Cooperative Extension time and interactive contact.

Time input: Estimated to be about 0.8 percent (1976 data).

Contact numbers: Estimated to be about 86,000 information contacts (not individuals) or some 0.87 percent of all Cooperative Extension contacts (1976 data).

(5) Environmental interests. Contemporary activist organizations are often addressed indirectly, or even as adversaries. Within agricultural science many environmental issues are treated from the perspective of producers' input needs. Producers of renewable resources products are becoming more common direct interactors with the public Extension system through Sea-Grant, Renewable Resources Extension Act, and other special programs. Most direct-interactive access for more activist environmental interests has come from nonagricultural research and selected nonprofit institutions. Overall, environmental interests represent a small portion of Extension time.

Time Input: Figures vary. 1981 annual Cooperative Extension reports show 4.6 percent.

Contact Numbers: Not estimable from available Cooperative Extension data.

(6) Rural development interests. Rural development audiences vary greatly, from elected and appointed officials, to organized action groups or agencies, to those individuals in rural areas sharing a need or problem. Generally, these interests have not been extensively addressed, and when they have been it is often in conjunction with or as a result of the interests of other groups. Cooperative Extension has a continuing tradition of both direct-interactive and information service to rural communities, bolstered in the last decade with special federal monies. Varied nonprofit organizations are somewhat more active, usually on the local level. Overall, a small portion of total Cooperative Extension time, despite special funding.

Audience: Cooperative Extension estimated 63 million persons potential (1980 National Extension Evaluation).

Time Input: Reported to be about 5.7 percent of total professional Cooperative Extension staff time, but declining (1980 data).

Contact Numbers: Reported at 6.1 million persons or about 5.5 percent of total reported Cooperative Extension contacts (1981 data). 1980 National Extension Evaluation reported the following breakdown: 16 percent local government, 12 percent service clubs, 12 percent state government, 9 percent federal government, 51 percent individuals.

(7) Human nutrition interests. Much work has been done outside agricultural science circles, most of which has featured direct-interactive access primarily for the medical community. Consumers have been addressed mostly via indirect informational mechanisms. Cooperative Extension has featured a nutrition program for low-

income consumers for a decade, but has emphasized information provision, rather than providing interactional development or strong applied research connections. Overall, a significant level of public Extension effort, due in large part to special federal funding. Also, significant levels of nonprofit activity at all levels.

Audience: Cooperative Extension estimates a potential audience of 226 million consumers.

Time Input: About 9 percent (1976 and 1981 data).

Contact Numbers: The 1980 National Extension evaluation reported 1.7 million contacts (not individuals) since 1968.

(8) Animal health interests. Production and production support interests have been addressed by both public and private sectors on a direct-interactive basis as well as via others. They are addressed and reported within production program data.

(9) Transportation interests. Very small amount of direct-interactional access, usually as a result of influence and demand on government performers, mostly outside the agricultural area. There is a small amount of reported effort. Little known, even within production or other related areas.

(10) Labor interests. Generally, very little access of any kind except through special projects and some nonprofits. Labor organizations themselves are often regarded adversarily by agricultural groups and so are difficult for agricultural science to cooperate with. There is no significant level of nationally reported effort beyond some economic assessment, farm safety programs, and a few pilot state efforts aimed directly at employers, supervisors, and employees themselves.

(11) World agricultural development interests. Substantial access to research overall, in part due to government and funding influence. Considerable direct-interactive access, focused largely on application and adoption of U.S. production practices, technologies, or products. There are no separately reported efforts. Mostly addressed and reported through and as adjunct of ongoing production work.

(12) Youth--4-H. A category not represented on the National Users Advisory Board, nor often thought of in connection with food and agricultural science. However, public Cooperative Extension programs have long expended significant effort on youth development programs associated with their main priorities. Represents a significant portion of total Cooperative Extension activity, no longer exclusively a rural or agricultural program.

Audience: The 1980 National Extension Evaluation estimated some 45 million potential youth participants.

Contact Numbers: 4-H members were reported in the 1980 evaluation report as 4.2 million (9 percent of potential audience), and total 4-H contacts (not

individuals) were reported in 1981 as 40.3 million or about 36 percent of all Cooperative Extension contacts.
Time Input: In 1981, 27 percent, declining slightly since 1975.

Although cursory and superficial, this brief review helps to demonstrate that far and away the greatest degree of direct interactive access to agricultural science afforded by various extension mechanisms is given to agricultural production interests. Further, given the assumptions made earlier regarding the mutuality of influence resulting from direct interactive relationships, it should not be surprising to find current agricultural science priorities and perspectives greatly influenced by contemporary production viewpoints.

SUMMARIZING SELECTED ISSUES

In summary, the four issues of express interest in this volume do not seem to be given time or resource priority by Cooperative Extension or other research access providers, in part, perhaps, because they are not expressed as high priority by the majority of current directly interactive users. This is also true, in part, because all four selected issues require cross-user group collaboration, which Extension-type programs have not emphasized and may not be well equipped to manage. Finally, partly because these issues are still relatively emergent in nature, they are not traditionally important to extension providers themselves. One significant result of the lower level of interactive attention to these issues is a lower level of demand for related research or development efforts.

An issue-by-issue summary suggests the following:

(1) Sustainability. Long-term sustainability of production potential has not been given a great deal of attention except as ancillary to contemporary production increase work. Given contemporary intensive production methods and our history of available land, water, energy, and other inputs, it is understandable why and how other concerns have been given higher priority. This is not to say that resource conservation and use have not been addressed at all. For example, significant effort has been expended to reduce dependence on chemical pesticides through the nationwide Integrated Pest Management program, and work has increased in the areas of minimum tillage, biomass, water use, and other resource conservation-related subjects. However, even considering these important programs, long-term sustainability as described here has not been given overall priority, and, where addressed, the emphasis has been and continues to be on marginal

cost saving modifications to contemporary resource intensive production practices with generally less attention to nonconventional alternatives. This has been so much the case that the National Agricultural Research and Extension Users Advisory Board (USDA-SEA, 1979-1981), after reviewing both public and private activities and investments nationally, put its highest priorities for three years in a row on sustainability related issues.

Given increasing concern for sustainability related issues, questions must be raised about the effectiveness of virtually all types of extension mechanisms with regard to their role in sensitizing the agricultural science establishment to emergent issues. "Sustainability" related issues have been somewhat regarded as issues of constraint by production increase interests. Given that the bulk of extension activity has been focused on production interests, it is not altogether surprising that extension mechanisms of all types have found it difficult to give more priority to sustainability issues as they have emerged. Recent profit threats posed by resource availability and cost are nonetheless making all parties more amenable to addressing these issues.

(2) Nutritional quality. In very broad terms nutrition has been given significant attention by virtually all extension mechanisms. One major national Cooperative Extension program has emphasized providing consumers with nutritional data on foods and dietary advice. There has also been substantial work leading to the development and adoption of conventional food processing technology, quality control, and preservation technologies. However, as noted elsewhere in this volume, more effort is needed on improving nutritional content of foods, foods availability, analysis of production practice effects on quality, marketing standards effects, and effects of preservation and processing practices. Complicating virtually all efforts is the overall lack of proven basic understanding and conflict of views among professionals, one of the results of which is that many nutrition related programs work at cross purposes to one another (e.g., a corporate sales unit vs. a nonprofit interest group). Secondarily, since most nutrition work has emphasized either consumer "education" or interaction with specific industries, little or no cross-interest interactive development effort has been undertaken. For the nutrition issues raised here to be addressed fully and effectively, multiple and occasionally opposed interests would have to be simultaneously and interactively addressed. To date, most nutrition efforts have not been so designed or coordinated.

(3) Food access equitability. This issue seems little addressed in its own right. Very likely, private sector marketing work constitutes much of what is done,

although for most of these increased sales and profits objectives supersede equitability of access. It can also be argued that equitability of access has been, and continues to be, increased through production increase work which keeps volume up and prices lower. Domestically, more direct attacks on equitability of access have had little priority because hunger has not been a critical problem for most of our population and food costs represent a small portion of living costs. Regarding world level needs, our prevailing view of the problem as one primarily of transportation and marketing has constrained approaches. Complicating all aspects of the issue is the concept of a "free market" which obfuscates the differences between need and demand. Additionally, there is very little public sector marketing, transportation, or distribution-related research or extension work. This may be due, in part, to the public sector's emphatic focus on interactive relations with production interests which do not see these issues as relevant to their needs and to private distributors' competition for market shares, which tends to discourage any activities which would make relevant information more widely known.

(4) Food system costs and benefits. Like equitability of access, some facets of these issues are well addressed, while others are not. Most extension mechanisms are oriented to interacting with and supporting the food system as it is. Hence, most cost/benefit issues addressed tend to be those dealing with increased efficiencies from the current system's perspective. As a result, there is little activity in this area regarding the broader scale, alternative issues being raised in this volume. There are numerous reasons why. First, at the operating level such cost and benefit issues are costly and difficult to deal with practically. Second, major actors in the current food system already know much about costs, and do not necessarily want them explicated without ready alternatives. Similarly, benefits accounting can raise public questions which may prove counter to some businesses' interests in withholding marketing performance related information for competitive reasons.

In summary, the four issues of express interest here have not in general been given high priority by the Extension Service or by other extension mechanisms. A point reinforced by reports of the National Agricultural Research and Extension Users Advisory Board, which in successive reports (USDA-SEA, 1979-1981) has given its most urgent priorities to: conservation and better use of agriculture's natural resource base; food quality and human nutrition; and the structure, competition, economics, and effects of the food system. This is not to say that the work given priority by most Extension units has been unimportant or ineffective. Indeed, it can be argued that the successes of extension units in food and agriculture have made possible the nation's current copious levels of food production and overall low level of hunger. Even the most severe critics of agricultural extension units do not call for dissolution of the work or system. Rather, questions are focused on the perceived lack of variety in interests given priority access and the slow pace of priority change and work redirection as new national concerns emerge.

A recent survey of redirection of efforts within state Cooperative Extension Services (Tanner, et al., 1982) suggests that redirection within broad work categories is taking place. Areas given some increased effort include crop and livestock production, farm business management and marketing, family resource management, conservation/minimum tillage, and food and nutrition. While the amount of effort reallocated was small, the redirection nevertheless represents some response to the emergent concerns addressed here. However, despite the fact that some Extension activity is and has been ongoing in these areas of interest, and there is evidence of some small level of redirection toward the broad areas, current levels of effort are still insufficient in light of the serious implications of these emerging national concerns. Why should this be so, and what, if anything, can or should be done? To ?wer these questions we must look at how extension mechanisms function, and how priorities are established.

Since it is by definition an extension mechanism's role to serve as an organization's most interactive interface with society, it follows that whether or not the research group sets overt goals, both the type and direction of its work as well as the type and extent of development or application resulting, will be much influenced by the interests reflected by user constituencies interacted with. As a result, there are two important perspectives from which to consider research uses and users. The first is one of equity: who tends to benefit and who does not. The second perspective is one of interaction: who influences research and development and who does not. Technological change is seldom neutral. Some issues and some participants must take precedence over others. Given that the process of development and use is not automatic or structurally preordained, this view is a positive one, since it means that uses of research are indeed influenceable.

Thus, one set of reasons why Extension mechanisms may not be giving more attention to the four issues described here has to do with the identity and interests of those users given direct interactive access as opposed to those who are not. Users heavily invested in contemporary production methods constitute the largest block of directly interactive users, but they are obviously not demanding, or even encouraging work on the four issues described. Furthermore, it seems reasonable that shared experience in previous development will lead to expectations of continued assistance. At local levels, where interactive development must of necessity be focused, Extension interactive contacts have emphasized prominent "influence leaders," a constituency which has not been easy to "de-invest" from as topical interests and needs change. Extension workers and these prominent clients know and trust one another, making change of interactive user groups difficult at best. Extension units need encouragement and both institutional and political support to provide more direct interactive access to broader constituencies, including some whose interests and views conflict with one another.

The second major basis for priority setting should be whether there is a body of relevant research work (or information) on hand or underway and accessible. To fuel its work, an Extension unit brings development interest based on perceived need or opportunity together with a relevant base of research information. Thus, another reason why Extension units are not giving more priority to these four emergent issues in their programs is that there is relatively little directly relevant current research information to draw on. Indeed, there is some evidence that this may be one reason for the current situation. Research units have tended to be influenced more by academic interests and structures than by external sources. The existence of separate extension units may, in fact, add to a research unit's sense of legitimate isolation from external interests and issues. Academic research units tend to emphasize information publication and dissemination as the natural conclusion of their responsibility. Thus, researchers have not always welcomed the demands of interactive relationships.

To what degree should various extension mechanisms give more priority to the four issues of this volume? This is not only a matter of observed need and knowledge base opportunity, but of opinion as well. Academic opinion is conservative, and conservatism has generally served it well. The emergent concerns suggested here represent serious concerns not only for current human welfare, but also for our future. A strong case for increased attention seems well justified, although not a wholesale reversal. Much current work is both necessary and productive.

Developing new priorities at any time, but especially in a period of economic constraint, will prove slow at best. It is helpful that there is evidence of some priority redirection already. Several steps can yet be taken which would help extension mechanisms of all types, especially those in public institutions, become more responsive to these and other emergent needs. Some steps which deserve consideration include:

- Administrators of research and extension units need to make more concerted efforts to improve linkage and cooperative planning between research and extension units, and developmental interaction with users themselves.
- Administrators and participants need longer planning and problem-definition time horizons, especially for problems which embrace a broader spectrum of perspectives and interests.
- Administrators need to put more premium on improving social science skills and resources to increase staff group organization and development abilities to the degree needed for sophisticated multiple and cross interest group interaction and for dealing effectively with social aspects of "technical " problems.
- Improved systemwide coordinated planning must be explored. The high degree of decentralized priority setting in the public sector is and has been a good thing, and should not be overly disrupted. However, this same decentralization coupled with little functional accountability seems to breed a standardization of process which discourages diversity in problem definition and approach.
- Administrators and funders must not accept the presumption that technology is neutral in the long term. They must not only tolerate, but inquire actively into, questions of the social and economic impacts of innovation. The point must be not to discourage selected research or innovation development but to better understand its impacts, real and potential, and to consider related policies and actions.
- Governments, funders, cooperators, administrators, and others need to express more aggressive support for public extension units' openness to divergent views. Political and other pressures on an extension unit can be intensely discouraging of openness to competing interests. Optimal innovation development depends on reciprocal gains by proponents of competing interests and ideas.
- Congress and state legislatures need to express clear and consistent support for these and other long-range issues, to help public providers avoid being predominantly directed by selective-interest constitutents.
- Serious consideration should be given to restructuring of public extension provider missions, and programs, to give clear emphasis to bringing about development and use of innovations from research on existing and emergent problems,

as opposed to roles of simple information dissemination. Such a shift should help differentiate more clearly between which users enjoy direct interactive access, whose influence is greater, and other users whose influence is less.

- Both research and extension performers and funders need to develop ways to improve and maintain a full food and agricultural system-perspective, to seek input from all segments, and to articulate explicit obligations to all segments. This must include objective feedback and checking procedures to help ensure that interaction is not dominated by only the most aggressive or sophisticated users.
- Greater support is needed in the public sector for cross-disciplinary and cross-user interest problems and issues which, because of their complexity--both technical and political--otherwise tend to be deferred.
- Both public providers and funders need to review jointly and on a regular basis the relationship between research efforts, Extension efforts, and reported needs from multiple sources.

Whether these steps will prove sufficient to the need is difficult to assess. But, extension personnel have already a great degree of commitment to human service and welfare, and great professional competency. The emerging needs of the future, like those of the past, can be met by a cooperative system tolerant of extreme diversity in thought, and collaborative in its approach to development.

REFERENCES

Rogers, E., and Shoemaker, F. 1971. Communication of Innovations: A Cross-Cultural Approach. New York: Free Press.
Sahal, D. 1981. Patterns of Technological Innovation. Reading: Addison-Wesley.
Tanner, B.; Bryan, J.; and K. Rygasewicz, comps. 1982. State Survey: Redirection of Program Efforts. An unpublished USDA-ES analysis and report, Washington, D.C.
U.S. Congress. 1977. Food and Agriculture Act of 1977. PL 95-113, Section 1408. Washington, D.C.
USDA. 1980. Inventory of Research and Extension Programs in the Food and Agricultural Sciences: Part I. An unpublished report to the National Research and Extension Users Advisory Board, Washington, D.C.
USDA-ES. 1982. National Summary of Extension Level of Effort for FY 1981. USDA, Washington, D.C.
USDA-SEA. 1980. Reports of The National Extension Evaluation Project. USDA-SEA, Washington, D.C.
______. 1981. Reports of the National Agricultural Research and Extension Users Advisory Board to the Secretary of Agriculture. USDA, Washington, D.C., October 1979, October 1980, October 1981.

International Implications

16
International Implications of U.S. Food Security Programs

C. Dean Freudenberger

The efforts of the United States to achieve national food security, in terms of adequacy and sustainability of the food supply, equity and access to food by the entire population, nutritional quality, and the social, economic, and health benefits of our food system, have had, by and large, a negative international impact. This is because the U.S. has not given sufficient attention to achieving a national food security system in broadly defined terms. Rather, its emphasis has been short term, with a narrow focus on high cost and high technology production. The negative international impact of this system has been particularly felt within the tropical and semi-arid food-deficit nations of Africa, Latin America, and Southeast Asia, even though the national intent, as expressed in a variety of food-aid and development assistance programs, has been otherwise. The purpose of this chapter is to substantiate this observation and suggest, in outline form, new directions for beginning the process of correction.

For purposes of brevity and clarity, the assumption is made that the reader is informed by the contributors to this volume about the nature of American agriculture: soil and water loss, conversion of prime agricultural land, fuel and fertilizer dependence, genetic vulnerability of our monocropping systems, food quality and safety, the narrowed focus of agricultural research, and associated problems in leadership training in our land-grant college establishment of research, education, and extension. My specific task is to point to the international impact of our own agricultural crisis and its associated food security system. This impact is largely unintentional and indirect. But, it is critical in terms of the immediate and future hope of many nations, as well as in terms of long-range projections about the health and stability of the whole biosphere.

I find it useful when considering the subject of "food security" to remind myself that within the relatively short history of agriculture itself, the fundamental problem of food production has yet to be solved by any generation, in any nation. This is to say, no significant progress has been made to overcome the problem of soil loss, and the wasteful use of water (Lowdermilk, 1978). A contemporary description of the process and magnitude of loss, set within this historical setting, can be seen in the relatively recent United Nations' final report of the World Conference on Desertification (1977). The maintenance of plant and animal species diversity, absolutely essential for sustainability in food systems, is in serious jeopardy (Myers, 1979, pp. 1-82). Overarching concerns about the general health of the biosphere, in terms of carbon dioxide levels, atmospheric particulate pollution, issues of stability in the ozone shield, the build-up of toxic chemicals, and problems related to global deforestation, are well documented in recent United Nations' Environmental Program Annual Reviews on the State of the Environment (1977-1983; see especially 1979 and 1980). Contemporary U.S. agriculture contributes to the tragic historical record of species loss, soil, forest, grassland and water loss, and desertification. No nation is exempt from responsibility for the record; almost all nations continue to add momentum to this historical record:

> More than a mere threat, desertification is actively at work. A great many people live in dry lands that are now undergoing the process, and their livelihoods are already affected. Estimates of present losses give rise to a pessimistic outlook, suggesting that the world will lose close to one-third of its arable lands by the end of the century. Such losses will take place while the food requirements of the human race are rising at least as rapidly as the human population is growing (UN, 1977:9).

Past and present policies and programs for achieving domestic food security on the part of all nations have yet to address these kinds of fundamental issues. Until they do, the hope for achieving an enduring security in U.S. food-producing systems, or for assisting other nations in overcoming food deficits, will be beyond reach(1).

PAST AND PRESENT U.S. POLICY PREOCCUPATIONS

One can summarize past and present U.S. food policy formulations in the following way. Policy has historically sought for adequate supplies of food, has attempted to place food in the marketplace at low prices, and has attempted to maintain (until recently) a viable farm economy and rural community (Lutz, 1980, ch.

2; Rodefeld et al., 1978). One can, of course, note that the overarching problem which has plagued U.S. agriculture for many decades is commodity surplus, particularly in small and coarse grains (Freivalds, 1976; Morgan, 1979). Research and development within our agricultural system has centered on these basic policies (Hadwiger, 1983). Agricultural extension activities, until recently, have been consistent with these policies and problems. Internationally, a heavy responsibility of our embassies and consular offices with their agricultural attaches has been that of market development for U.S. agricultural surpluses (Freivalds, 1976). For decades, the problem of surplus commodities has been a part of U.S. food-aid programs(2).

Since the end of World War II there has been a constrained discussion about supports, markets, small and large farm interests, and social philosophy in the national politic. The discussion of agricultural policy has been governed by national and world demands. These discussions take place in the House and Senate agricultural committees, subcommittees, the lobbying core, and interest groups. The discussion ultimately takes the form of agricultural, food, and farm policy. There is a double emphasis: first, upon small farms, the rural community, social welfare, and the democratic society; and second, upon the market needs of large producers and crop production efficiency. Efforts are constant in the area of stabilized farm income, commodity production, and surplus, with a peripheral struggle about who farms, what is farmed, and how much is farmed. A contemporary illustration of this is the "PIK" (Payment in Kind) program in effect during 1983.

Each presidential administration has different policy emphases. In the Truman years, the emphasis was upon flexible but permanent controls on parity, with an eye on the welfare of the small family farm. The defeat of the Brannan plan by powerful interest groups shifted the emphasis away from the small producer. During the Eisenhower years, emphasis was placed upon the soil bank and acreage reserves. This was undone by the international food crisis of the late '60s and early '70s. During the Kennedy and Johnson years, the emphasis was upon supply management, a flexible version of parity policies. The Nixon years instituted an emphasis upon a market-oriented farm program (Friedland, 1982). Small farmer protection legislation was weakened to the advantage of large producers (Paarlberg, in Rodefeld, et al., 1978). The Ford administration felt the decline of worldwide prices for agricultural commodities and the beginning of the farmer strike movement. Policy was made around commodity interests in cotton, wheat, soybeans, corn, meat, and dairy products.

Thus, during the 1950s and '60s, the problem was overproduction. During the 1970s to 1980, the global situation was one of food deficits. Of course, the

Rome World Food Conference of the United Nations brought the critical situation to the attention of the world community (1974). The shift was made from small farmers to agribusiness activities(3). During this time, budgets and grants were cut for agricultural research for the benefit of the public sector, while the private corporate sector did its own research and development (Hadwiger, 1983). Also during these years, the Poor People's Campaign for school lunches emerged under the leadership of Ralph Abernathy. The Senate Select Committee on Nutrition and Human Need was developed under the leadership of Senator George McGovern. Now, during 1983, there is a growing concern in the field of consumer protection (food purity) (see chapters 10 and 11 in this volume). Equal concern is being given to agricultural land loss and nonsustainable rates of both surface and ground water use (NAL Study, 1980d,e; Fisher, 1978).

Although all of these policy issues are important, they do not address the critical problems confronting U.S. agriculture today (Ebeling, 1976). Furthermore, a preoccupation with these issues leads to international resource and social stress within the agricultural sectors of many food-deficit nations. An explanation of this point is essential.

FOUNDATIONAL ISSUES NEVER ADDRESSED IN THE FOOD SECURITY FORMULA

Several issues critical to U.S. food security never entered our imaginations 20 years ago (Ward and Dubos, 1972; Brown, 1978; Eckholm, 1982). I refer to these issues only in outline form. It will be seen that because of this negligence the issues facing the international community of nations have likewise received little attention, because in many ways what does or does not happen in the U.S. can be observed internationally. Many of the negative aspects of the so-called "Green Revolution" illustrate this point (Ruttan, 1971; Borgstrom, 1973; Hunter, 1972; Ceres, 1973).

Of major priority is the need to develop a new social ethos (the total complex of attitudes that predetermines collective behavior, attitudes about ultimate meaning, ends, claims that justify actions) and policy about land and land use. To date, in our own national history, the norm has been freedom to use the land. The question today is, "What is the nature of responsible land use?" (NAL Study, 1982). Following this issue is the need for developing a new social understanding about the nature and essence of the concept of agriculture: Is it a science, an art, a technology, a way of life, or does it have something to do with all of these things? (Dubos, 1980).

This issue of social ethos, and understanding about agriculture, in the context of the destructive history of agriculture everywhere, is discussed in Donald Worster, Dust Bowl: The Southern Plains in the 1930s:

> As the world's population moves increasingly onto marginal land . . . and already more than half a billion people live in deserts or semi-arid places . . . and as unfavorable shifts in climate appear likely, even in temperate zones, the need for ecologically adaptive cultures becomes all the more crucial. Capitalism cannot fill that need; all its drives and motives tend to push the other way, toward overrunning a fragile earth. Man, therefore, needs another kind of farming by which he can satisfy his needs without making a wasteland. It would be fitting if we should find this new agriculture emerging someday soon in the old Dust Bowl (1979:243).

Every nation faces the need for a new social understanding about relationships of people to the environment and ecosystems in their physical and biological context (Brown, 1981). Rene Dubos (1980) observes with hope and anticipation that societies are now beginning to learn to anticipate the dangers of failing to deal with the issues of environmental and human exploitation. He stresses that modern human beings have become destructive because they have lost their sense of quality and level of high responsibility for maintaining harmonious relationships to the earth. The "wooing of the earth" is possible only if a relationship of respect, in contrast to exploitation, prevails. He is quick to add that "the recycling of degraded environments is one of the most urgent tasks of our age." It is in this reclaiming process that our security and hope for the future resides.

Within our national debate, as outlined in the previous section, and in comparison to the magnitude of the problem being addressed, very little has been said about soil loss, soil compaction, reduction of humus, water loss, water and atmospheric pollution, ground water overdraft, history of weather, and weather change. Only recently have these issues, of a critical nature, begun to surface in any kind of general public way (Sampson, 1981). Similarly, few policy debates have focused on the vulnerability of monocultural cropping systems, patterns of crop diversity and environmental stability in forests, surface water, and ground water management (Brown, 1978). The issue of the welfare of future generations in relation to prime agricultural land use has received minimal attention (NAL Study, 1982), as have questions about the patterns of resource use and dependencies in agricultural production, processing, and distribution (i.e., fossil fuels and related agro-chemicals) (Pimentel and Pimentel, 1979). Only in the past decade has the concern about agricultural product safety and purity levels become a focus of food security policy issues (Marine and VanAllen, 1972). Nothing substantial has been discussed at the congressional level about the impact of the world food crisis on U.S. agriculture. It was during the Rome World Food Conference, called by the United Nations in 1973, that this issue really emerged. Former Secretary of Agriculture Earl Butz declared in Rome that the U.S. would do all it could to move to levels of full production in response to the world food crisis. Today, we experience negatively the environmental, natural, and human resource impact of this policy. We observe clearly the impact of the world food crisis on U.S. agriculture, involving food deficits in more than 100 nations. The needed response is not in the area of increasing U.S. productivity. Rather, it lies in the area of U.S. as well as international agricultural development assistance to food-deficit nations in common efforts to create sustainable and self-reliant food production systems as replacements for the continuing existence of the old colonial cropping and export structures and systems involving coffee, tea, tobacco, groundnut, palm oil, rubber, cacao, and pyrethrum production.

All of the above issues need attention. The massive groundwork needed for conducting a high-level debate on these foundational issues is only beginning to be observed in the form of Congressional debate and appropriate legislative response. The National Agricultural Lands Study, which was developed during the Carter administration, and languished during the Reagan years, best illustrates this point. The question that must be raised here is, "What are the international implications of this neglect?" An associated but more specific question is, "How do our assumption patterns and the technical, economic, and value structures of our prevailing agriculture get transferred to the international community of nations?"

INTERNATIONAL IMPACT OF DOMESTIC FOOD SECURITY POLICIES

The crucial point is that short-term success in U.S. agriculture has blinded the nation and the international community of nations. Spectacular yields, as a consequence of heavy capital investments and uses of petro-chemicals, have become the vision of international productivity, particularly in the food-deficit nations of the semi-arid and tropical world. Related to this record of high yields are questions that have not been examined about soil and water loss, costs of inputs of exhausting resources (fossil fuel-based fertilizers), and costs of rural community dislocations and farm debts. Despite this knowledge gap, many of the basic assumptions about international agricultural development, as well as technologies from the northern temperate regions of the world, have been exported. The technologies that have been developed in temperate climates and upon rich, deep and youthful soils, are more often than not transferred unquestioningly into the tropical and semi-arid world.

One of the best observers of this phenomenon is Rene Dumont. Predominant technologies undergoing transfer are conveyed through development assistance programs sponsored by the U.S. Agency for International Development (AID), the U.S. Peace Corps, the many nongovernmental assistance organizations, and through the lives of many national leaders trained in our agricultural centers. This problem is being more aggressively addressed by the international network of agricultural research programs, yet there is a long way to go before some of the most urgent problems, such as tropical soil fertility maintenance and the development of permacultural and agro-forestry systems, are adequately addressed. The assumptions about the high productivity of hybrids, annual cropping systems (mainly in cereal and coarse feed grains), and livestocking, have resulted in disaster in many nations of the world where adaptation has been attempted (Dumont, 1969; Dumont and Cohen, 1980). This observation is made even outside the consideration of social problems involving labor and land displacement, the destabilization of rural/urban settlement patterns, and problems of rural migration to the cities. Methods of measuring the value of an agricultural system in terms of the maximization of yields at the expense of soil loss and petro-chemical inputs, the environmental, social, and economic impact of these inputs, and concepts of comparative advantage in the international marketplace have resulted in heavy losses of soil, forest, and grassland reserves (Eckholm, 1976; Brown, 1983). A further dimension of the international impact of our national attempts at food security has been the development of capital-intensive, large-scale agricultural enterprises in nations with high percentages of rural populations. In many places, the introduction of such systems has resulted in social upheaval and the movement of rural people to marginal highlands, forests, and deserts, with irrevocable resource loss and growing poverty (Burbach and Flynn, 1980; Lappe and Collins, 1977). Generally, the patterns of negative impact of Western international agribusiness have resulted in greater in-country food deficits and the institution of nonsustainable agricultural industries. One can see this phenomenon in Zambia and Zimbabwe in tobacco, in the tea plantations of Kenya, in the rubber production of Liberia and Indonesia, in groundnut and cotton schemes in West Africa (Benin,

Togo, Niger, Mali), in beef cattle production and bananas in Central America, and in pineapples on some of the islands in the Philippines.

There are several major issues related to the expansion of worldwide agricultural developments which, to date, have no solution. They must soon receive priority attention by all of the nations. Emerging from the several United Nations State of the World Environment Reports (between 1977 and 1983) is the growing concern about the stability of the ozone shield and the buildup of atmospheric carbon dioxide, two issues deeply related to prevailing agricultural technologies utilizing manufactured nitrogen fertilizers, and the impact of annual tillage and expanding farmlands on forestlands. These are of growing concern, in reference to the issues of maintenance of atmospheric transparency and the balancing of incoming infra-red light (Great Britain, 1976). The question about the maintenance of stability in climatic cycles and histories is also raised.

On a parallel magnitude of concern within the international agricultural community are the possible consequences of global fallout of radioactivity (resulting from an outbreak of limited nuclear warfare) upon insects and patterns of crop pollination. These are awesome issues which must be resolved in our generation, yet they are neglected in the national debate and legislative chambers where food security issues are discussed and programs formulated. Only rarely does the question of the impact on national and worldwide agriculture get raised in the context of discussions about effects of nuclear warfare (Sagan, 1980).

There are two additional issues of international impact involved in our food security policies. The first involves the sale and grants-in-aid of U.S. surplus food. The second is in reference to dominant trends in U.S. agricultural research, development, the education of agricultural leaders, and how international leaders, as well as U.S. citizen development assistance personnel, are orientated in approaches to international agricultural development. The need for review is long overdue. In many instances, the present need for locating markets for U.S. agricultural commodities works against efforts to resolve the issues of food assistance and relevance in leadership training for the development of international food self-reliance. These issues are hardly documented. Against the Grain (Jackson and Eade, 1982) is one of the most concise publications which addresses the so-called dilemma of projects involving food aid, projects of both federal and nongovernmental origins. Also, examining the 3,000 bibliographic citations in Nicole Ball's World Hunger: A Guide to the Economic and Political Dimensions (1981), a first of its kind, one is immediately impressed by the Western bias in conceptualizations about the problem of food production and distribution in food-deficit nations, and, in relation to the second point of this paragraph, the very narrow focus in the fields of agricultural research, development, and agricultural leadership education and training.

What is being suggested is simply that U.S. domestic food security is interrelated with the question of international food security. It is difficult to envision one without the other. To complicate the situation, no one nation can provide a model for another, beyond methodologies for approaching the challenge of sustainability in agriculture itself.

STEPS TOWARD A MORE SECURE FUTURE

Note that what is being said in this chapter has been said before. The point is that it must be repeatedly stated in every symposium dealing with food policy matters, and in research and development proposals. These issues must be reiterated until appreciable response to them can be observed.

In summary, I will identify the agenda that is before us:

(1) We face a critical need to generate public awareness of problems regarding clarification of agricultural technological assumptions. The level of understanding about these matters is too low for one to expect adequate, let alone inventive, discussion.

(2) Similarly, we need to be more realistic about our simplistic approaches to food security. The nation thinks primarily in terms of increased food production, storage, and adequate distribution, stockpiling in "rural interiors" in anticipation of the need to move populations from the massive urban centers to safety in the rural interior in time of international military threats.

(3) We need to understand and more clearly analyze the problem of fossil-fuel dependence in our food production technologies, which has emerged in our relatively new industrialized and integrated agricultural system. Within an historically short period of time, oil and gas will be depleted to the point where, because of increased price and availability, their use will be economically impossible for agricultural production (Brown, 1983). Unfortunately, this fossil-fuel dependent technology has become a reality throughout the global food system (Smil and Knowland, 1980). We are running out of time in our race to develop a solar and biologically intensive agricultural system.

(4) Another serious problem is the low level of federal funding for agricultural research and

development targeted at the massive array of problems associated with the food security crisis. Related to this priority is the hope for the development of a post-petroleum or solar/biological intensive international system for achieving sustainable self-reliance in agriculture. As I observed earlier, as the Western world goes in its agriculture, so, by and large, goes the rest of the world.

(5) Finally, our food security depends upon both the security of the nation's farming community and the level of its management skills. When soil loss and the conversion of prime agricultural farmland is at historically unprecedented levels, and the national farm debt is totally out of proportion with annual income ($200 billion debt, with an expected income of about $20 billion), it is obvious that our food system is in jeopardy. But a related issue of a more subtle nature is that of the displacement of massive numbers of people from the farming sector. As history and the global resource base of fossil fuel presses in upon us (and as the problem of atmospheric stability increases, partly as a consequence of increasing combustion of these fuels), the need for a more biologically intensive agriculture becomes clear. To meet the challenge there is need for adequate numbers of highly skilled farm "managers," numbers of skilled people who are able to manage the numerous and diverse micro-ecosystems of our nation. Our food security, as well as the world's food security, depends upon vast numbers of secure and skilled farmers. At the present moment in global and our own national history, the challenge has yet to be addressed. Recent migration of rural people to the cities across the world has created a large vacuum, and the future of all food systems is now in question.

Leaders must be recruited and trained to meet the above-identified needs. Agricultural studies need to be undertaken within the liberal arts colleges of this nation, and in the departments of the humanities of the great state universities, where tomorrow's leaders will be prepared for assuming responsibility in government and policy making for tomorrow's agriculture. Through the auspices of the Kellogg Foundation, significant effort has already been made in this area. Steps need to be taken to develop a sound agricultural land-use policy which addresses the loss of soil, water, grassland and forest resources, and prime farmland preservation. In this same category of needed action, greater effort is required for the development of policy and law to stabilize the farming community, to bring it out of bankruptcy and to develop essential rural community amenities. Several chapters in this volume have addressed the need to deal with the problems of toxicity in our food production and processing systems. Required longer term steps point to the need for the development of a normative conceptualization of agriculture--of its nature, role in society, and the future (Dahlberg, 1979; Merrill, 1976). The concept of long-range sustainability (a notion quickly dismissed in contemporary discussion), and the impact of agriculture upon the whole biosphere must be addressed (Daly, 1977). Research and development, including the development of appropriate leadership for the emergence of an agriculture that is sustainable in long-range perspectives of decades and centuries, needs to take place (Jackson, 1980; Berry, 1981). Research and development will need to be done with an awareness of the fragile nature of the earth's various bio-geographical provinces (Udvardy, 1975). It must search for time references and technologies which are consistent with new realities of resource limits, particular fossil fuels, soil, forest and grasslands, human population numbers and their projections, and maturing insights about the nature of biospheric equilibrium (Schneider, 1976).

In terms of food security, this all adds up to the need to reduce U.S. agricultural production to a sustainable basis (with regard to the way that sustainability has been defined), to preserve agricultural lands for future generations (as is generally the pattern in Western Europe), to maintain essential forest, water, and air resources, to reduce international dependence on U.S. agricultural production for basic foods, fibers, and vegetable oils, and to work for the emergence of commodity price policies and mechanisms which are in line with the costs of agricultural production on a sustainable-yield basis (Strong, 1979; Sampson, 1981, ch. 13-15). Every effort must be made to cooperate with the international community of agricultural leaders charged with the development of reasonable sustainability in national self-reliant food and fiber production systems, and for the development of a biologically sophisticated post-petroleum agriculture.

CONCLUSIONS

We need (a) global perspectives brought to bear on U.S. agriculture and its quest for domestic food security, (b) a profound discussion about agricultural policy and a program within the national community which will be comprehensive in consideration (to include the issues of atmospherics, essential resource limits), (c) a probing of "foundational issues" that have yet to be addressed adequately in the public forum, (d) a careful look at the international impact of our agriculture upon the nations struggling for sustainable self-reliance in food and fiber, and (e) an articulation of strategies and sustainable guidelines for taking steps toward a more food-secure future for humankind.

To be responsible in our quest for domestic food security, it is essential that we think globally and act locally, and that we act according to our growing knowledge about the nature of global interdependencies. This is a task which cuts across every aspect of the vast complexity of agriculture--its research agenda, educational establishments, technology, and associated industries and values. It is an agenda with little historical precedent. The question of national food security, and the global impact of attempts to build security in our food and fiber supply, points to issues which, twenty years ago, were hardly in our imagination, yet which must be resolved in our lifetime if there is to be even minimal quality in the lives, and the environment, of most of the world's children who are already born.

NOTES

1. Tolba, Development Without Destruction: Evolving Environmental Perspective (1982). This collection of major addresses by the Director General of UNEP provides good background on this subject. Sheridan, Desertification of the United States (1981), written for the Council on Environmental Quality, provides a focused portrayal of U.S. conditions.
Fletcher and Little, The American Cropland Crisis (1982), describes the magnitude of loss and citizens' and government's attempt to save what is left.
Worster, Dust Bowl (1979), describes fundamental issues on the prairie lands of the United States.

2. See Jackson and Eade, Against the Grain (1982), for a full description of the impact of "Food for Peace" and nongovernmental organizations' programs for food assistance, and the impact this has on indigenous markets and food producers.

3. For a careful analysis of the depth and significance of this shift, using two rural communities in California as a basis for comparison, see Goldschmidt, As You Sow (1978). The Preface and Introduction should receive careful attention, as these sections give reference to "Agriculture and the Social Order." Also, see Kramer, Three Farms (1977).

REFERENCES

Ball, N. 1981. World Hunger: A Guide to the Economic and Political Dimensions. Santa Barbara, Oxford, ABC-Clio.

Berry. W. 1981. The Gift of Good Land. San Francisco: North Point Press.

Borgstrom, G. 1973. Focal Points: A Global Food Strategy. New York: Macmillan.

Brown, L.R. 1970. Seeds of Change: The Green Revolution and Development in the 1970s. New York: Praeger Publishers.

______. 1978. "The Worldwide Loss of Cropland." Worldwatch Paper #24. Washington, D.C.:Worldwatch Institute.

______. 1981. Building a Sustainable Society. New York and London: W.W. Norton and Company.

______. 1983. "Population Policies for a New Economic Era." Worldwatch Paper #53. Washington, D.C.: Worldwatch Institute.

Burbach, R. and Flynn, P. 1980. Agribusiness in the Americas. New York and London: Monthly Review Press.

Ceres. March-April, 1973. Special issue on "The Costs of Science and Technology." Rome: FAO.

Dahlberg, K.A. 1979. Beyond the Green Revolution: The Ecology and Politics of Global Agricultural Development. New York and London: Plenum Press.

Daly, H.E. 1977. Steady-State Economics: The Economics of Biophysical Equilibrium and Moral Growth. San Francisco: W.H. Freeman and Company.

Dubos, R. 1980. The Wooing of Earth. New York: Scribner's.

Dumont, R. 1969. False Start in Africa. 2d revised ed. New York: Praeger.

Dumont, R. and Cohen, N. 1980. The Growth of Hunger, A New Politics of Agriculture. London: M. Boyers.

Dumont, R. and Mottin, M.F. 1983. Stranglehold on Africa. (Translated from French by Vivienne Menkes). Deutsch.

Ebeling, W. 1976. The Fruited Plain: The Story of American Agriculture. Berkeley: Univ. of California Press.
Eckholm, E.P. 1976. Losing Ground. New York: W.W. Norton and Co.
______. 1982. Down to Earth. New York and London: W.W. Norton and Co.
Fisher, J. 1978. From the High Plains. New York: Harper and Row.
Fletcher, W.W. and Little, C.E. 1982. The American Cropland Crisis. Bethesda: American Land Forum.
Freivalds, J. 1976. Grain Trade: The Key to World Power and Human Survival. New York: Stein and Day.
Freudenberger, C.D. 1984. Food For Tomorrow? Minneapolis: Augsburg Publishing House.
Friedland, W.H. 1982. "The End of Rural Society and the Future of Rural Sociology." Rural Sociology, 47(4):589-608.
Goldschmidt, W. 1978. As You Sow: Three Studies in the Social Consequences of Agribusiness. Montclair: Alanheld, Osmun and Co.
Great Britain. 1976. Chlorofluorocarbons and Their Effect on Stratospheric Ozone. Department of the Environment. Control Unit on Environmental Pollution. Pollution Paper No. 5. London: Her Majesty's Stationery Office.
Hadwiger, D.F. 1983. The Politics of Agricultural Research. Lincoln and London: Univ. of Nebraska Press.
Hayes, D. 1977. Rays of Hope: The Transition to a Post-Petroleum World. A Worldwatch Institute book. New York: W.W. Norton.
Hunter, G. 1972. "Some Western Transplants Yield Strange Fruits." Ceres, Issue 25. Rome:FAO.
Jackson, T. and Eade, D. 1982. Against the Grain: The Dilemma of Project Food Aid. Oxford: OXFAM.
Jackson, W. 1980. New Roots for Agriculture. San Francisco: Friends of the Earth. In cooperation with the Land Institute.
Kramer, M. 1977. Three Farms: Making Milk, Meat and Money from the American Soil. Boston and Toronto: Little, Brown and Company.
Lappe, F.M. and Collins, J. 1977. Food First: Beyond the Myth of Scarcity. Boston: Houghton Mifflin Company.
Lowdermilk, W.C. 1978. Conquest of the Land Through Seven Thousand Years, rev. ed. Agricultural Information Bulletin No. 99, USDA, Soil Conservation Service. Washington, D.C.: U.S. Government Printing Office.
Lutz, C.P., ed. 1980. Farming the Lord's Land: Christian Perspectives on American Agriculture. Minneapolis: Augsburg Publishing House.

Marine, G. and Van Allen, J. 1972. Food Pollution: The Violation of Inner Ecology. New York: Holt, Rinehart and Winston.
Merrill, R., ed. 1976. Radical Agriculture. New York and San Francisco: Harper Colophon Books.
Morgan, D. 1979. Merchants of Grain: The Power and Profits of the Five Giant Companies at the Center of the World's Food Supply. New York: Viking Press.
Myers, N. 1979. The Sinking Ark: A New Look at the Problem of Disappearing Species. New York: Pergamon Press.
National Agricultural Lands Study. 1980a. Interim Report No. 1. The Program of Study. Washington, D.C.: National Agricultural Lands Study.
______. 1980b. Interim Report No. 2. Agricultural Land Data Sheet. Washington, D.C.: National Agricultural Lands Study.
______. 1980c. Interim Report No. 3. Farm Land and Energy: Conflicts in the Making. By W. W. Fletcher. Washington, D.C.: National Agricultural Lands Study.
______. 1980d. Interim Report No. 4. Soil Degradation: Effects on Agricultural Productivity, prepared by the National Association of Conservation Districts. Washington, D.C.: National Agricultural Lands Study.
______. 1980e. Interim Report Number 5. America's Agricultural Land Base in 1977. Washington, D.C.: National Agricultural Lands Study.
______. 1982. The Protection of Farmland: A Reference Guidebook for State and Local Governments. Washington, D.C.: National Agricultural Lands Study.
Pimentel, D. and Pimentel, M. 1979. Food, Energy and Society. Resource and Environmental Science Series. New York: Wiley.
Rodefeld, R.D.; Flora, J.; Voth, D.; Fujimoto, I.; and Converse, J., eds. 1978. Change in Rural America: Causes, Consequences, and Alternatives. St. Louis: C.V. Mosby.
Ruttan, V. W. 1971. "Perspectives on the Green Revolution." Univ. of Minnesota: Staff Paper, Dept. of Agriculture and Applied Sciences.
Sagan, C. 1980. Cosmos. New York: Random House.
Sampson, R.N. 1981. Farmland Or Wasteland: A Time to Choose, Overcoming the Threat to America's Farm and Food Future. Emmaus: Rodale Press.
Schneider, S.H. 1976. The Genesis Strategy: Climate and Global Survival. New York and London: Plenum Press.
Sheridan, D. 1981. Desertification of the United States. Written for the Council on Environmental Quality. Washington, D.C.: U.S. Government Printing Office.
Smil, V. and Knowland, W.E., eds. 1980. Energy in the Developing World: The Real Energy Crisis. Oxford: Oxford Univ. Press.

Strong, A.L. 1979. Land Banking: European Reality, American Prospect. Baltimore: Johns Hopkins Univ. Press.
Tolba, M.K. 1982. Development Without Destruction: Evolving Environmental Perceptions. Dublin: Tycooly International Publishing, Ltd.
Udvardy, Miklos D.F. 1975. A Classification of the Biogeographical Provinces of the World. Prepared as a contribution to UNESCO's Man and the Biosphere Programme, Project No. 8. Morges, Switzerland: International Union for the Conservation of Nature and Natural Resources.
United Nations. 1974. Assessment of the World Food Situation: Present and Future. Item No. 8 of the Provisional Agenda, E/Conf. 65/3, Rome.
United Nations Conference On Desertification, Nairobi, 1977. Desertification: Its Causes and Consequences. Compiled and edited by the Secretariat of the United Nations Conference on Desertification, Nairobi, Kenya, August 29 to September 9. New York: Pergamon Press.
United Nations Environment Programme Annual Review. Nairobi: United Nations Environmental Program, reports from 1976 to 1983.
Ward, Barbara and Dubos, Rene. 1972. Only One Earth: The Care and Maintenance of a Small Planet. An unofficial report commissioned by the Secretary-General of the United Nations Conference on the Human Environment. New York: W.W. Norton and Company.
Worster, D. 1979. Dust Bowl: The Southern Plains in the 1930s. New York: Oxford Univ. Press.

17
International Trade and American Food Security

Bruce Koppel

INTRODUCTION

In the 1890s, one-third of the American wheat crop and one-half of the country's cotton crop were sold to foreign markets. By the 1950s, one-third of the wheat and cotton produced by American farmers and one-fourth of the soybeans were sold to foreign markets. In recent years, three-fifths of the wheat, one-third of the corn and two-fifths of the soybeans produced in the United States were consumed abroad. In fact, in the 1970s fully 85 percent of the growth in consumption of American farm products was a result of an export volume which tripled between 1971 and 1981. While international trade has been important to U.S. agriculture for many years, the 1970s witnessed both the strengthening and added recognition of this international dimension. Sandra Batie and Robert Healy spoke for many when they wrote:

> As the preeminent exporter of wheat, wheat flour, coarse grains, and soybeans, the United States both influences and is influenced by world markets and world events to an extent unprecedented since the United States' colonial years (Batie and Healy, 1980:13).

Just as the U.S. was proclaimed the breadbasket of Europe and Asia (e.g., Brown, 1975), a series of large Russian grain purchases and a serious inflation in consumer food prices heralded the ways in which America's food producers and consumers were sensitive, even vulnerable, to supply and price variability associated with international trade. The 1980s have already revealed that deeper participation in international trade can be associated with volatile trends in demand, supply and prices; trends which have potentially serious consequences for the economic viability of many agricultural producers, the ecologic sustainability of the agricultural resource base, the nutritional status of low income consumers, and the financial integrity of the public treasury. Such consequences are associated, in turn, with systematic and systemic changes in the way the American food system is organized, how vulnerabilities and risks are distributed in that system, and how policy institutions are pressured to protect and promote specific interests within that system.

As is argued elsewhere in this volume, the overall food security of the United States is not at risk in the short or medium term. However, the relationship of the American food system to international trade in agricultural commodities does suggest that food insecurity for specific groups and regions within (as well as outside) the United States may be a periodic near-term risk. That possibility is not based on an analysis of recent developments alone. Instead, it is a concern that has repeatedly surfaced in policy discussions about the American food system at least since the Civil War. In this century, the issue has arisen several times including after the First World War, during the depression, after the Second World War, in the early and mid-1970s, and in the early 1980s.

This chapter will examine the relationship of American agriculture to international agricultural trade. While there are other international dimensions of U.S. agriculture which can be considered in understanding American food security prospects, such as aid, investment, and transnational environmental change, the major emphasis will be placed on international trade, particularly American agricultural exports, and the relationship of that trade to structural change in the American food system. The major conceptions of the problems in that relationship, the typical policy responses that have evolved to meet those problems, and how the U.S. food system has reorganized to participate more broadly in international trade will be reviewed. The chapter will conclude with a number of questions about the future relationship of the American food system to international trade, how that relationship is understood within the policy system, and future implications for the stewardship of American food security.

INTERNATIONAL TRADE AND AMERICAN AGRICULTURE: CONCEPTIONS OF PROBLEMS

Three major themes have dominated policy and academic discussions of the relationship of international trade to the development and future of American agriculture. They are: (1) excess capacity, (2) instability, and (3) change in the structure of international agricultural commodity markets. This section will review each of these themes.

Excess Capacity

The idea that the problem of American agriculture is excess capacity has the longest and most continuous presence of any of the three themes. Domestically, excess capacity as too much production has been addressed by commodity-specific acreage control plans. These policies seek to reduce production and protect floor prices by reducing the area cultivated. In addition, there was, and is, a belief that American agriculture has too many people directly dependent on it for income and employment. Excess capacity, in other words, means having too many unproductive people in agriculture. The dilemma of low labor productivity can be resolved by increasing production and decreasing the number of people directly involved in the production process. Both strategies have been pursued through a combination of technology and economic policies that encourage what have been viewed as "natural trends" in farm size growth, mechanization, and out-migration from agriculture. However, while capitalization levels on American farms have increased, along with labor productivity, thereby "solving" the problem of excess capacity as too many people on the farm, increases in production have prolonged the problem of excess capacity as too much output.

Closely related to excess capacity as too much production is the idea that there is too little consumption. A solution to the problem of inadequate consumption is the international market. Hundreds of millions of people are potential consumers of American agriculture. The growth potential abroad significantly exceeds any growth that can be expected from the domestic market. In fact, significant expansion of domestic consumption has seldom been seen as a viable option for dealing with the excess capacity problem. Recent declines in domestic per-capita consumption of sugar and red meat support the belief that domestic food consumption relationships are inelastic. Consequently, foreign markets are viewed as an attractive solution. Two major exceptions to that view received serious policy attention, but did not weaken the basic belief that underconsumption was a foreign problem. One exception, the discussion surrounding the Brannan Plan in 1948, sought to maintain smaller average farm units through an incomes policy that included support prices to encourage production shifts to commodities which were underconsumed in the domestic market rather than production cutbacks for commodities that were over-produced. The plan was never adopted intact, although elements of the plan were incorporated into the farm policy portfolio. The second exception, the food stamp program, began as an effort to reduce storage costs for surplus production rather than to correct underconsumption. The program was linked to consumption and welfare objectives in the 1970s, and it developed a strong constituency among producers who viewed it as a valuable demand enhancement strategy. In recent years, the food stamp program appears to be returning to its roots, reducing storage costs (rather than curbing overproduction or correcting underconsumption).

During the mid and late 1970s, a belief developed that the excess capacity problem had been solved by the rapid expansion of foreign demand for American agricultural products. G. Edward Schuh, for example, concluded in 1976 that the emerging equilibrium in agricultural labor markets, the decline in the rate of productivity growth, the release of most idled agricultural land, and an increased demand for agricultural exports together meant "an almost total disappearance of the excess capacity that existed at prevailing price ratios for such a long period of time" (Schuh, 1976:805-6). In a widely circulated USDA volume, Lyle Schertz concluded that the expansion of agricultural exports in the 1970s provided an opportunity to realize "politically acceptable prices and farm income with only modest restraints on production" (Schertz, 1979:2). A Time to Choose, the Carter Administration's effort to stimulate debate on the structure of American agriculture, concluded: "Our agriculture is today at a crossroads. The time of chronic surpluses is behind; a time of growing demand and tighter supplies lies ahead" (USDA, 1981a:152).

By the early 1980s another variation on the excess capacity theme began to receive wider attention: should the export boom of the 1970s be seen as a solution to an excess production problem or was production accelerated precisely to meet expanded foreign demand? If the latter were the case, then some questions could be raised about whether the level of intensification associated with meeting the export boom was simply fuller utilization of American productive capacity or an over-utilization of that capacity, especially in environmental terms. Lauren Soth (1981:898) stated this concern succinctly: "The crux of the food-agriculture problem facing America is soil resource maintenance versus unrestrained grain exports." Others (c.f. Batie and Healy, 1983; Crosson and Brubaker, 1982; Hitzhusen, 1982; USDA, 1981b; Nelson, 1983; Winsberg, 1982) raised questions about erosion, mining the Ogalala aquifer, the destruction of shelterbelts, the demands on marginal agricultural land, and the growth of single-commodity specialization across large areas (increasing vulnerability to market fluctuations and plant disease). However, questioning whether the U.S. was exporting its own agricultural future was difficult when agriculture was a bright spot on an otherwise dim international trade balance sheet for a faltering U.S. economy.

An important point worth noting in a review of the excess capacity theme is the almost complete disassociation between a policy attempt to meet the problem through reducing cultivated areas and the successful efforts of the publicly-financed agricultural research system to provide technology which would increase operating-unit productivity. At the Executive and Congressional levels, excess capacity has generally meant surplus production or a strong tendency in that direction. It has been met by a combination of area reduction programs (to reduce production), price support programs (to reduce the economic losses specific commodity producers might suffer in years when production is high and prices are low), and export promotion programs (to increase foreign consumption of surplus American production and thereby protect the floor on agricultural commodity prices). At the same time, the public agricultural research system has tended to view excess capacity as underutilized resources, typically measured in returns to investment in farm enterprises and productivity per operating farm unit. To address this version of the excess capacity problem, both the public and private agricultural research systems have developed a range of mechanical, chemical and biological technologies which have generally increased labor productivity (essentially by a significant substitution of capital investments for labor investments), increased the "optimal" scale of an operating farm unit, and increased operating unit productivity.

In this context, it becomes clear that the excess capacity problem is not simply a relationship between domestic production of agricultural products and global demand for those same products. It is also a relationship between specific producer and consumer interests in acceptable price and supply levels and the capabilities of those same interests to mobilize various parts of the policy and research systems to support their specific interests (Perelman, 1977). The export boom of the 1970s did less to solve the excess capacity problem than it did to facilitate a reformulation of that theme in terms of the relationships between the short-run demands of the economic system and the long-run requirements of a biologically based production system. Whose interests are at risk in such relationships? Rapid developments in the biotechnology field suggest the inclination is stronger to believe that the relationship can be arbitrated through a continuation of past strategies (policy adjustments, technological intervention) than it is to reconsider the premises of a system that generates such choices.

Instability

Since the Russian grain purchases in 1973 and the ensuing food price inflation and export boom, the theme of instability has been widely discussed. The analysis underlying this theme is quite straightforward. While there are few food exporting countries, there are many food importing countries. There are more than 50 grain importers, for example, and not one of them purchases more than 10 percent of total international grain shipments. For the vast majority of importers, the specific amounts they will purchase from international markets in any year will be a function of domestic production (which typically will provide between 95 and 100 percent of the food consumed), availability and costs of foreign exchange needed to finance imports, and a range of domestic and international political considerations. Focusing on the first two, there is considerable year-to-year variability in the import requirements of specific countries as well as considerable variability in the financial feasibility of actually importing specific amounts of food. The agricultural supply system that provides commodities for international trade, however, cannot turn on and off that quickly. In the absence of significant reserve stocks, the result can be substantial gyrations in international prices.

By the 1970s, U.S. agriculture had come into closer balance with the world market in terms of the association between international and domestic prices for agricultural commodities. Moreover, during most of the 1970s, the value of the U.S. dollar depreciated in relationship to other major currencies, making the cost of importing American products cheaper for those currencies. This meant that year-to-year variation in international demand for American agricultural products could be substantial. For the U.S., producing at full capacity to meet that demand and not having any major commodity reserves or unused areas, international demand variability could really reverberate through the American food system. Cochrane summarizes the instability theme:

> Without doubt, world integration has resulted in an expansion in the demand for the basic agricultural commodities of the United States and so has contributed to a more prosperous farming community than otherwise would have been the case. But world integration...has opened the trade doors and exposed U.S. agriculture to the vicissitudes of the international market. Supply instability in the world market leads to--in fact generates--domestic price instability through the wide open trade doors (Cochrane, 1979:153).

However, two points should be made. First, instability or fear of instability did not begin in 1973. Concern about prospective postwar instability generated a vigorous international and domestic debate in the 1920s (Dawell and Jesness, 1934; Koerselman, 1971; Nourse, 1924; Schideler, 1957) and again beginning in the late 1940s (Matusow, 1967). That debate included discussion (and defeat) of a plan for internationally coordinated buffer stocks to reduce instability and ensure food security globally; discussion (and implementation) of an international agreement to establish mechanisms for stabilizing prices of wheat; and discussion (and partial implementation) of programs to reduce the exposure of American farmers to downward shifts in farm prices and the exposure of American consumers to upward flights of food prices. Much of the postwar concern was based on the experience of the 1921-23 farm depression in the U.S., a depression that was linked to the collapse of European (especially British) markets for American agricultural products.

Second, instability turns out to be really three different themes: variance, uncertainty, and vulnerability. The issue of instability as variance is seen by many as the issue for the 1980s, more so than simply the levels that prices will occupy. The basic question involved is: how can American consumers face tolerable prices and American farmers anticipate income stability when the American food system is integrated into world markets through major food commodities? Economic theory advises the policy system that the answer is to set domestic food prices at world levels. The problem, of course, is that setting prices at world levels increases the variability of prices that Americans will face. Another answer, proposed in a string of publications on the food security theme, has been a reserve. By maintaining a reserve, it is argued, the United States can modulate the amount of price variance within defined limits. This was essentially the policy followed through the 1960s. During the 1970s, the reserve diminished substantially, but with demand high, nobody seemed to notice.

In the early 1980s three major problems with the "reserve" strategy are receiving attention. First, the cost of maintaining reserves can be very high, particularly if a collapse in foreign demand is not matched (as it cannot really be) by an immediate proportionate reduction in production. Second, in addition to financial costs, there are political costs. Maintaining reserves was joined to the other instruments of American farm policy. Those instruments have significant political constituencies. Removing or lowering the scope of a reserve program is politically difficult. Third, some analysts have argued it is not reasonable to expect that the U.S. can, by itself, stabilize international agricultural commodity markets by trying to stabilize only its own domestic prices. Exactly the opposite happens--instability that would otherwise face domestic consumers and producers is exported (Josling, 1980). Ultimately, that instability is reflected in variability in international demand for American (and other exporters') agricultural products. The policy, in other words, is self-defeating. One response has been to argue for an _international_ reserve. Efforts to develop international buffer-stocks, however, have essentially gotten nowhere. There are small regional buffer stocks (e.g., in the Association of Southeast Asian Nations), but it appears unlikely that the many proposals made and debated by economists and others will ever be implemented.

Instability as uncertainty is closely related to the variability problem. The thrust of the uncertainty theme is that not knowing what to expect severely complicates the problem of resource allocation for producers and processors and substantially weakens the ability of the policy system to design and implement policies that anticipate conditions and consequences. In this light, it is not surprising that enthusiasm for buffer stocks is waning and commitment building for comparative advantage, free trade and forward contracting. That becomes an uncertainty issue as distinct from a variance issue when the focus turns to income levels. Proponents of international trade theory are arguing for an approach to reducing instability of prices, but at what average levels? Modulating price variability at levels that are not consistent with the economic viability of large numbers of producers introduces uncertainty across a broad swath of the American agricultural system, something that can be seen as U.S. agriculture has, in fact, become more open. Being export competitive, moreover, something that has become extremely important to U.S. agriculture, does not mean stability in terms of prices or incomes (Groenewegen and Cochrane, 1980). In a real sense, it means just the opposite, in large part because of variability in global production and demand.

When instability means uncertainty, the value of information and control are dramatically increased. Information which permits gaining some edge on price movements and price levels can make the difference between economic success and failure. Where does such information come from and who are in the best positions to appropriate that information? The executive director of the North American Export Grain Association provided insight in rather stark terms on both the nature of the problem and the outlines of the answer when he said: "Real merchandising ability...is to buy a bushel of grain and sell it at a price a little under the cost to the seller and still make a profit" (Harlow, 1980:896). As will be discussed later, instability as uncertainty has probably accelerated processes of "rationalization" within the agricultural production and marketing systems. These processes take the form of vertical contracting, production contracts, patterns of informal coordination, enterprise consolidation and the like

(Martinson, 1978; Martinson and Campbell, 1980). The effect of such processes is to reduce uncertainty in some cases and internalize its benefits in others. One effect, however, that is less apparent is any increase in the efficiency of the enterprises involved in such activities (Polopolus, 1982).

Instability as vulnerability has been discussed by both academics and those in policy circles largely in terms of the economic and environmental consequences of broader and more intimate participation in world markets. The U.S. is vulnerable, for example, to various forms of protectionism and non-tariff trade restrictions (e.g., quotas, consumption taxes). The U.S. is vulnerable to such actions insofar as its agricultural production system relies on markets in many countries where government efforts to protect local producers extend to various trade restrictions and import levies. That lesson was learned indelibly in the 1920s and 1930s and it is being learned again in skirmishing between the U.S., Japan, and the European Economic Community. American agricultural producers have also learned they are vulnerable to embargo actions taken by the U.S. itself. This subject will be considered further when the discussion turns to changes in the international market, but it is appropriate to note here that the grain embargoes of 1980 and 1982 contributed to depressing farm prices in the U.S. and weakened, in Soviet eyes, the reliability of the U.S. as a grain exporter. The USSR has successfully diversified its import sources and the United States currently holds a 19 percent share of a market in which as recently as 1978 it held a 78 percent share.

The vulnerability of American producers to embargoes initiated by their own government illustrates a much broader point. U.S. agriculture is tightly incorporated into the broader American economy. While agriculture still has political influence that is disproportionate to the numbers of people directly engaged in agriculture, the nature of its political influence has changed and the types of trade-offs and issue arenas it confronts have altered. Today, agriculture is represented more strongly in lobbying circles by specific commodity interests rather than by the farm groups of several decades ago. Similarly, as many people have noted, contemporary U.S. farm policy cannot be clearly and consistently distinguished from foreign policy, economic development policy, and welfare policy (c.f. Destler, 1980; Lyons and Taylor, 1981; Paarlberg, 1980; Schuh, 1981). Given the broader internationalization of the American economy, it follows that general economic policies (e.g., monetary and fiscal policies, exchange rate regulations, tax policies, and broader trade policies) are often more important to agricultural welfare than agricultural or commodity policies as such.

The broader point is that trade in agriculture is supported, constrained, promoted, and hindered more by factors outside agriculture than within it. Incorporation of agriculture into a broader industrial economy means that international trade policy has acquired a closer relationship to internal trade policy (Paarlberg, 1980:247). Similarly, growing interdependence between financial markets and commodity markets, growing internationalization of capital markets, and growing integration of agricultural labor markets with nonagricultural labor markets all imply that government has less ability in the 1980s than it did in the 1950s to autonomously manage the American agricultural economy (Gardner, 1981a). It is against this background that Senator Percy's defense of agricultural exports, as a necessary part of financing American energy imports, needs to be seen as an indicator of the vulnerability side of instability:

> If we are to continue paying those [oil import] bills, if we are to continue to keep the dollar strong, if we are to keep a surplus in our balance of payments, as we must have over a period of time, it's going to depend upon agriculture more than any other single thing, and certainly anything that impedes or looks as though it's going to stand in the way of our exporting agriculture to the world is a matter of deep concern (U.S. Senate, 1982:2)

There is another international aspect of instability as vulnerability which cannot be discussed at length here (in part, because it is not a question related to trade), but one that requires mention as a window on the broader issue of international sources of vulnerability in the American food and agriculture system: the international sources of climate change. During the last 30 years, U.S. agriculture has accelerated and deepened its commitment to a form of production that is dependent on maximum utilization of natural resource endowments. Limitations in those endowments are overcome through strategies of intensive single-commodity cultivation, large-scale irrigation, heavy fertilization, deep plowing, hybridization, and plant breeding to specific agroclimatic niches. What would happen to this system if annual average temperatures were to rise by 3°C, precipitation belts were to shift, and carbon dioxide in the atmosphere were to double? Nobody really knows the answer to such questions, in part because regional predictions from global climate models are not yet reliable. However, there are certain points worth considering:

1. If the growth rate in fossil fuel consumption in the third world is 6 percent/year and the growth rate in developed countries is only one

percent, there is general agreement that the global concentration of carbon dioxide in the atmosphere would double in the first half of the 21st century. When that happens, there is also wide agreement that global average temperatures will increase by 3°C $\pm$ 1.5°C with higher increases in more northerly latitudes. That could mean a global shift of precipitation belts and possibly a significant intrusion of the Sonoran Desert into Midwestern wheat and corn areas (Bernard, 1980; Kellogg and Schware, 1982; National Research Council, 1979; 1982). If the temperature rises, precipitation falls, and evapotranspiration increases, the most desirable climate for corn and soybeans will be more than 200 kilometers north and east within the lake states region (Tucker, 1982). An increase in carbon dioxide can benefit plant growth (Kellogg and Schware, 1981), but it can also alter the balance that allows crops to dominate weedy species. Many noxious perennials that use the C_3 pathway for photosynthesis will become more competitive with crops such as corn which use the C_4 pathway (Newman, 1982)(1).

2. Any evaluation of the consequences of an increased carbon dioxide load in the atmosphere needs to consider the interactive effects of carbon dioxide with other gases and pollutants that may have high local concentrations. For example, one group has explored the interactions of carbon dioxide growth rates with growth rates in atmospheric chloroflurocarbons and nitrous oxide. They estimate, under realistic assumptions of growth rates, that American corn production would be vulnerable to production decreases exceeding 40 percent of 1980 levels (Kelejian and Vavrichek, 1982:224-268).
3. The last 50 years, during which dramatic increases in production have been made possible in part by breeding and management strategies designed to optimize average <u>weather</u> conditions, have been an abnormally benign period in terms of global <u>climate</u> (Schneider, 1976). How serious a vulnerability problem does that imply? At a conference on "Climate Change, Food Production, and Interstate Conflict" sponsored by the Rockefeller Foundation in 1975, participants concluded that weather had reemerged as a major destabilizer. F. Kenneth Hare concluded that "climatologists and plant geneticists have a clear obligation to try to cope with the root causes: no one else is competent to do so" (Rockefeller

> Foundation, 1976:ix). Dean Haynes points out that many plants are already tailored to grow at the limits of their range. In other words, climatic stress is a problem even now. The job of "redesigning" the technological basis of agriculture is not simple: "It would be overly simplistic to think that crop species could simply be moved in response to temperature changes. Massive plant breeding programs would be required to 're-engineer' the environmental triggers of the crop cultivars and then complement them with new resistance to pests or other management strategies" (Haynes, 1982:26).

All of this illustrates that instability as vulnerability can have an international origin the impact of which will be conditioned by the type of agricultural system the U.S. has. If the U.S. intensifies to compete internationally, how much more vulnerable will its agriculture be to what appear to be inevitable changes in the earth's climate?

The focus on instability as a theme in understanding the relationship of international trade to American food security represents a more global perspective on the factors influencing American agriculture. However, policy discussions within this theme have avoided deeper questioning of how the organization of agriculture in the U.S. itself exacerbates the instability. The policy system also has generally avoided confronting a basic structural issue on the relationship between price stability and income stability: The two are not directly related because the latter, income stability, is very sensitive to the real terms of trade between agriculture and the rest of the economy (Grennes, Johnson and Thursby, 1978).

Change in the Structure of International Agricultural Commodity Markets

While international markets have long been important for American agricultural development, export growth in the 1970s represented such a sharp increase from the pattern of the prior 20 years that it is appropriate to look for major changes in the international market and ask whether such changes are durable and what consequences will follow. Several such changes can be identified and together they lead to a more fundamental question: Is the American consumer now in some kind of confrontation or competition with foreign consumers for the products of American agriculture? This section will review some of the major changes in the international market for agricultural commodities and the discusison that has developed on whether American consumers are in competition with foreign consumers.

In the 1950s, less developed countries and the centrally planned economies of Eastern Europe and the Soviet Union accounted for 18 percent of world agricultural trade. By 1976, that percentage was up to 28 percent. By 1983, it was close to 43 percent. Sixty percent of the increase in global grain trade in the 1970s is increased demand from the centrally planned economies. Especially important in this context was the entry of the Soviet Union into international agricultural markets in the 1970s as a major grain importer.

The volume of agricultural trade globally in the 1970s grew faster than agricultural production (5.5 percent/year versus 3.5 percent/year). Since the 1950s, the share of global grain production which is destined for livestock consumption has doubled to 40 percent. This is reflected in a change in the composition of agricultural trade. Prior to 1945, agricultural trade was dominated by high quality wheat destined for human consumption. Since 1945, high quality wheat has been replaced by coarse grains and oilseeds destined for livestock consumption as the dominant components of international agricultural trade. The increased role of livestock consumption in international grain trade means, for example, that growth in demand from less developed countries is really growth in demand from what are today called the newly industrialized countries such as Korea, Taiwan, and Brazil--the wealthier of the so-called developing countries.

During the early 1960s, concessional exports constituted 90 percent of the American agricultural products moving abroad. By the late 1970s, that proportion had declined to 16 percent. During the 1982-83 period, it declined to 3 percent with the vast majority going to one country, Egypt. What all this represents is a major growth in commercial sales and a policy decision by the United States to seek and support growth in exports through commercial rather than concessional channels. Here again, what is being reflected is the expanded significance of customers able to pay--Japan, the European Economic Community, the Soviet Union, Canada, and the newly industrialized nations of Asia and Latin America--and the diminishing commitment by the U.S. to support concessional exports to all but a handful of other countries.

During the period 1947-1968, international wheat trade was more or less governed by an international wheat agreement which restricted dumping, tariffs and nontariff restrictions, and sought to keep wheat prices moving within a narrow band through government (particularly Canadian and American) measures to maintain a price floor and, via reserves, to maintain price ceilings. In 1968, the agreement broke down and efforts to renegotiate were unsuccessful. Subsequently, the international wheat market became more volatile.

Exporters moved to reduce stocks through restrictions on production. World acreage planted to wheat dropped 8 percent between 1968/69 and 1970/71 and global wheat reserves dropped 35 percent. Wheat prices moved in a much wider band.

Thus, by the early 1970s, the U.S. had abandoned its large grain reserves. Those reserves had served as a global buffer stock in many ways and had helped to keep grain prices down. The shortages which developed in the early 1970s and the price escalations and supply problems that followed were exacerbated, in some views, by the failure of American policy to maintain a larger reserve (Bosworth and Lawrence, 1982). American consumers paid a price for that. Another view, however, is that consumer food price inflation was not simply a result of tight supplies in the face of Russian demand and unfavorable growing conditions in many parts of the world. Farm prices were declining when consumer prices took off. The supply was not as tight as control over disposition of the supply.

Finally, two trends that accelerated in the late 1970s and into the 1980s were the increase in nontariff barriers, particularly by Japan and the European Economic Community, and the use of agricultural export embargoes by the United States as an instrument of foreign policy. Both created additional downward pressures on producer prices for American farmers. The growth of nontariff barriers increased tension between the U.S. and those countries raising such barriers. The U.S. government found itself drawn into a more active role in combatting such measures (Lewis, 1983). The grain embargoes of 1980 and 1982 had little impact on the Soviet Union, but complicated the relationship between the U.S. government, agricultural producers, and food consumers in the United States.

Taken together all these changes implied a larger, more commercial, more competitive, more variable, and potentially, a more lucrative international market. In the face of such changes, the questions began to surface: Is the American consumer on a confrontation course with foreign consumers? In a shortage, would the American consumer have "first" access to the products of American farmers? In an international political crisis, would the economic welfare of American farmers be sacrificed to foreign producers as a price for the implementation of certain U.S. foreign policy strategies?

Some have argued that a consumption confrontation is possible, indeed likely, if a tight supply situation should arise (Breimyer, 1982). The argument is made that the U.S. can best establish its reliability as a source of agricultural goods when supplies are short. Moreover, with other grain exporters such as Canada, Argentina, France, and Thailand becoming more effective competitors, the risks the U.S. would court by only being a fair-weather exporter are substantial. Two confrontations of this sort are possible. One is between domestic and foreign consumers. A second is between domestic and foreign livestock (Raup, 1980). These views are reinforced by a wide consensus that projects expanding and large markets for U.S. agriculture in Japan, the Soviet Union, many developing countries, and even the European Economic Community. In the light of such projections, there is thought to be little question that foreign demand for U.S. agricultural products is secure. That leaves open what would happen if global production veered below trend.

Another view, however, can be presented. This view does not deny the possibility of a domestic-foreign consumption confrontation, but the probability is treated as contingent on a less certain assumption: Is foreign demand as secure and expansive as advertised? Several points can be made which together suggest that the boom of the '70s may not necessarily be the initiation of a long era.

First, food needs and the ability to finance those needs are two quite distinct subjects (Burki and Goering, 1977; Sarris and Schmitz, 1981). As events in the early 1980s have amply demonstrated, appreciation of the American dollar, mounting debt problems for many countries, and a sluggish global economy are all quite sufficient to dramatically reduce effective demand for American agricultural exports.

Second, it is important to recognize that the U.S. share of global agricultural trade has declined since the mid-'70s while the shares of some major competitors, particularly the European Economic Community, have increased. Thus, during the period 1976-1981, U.S. agricultural exports increased 70 percent to $42 billion. During that same period, the value of farm exports from the EEC increased 156 percent to $27 billion. The EEC's share of global trade increased for milk, butter, and sugar. Australia, Canada, Argentina, Thailand, and Brazil have increased their share of internationally traded grain. The EEC currently represents a $6-7 billion agricultural trade surplus for the U.S. Prospective initiatives under the Community's common agricultural policy which would impact imports of soybean oil, corn gluten, and citrus pellets, and the probable effects of Spain's entry into the EEC will combine to reduce that surplus. Similarly, the U.S. had a $1 billion agricultural trade surplus with Canada in the late 1970s. By 1983, the surplus was essentially wiped out.

Japan is American agriculture's best single customer and there is good reason to believe that situation will remain. But is there equal reason to believe (e.g., Sanderson, 1978; 1982) that Japan's reliance on American agriculture will increase in the future at rates comparable to those in the 1970s when the overall value of Japanese agricultural imports increased sixfold? Japan's food system has changed from an earlier dependence on staples to increased reliance on high value livestock. Consequently, the 1970s saw a rapid increase in the import of feed grains and soybeans. In addition livestock imports now account for almost one-fourth of all agricultural imports. Sixty percent of those agricultural imports are from the United States, Canada, and Australia. Looking forward, two factors have to be weighed. First, will Japan's import requirements continue to grow rapidly or will a slower growth pattern develop? Second, will the sources of those imports remain as they have been, or will diversification occur that will have the effect of significantly reducing growth in real demand for American exports?

The answer to the first question would appear to be that a slowing of Japan's growth rate in imports seems inevitable. Some evidence of that can already be seen in surpluses of some products, including some livestock products. The same trajectory that Japan has followed in the last 20 years may lead to a conversion of Japan's rice fields to other crops. Finally, and perhaps least understood by American writers, is the salience of the self-sufficiency issue in Japan. Unlike the U.S., there is considerable political support for Japan's farmers by Japan's consumers, the economic inefficiencies of that support notwithstanding (Houck, 1982; Yamada, 1982; Yoshioka, 1982). The answer to the second question may be that Japan's rate of growth specifically for American agricultural exports may decrease. There has been evidence of that since late 1981 (USDA, 1982b; 1983a; 1983b). One consideration is the degree to which the rapid expansion in Japanese demand for U.S. agricultural exports was facilitated by the depreciation of the dollar relative to the yen through the 1970s. Similarly, as the dollar has appreciated relative to the yen since 1979, the rate of growth in demand for U.S. agricultural exports has slowed down.

Two points are illustrated. First, U.S. agriculture's fate is closely linked to factors outside of agriculture, such as international capital markets, monetary policies, and exchange rates. Second, Japan's interests in self-sufficiency do not have to be understood as independence from imports, but rather as insulation from excessive reliance on any single source. The so-called soybean embargo really embargoed very few soybeans, but Japan's reaction to the possibility of an effective embargo was sharp and long-term: investment in Brazil as a soybean source. As will be discussed shortly, Japan has an increasing investment in the marketing of American grain. In effect, they are storing their grain in U.S. elevators. That means that the U.S. producer will continue to be a significant trading partner. But Japan is doing with food what it is doing with another vital resource, oil. It is diversifying its sources and encouraging, through direct investment and the prospect of a Japanese market, expansion of production in diverse locations. It may well be the case that the U.S. is at least (if not more) dependent on Japan as a market than Japan is on the U.S. as a source. An economic slowdown in Japan, complications in U.S.-Japan overall trade relationships, and a number of other quite plausible developments could significantly modify the high optimism that many in the U.S. hold for the future of the Japanese market. In that case, the confrontation may not be between American and Japanese consumers, but rather between American producers and producers in Canada, Australia, Brazil, Thailand, China, Indonesia, the Philippines, Korea, etc.

The Soviet Union is a second example of a relatively recent market that has acquired, in many quarters, the status of a long-term, ever-expanding destination for American agricultural products. Since the famous grain purchases in 1973, Russian-American grain trade has gone through (now) two agreements, two embargoes and a significant reduction in America's role as the supplier of Soviet agricultural import needs. The 1983 grain agreement includes a commitment by the U.S. not to engage in embargo actions except in the event of a national emergency or war and represents a serious effort by the U.S. to recover what is perceived to be a very significant market. Between 1983 and 1988, the USSR is committed to purchase at least 9 million metric tons of grain per year from the U.S. They will need to do that and more as part of their efforts to rebuild depleted domestic feed reserves and strengthen their livestock herds.

In 1978, the Soviet grain harvest was 237 million tons. Since then, it has been problematic and has hovered around 200 million tons. Such figures are approximately 15 percent below where they wish them to be, resulting in substantial imports. However, investment in Soviet agriculture and reconsideration of management strategies is proceeding at a high and continual pace. Soviet agriculture is seriously inefficient, but that only means that there is considerable room for improvement. A simple example illustrates the point. To get 100 pounds of liveweight pork, Japan and West Germany need 350-400 pounds of corn. The USSR needs 700 pounds and additional protein feed. With Western feed conversion ratios, Soviet grain imports could be halved (Wadeken, 1982). The Soviet Union will be an important customer for American agricultural exports for many years. That demand will be quite high in some years given the especially difficult conditions Soviet agriculture faces. However, such a situation is not likely to continue indefinitely (Csaki, 1982).

The experience of the two embargoes was that U.S. producers suffered as well as U.S. taxpayers. The

Russians simply bought their grain elsewhere. Canada and Argentina are now significant suppliers and the European Economic Community is becoming one. Secretary of Agriculture John R. Block recently stated that "when the Soviet Union buys grain from us, they transfer some of their resources here, and it's that much of the resources that aren't available for military hardware" (New York Times, August 22, 1983). Here again, who is more dependent on whom? The new Soviet-American grain agreement, unlike the first agreement, does not contain any short-supply provision. In the first agreement, the U.S. claimed the prerogative to ship less than the agreed minimum (6 million tons) in the event of a shortage (Farnsworth, 1983). With 1983 American corn yield declines of more than 30 percent likely because of drought conditions, what will be the broader consequences if the U.S. is to meet Soviet demand for several million tons? Is the plausible confrontation between livestock consumers in New York and Moscow or is it between coarse grain producers in Illinois and Thailand?

Similar arguments can be made for the hungry nations of the Third World. At the same time that the prices American farmers were receiving for wheat were declining to record low levels, the real costs of purchasing that wheat among the 23 developing countries that import 78 percent of America's wheat exports were increasing. This is a result of the very significant strengthening of the U.S. dollar and the very signficant weakening of the economies of many of those importing countries. In contrast, 75 percent of American soybean exports go to developed countries where the real price of buying those soybeans has increased substantially less than has been the case for wheat.

Moreover, it is important to understand what the role of such imports often is. For example, China has, since the mid-1970s, imported 2-3 percent of the amount of grain it produces. But a major reason China does this has been to feed its large cities, thus relieving strains on internal transport (Barnett, 1982). In the future there may be other reasons for increased demand for agricultural imports, but it is important to recognize that imports in the Third World are anything but the simple filling-in of production shortfalls, as many projections assume.

The theme that the international market changed in the 1970s was an understandable reaction to the export boom that the period witnessed. The same theme was only moving to logical conclusions (and perhaps succumbing to popular interpretations of the 1973-1975 food price inflation in the U.S.) by hinting that domestic-foreign consumer confrontations were in the offing. The major question raised here has been whether the 1970s have been over-interpreted or, to put it differently, whether redefining the 1970s as "normal" has skewed expectations for the 1980s and 1990s. Some investment economists are predicting the annual growth rate for U.S. agricultural exports in the 1980s will be 2 percent, down considerably from the 8 percent annual rate of the 1970s (Wall Street Journal, February 1, 1982). Developments in international agricultural markets, especially since 1981, can be over-interpreted also. That just illustrates the uncertainties associated with linking the American agricultural system to significant levels of participation in international markets.

The Themes: An Epilogue

Three major themes in conceptions of the relationship of U.S. agriculture and international trade have been reviewed: excess capacity, instability, and changed international markets. It is important to reemphasize that all three themes have long histories. At any given time during the last several decades, one of the themes has dominated academic or policy discussions. However, all three themes have been developed and elaborated as the U.S. grapples with understanding what international markets mean for its agriculture; what, if anything, needs to be done to ensure that the opportunities presented by international markets are realized and the pitfalls avoided. The dramatic decline in net farm income in the early 1980s and increasing awareness that a soil erosion problem of substantial magnitude is building, illustrate what these themes have tended to downplay: that U.S. agriculture is changing in ways which make it both more capable of participating in international agricultural trade as an effective competitor *and* more exposed to the many pitfalls that such participation entails. Essentially, the same group of policy instruments and the same vocabulary for policy discussion have been in place for 50 years. They have their constituencies. The discussion of excess capacity, instability, and the changing international market continues.

THE STATE AND THE MIDDLE SECTOR: SOME ASPECTS OF HOW THE U.S. IS ORGANIZED TO PARTICIPATE IN INTERNATIONAL TRADE

Although agricultural exports from the U.S. move through private channels, the government has for many years been an active participant in establishing the conditions under which that business functions as well as promoting the trade itself. Here it is important to recognize that such involvement predates the New Deal. The Bureau of Foreign and Domestic Commerce, for example, was established in 1921 in the Department of Commerce to engage in agricultural promotion activities that were the forerunner of those currently performed by the Foreign Agricultural Service (FAS) of USDA (McKinna, 1978).

Today, under the rubric of the FAS, a large number of domestic and foreign groups promote the sale of branded U.S. products in foreign markets. Activities include efforts to promote changes in taste compatible with the products the U.S. is trying to sell. A second area of government involvement in trade promotion also has a long pedigree--government financing or subsidization of foreign purchases of U.S. grain, from the War Finance Corporation of the post World War I period to PL480 to the current blended credit and export credit guarantee programs. The blended credit program, established in 1982, provides $1.5 billion of interest-free export credits to foreign buyers and blends that with commercial bank credits to achieve a lower interest rate. In 1981, an Export Credit Guarantee Program (GSM-102) replaced the Noncommercial Risk Assurance Program (GSM-101). GSM-102 enables foreign buyers to get commercial credit at American banks for terms up to three years by covering payment defaults due to commercial or noncommercial reasons. Under the older Commodity Credit Corporation Export Credit Sales Program (GSM-5), the government provided credit at commercial bank terms. The major destinations thus far for GSM-102 assistance have been Korea, Portugal, Brazil, and the Dominican Republic (USDA, 1983b). These programs need to be seen in part as the policy system's approach to competing with the European Economic Community for foreign markets and in part as the way in which the public and private sectors share the risks involved in financing export expansion.

What parts of the private sector are best organized to manage participation in and extract benefits from both international agricultural trade generally and the dispensations of government policy that relate to international agricultural trade specifically? As Dan Morgan's _Merchants of Grain_ (1979) suggests, the logistics, financing, contracting, and information aspects of international grain trading put transnational private grain traders in an especially good position. What about agricultural producers? American agricultural producers have organized themselves into cooperatives for a number of marketing purposes. The Dairy cooperatives are relatively well known features of the domestic landscape. However, cooperative organization for foreign marketing purposes traditionally has been an important feature as well. The California Rice Cooperative, for example, was organized for this purpose in the early 1920s under stimulation of the Bureau of Foreign and Domestic Commerce. Producer cooperatives are represented on the Agricultural Cooperators Council of the Foreign Agricultural Service (Foreign Agriculture, 1982a;

1982b; 1982c). The Cooperator Program of the FAS is an offshoot of PL480 (Title I), except that in recent years, American dollars are used rather than foreign currency proceeds of PL480 sales alone. The Agricultural Cooperators Council can be viewed as an interest group, pressing the interests of commodity groups to Congress and the USDA (Talbot and Kihl, 1982). However, what about direct involvement by producer cooperatives in agricultural exports?

Sixty-three agricultural cooperatives had direct exports worth $3.2 billion in calendar year 1980. Grains constituted the largest dollar volume followed by oilseeds, cotton, and fruit. About half of the shipments went to Asia (principally to Japan), a quarter to Europe (principally the EEC), and most of the rest to Canada and Latin America. In 1976, 73 cooperatives engaged in direct export activities with a total dollar volume of $2.3 billion. However, as shares of overall American agricultural exports, the role of agricultural cooperatives declined from 9.2 percent in 1976 to 7.8 percent in 1980. For rice, fruits, nuts, and cotton the cooperatives have maintained at least a 20 percent share. But for grains other than rice and for oils, the two major categories of U.S. agricultural exports, the share of direct exports has remained under 8 percent.

The agricultural cooperatives also engage in <u>indirect</u> export sales. That means that transactions are made through American export agents to American export merchants, international grain trading companies, and foreign government purchasing agents in the U.S. Indirect sales in 1980 totaled $3.1 billion. They exceeded dollar volumes of direct sales for grains and oilseeds. Between 1976 and 1980, the percentage of direct sales declined from 61 to 50 percent. Together, direct and indirect sales accounted for about 15 percent of total U.S. agricultural exports in 1980. However, an examination of the size distribution of direct export cooperatives based on the dollar volume of such activities indicates that sales are highly concentrated. Thirty-five of the 63 active cooperatives in 1980 account for 2 percent of the total dollar volume. Ten of the 63 account for 82 percent of the total dollar volume (USDA, 1982a).

What all this means is that agricultural producers in the U.S. are not, for the most part, closely involved in the direct or indirect marketing of their products in foreign markets. More commonly, they sell their products at prevailing market prices. The gains and losses from international trade and the forward contracting it involves are primarily in the hands of a middle sector--the traders, transporters, and processors. Input, finance, service, processing, transport, distribution, and storage have become the dominant components of the American food system generally. Increases in food marketing and distribution costs contributed substantially to post-1972 inflation in consumer food costs. Productivity in this sector, to the extent it can be measured, has remained low and grown slower than productivity in primary production. This characterization, once considered "radical" (cf. Hightower, 1975) has now become conventional wisdom (cf. Polopolus, 1982). Where debate remains is on the competitiveness of the middle sector. That issue has special relevance for international trade since the absence of competitive pressures might suggest that inordinate risks associated with trade could be shifted to producers and consumers. Several points can be noted.

During the 1970s, commercial storage facilities and the covered barge fleet both expanded significantly. The railroads, traditionally an important feature of grain marketing (52 percent of all grain is handled by railroad and 58 percent of all grain sold moves via a grain elevator served by a single railroad), expanded their hopper car fleets and reorganized contracting, requiring the leasing of a minimum of 48 cars. More recently, the Staggers Act is likely to foster consolidation of railroads and reduce interline transfers via railroad gateways--with effects on rates and farm prices that remain to be seen (USDA, 1983c; Fornari, 1982; O'Donnell, 1982). Export merchants, international grain trading firms, and agricultural processors are actively involved in the financing, management, and ownership of the various facilities through which grain passes from field to ship. Some writers are concerned that the middle sector is expanding its use of vertical contracting and even moving into the purchase of agricultural land (Martinson and Campbell, 1980; Macleans, 1981). However, an equally significant development, changing patterns of ownership of the infrastructure without which marketing, especially international marketing, is impossible, has received less attention.

For example, approximately 20 percent of American grain which is exported is exported under control of Japanese firms. Companies such as Mitsui and Mitsubishi are buying and building grain elevators and selling grain internationally. For now, Japanese affiliated grain merchants in the U.S. tend to pay farmers a bit more via basis improvement (the difference between cash bid and the Chicago Futures Market) (Successful Farming, January, 1982). However, as suggested earlier, this should probably be viewed as a mixed blessing, the costs of which to American consumers may materialize in the event of a grain shortage.

Between 1974/75 and 1980/81 the market shares of the 5 largest grain exporters declined 5.3 percent; and the shares of Japanese owned or affiliated firms increased 4.7 percent. A recent study of competitiveness in the grain trade industry concludes the industry is contestable and that concentration may be less than generally thought (top 4 or 5 do not control 80-90 percent of the grain trade), although it is certainly higher than levels of concentration for domestic grain trade. The conclusion drawn is that

> grain-exporting firms are not vertically integrated in the conventional sense: their operations do not require that grain pass in sequence from controlled county elevators to or through their terminal or export facilities (Caves and Pugel, 1982).

That conclusion depends heavily on evidence that leading export firms hold smaller shares of inland terminal elevators than of export elevators (18 versus 53 percent of capacity) with no data available on county elevators. The problem with that conclusion is that it is not necessary for export firms to own substantial numbers of county or even inland terminals in order to protect their market positions. Who else are the railroads going to ship to? It is also not necessary to own county elevators where a capacity exists to operate actively on the Chicago Futures Market, a lesson that is being well illustrated by Japan's Public Food Agency as well as by the five biggest international grain traders. The significance of the Staggers Act and of similar legislation currently circulating which would require more ocean cargoes to use American ships and which would introduce user fees for ports and inland waterways is that it reduces uncertainty on the domestic side associated with assembling commodities for export.

In that context, the American producer can really do little more than produce grain and attempt to be smart about when to sell that grain. That raises the issue of the role of government-sponsored commodity storage programs. Several USDA studies, the most recent in 1982, sound the same theme: farmers who store grain in the reserve program tend to be large, well-equipped, and specialized. They use it and other related government programs for both risk management and income support. While smaller producers are the majority of participants, a small minority of participants constitute the majority of enrolled acreage and production. The ratio of the share of benefits to the share of production increases with farm size (Johnson and Clayton, 1982). Although at this writing it is too soon to tell, it is unlikely that the government's Payment-in-Kind (PIK) program will work any differently. Larger farmers seem better able to shift a significant portion of their production risk via use of the commodity programs. However, that is not because larger farmers are smarter but rather because it is for the larger farmers that those programs operate (Gardner, 1981b).

There is more involved than the commodity programs

alone. Some observers now argue, in fact, that farm income is primarily a result of returns from tax planning and real estate investment, not just agricultural production. Depreciation allowances permit income exclusion. Because marginal tax rates are regressive, higher-income purchasers of land have access to greater proportional income retention. Strickland (1982) argues that an after-tax farm income series is needed. The problem for such a series is quantifying the current income effects of real estate investments. National policies directed toward taxation, inflation and other macroeconomic concerns significantly influence the level and distribution of returns to agricultural resources. That process can be seen in the domestic farm income problems that have troubled American agriculture in the early 1980s. What can also be seen is that it is the same farmers who expanded through leveraging the appreciated values of their assets who, in the ensuing and now unfolding collapse of land market prices, can best utilize the commodity programs to cover that downside risk. As Johnson and Clayton conclude: "In this way, the commodity programs operate jointly with other national policies to promote increasing scale and concentration with the farm sector, each counteracting the adverse effects of the other" (Johnson and Clayton, 1982:956).

Macroeconomic policies generally and government commodity programs in particular serve the explicit purpose of helping farmers overcome some of the risks inherent in agricultural marketing. At the same time, those programs assist the middle sector of grain traders by making it unnecessary for them to bid grain prices up to ensure adequate supplies--the commodity programs ensure that. Similarly, it is important to note that the Federal Land Banks have become the leading lenders for farm real estate financing, displacing individual sellers (until 1977 the largest single source), insurance companies, and private banks. In effect, the public treasury is backstopping the risks of real estate leveraged expansion, relieving the private financial system of a good portion of that same risk (USDA, 1982c; also Robinson and Leatham, 1978). Consider the background against which this change has unfolded. In 1979, 63 percent of all farmland purchases were for farm enlargement. In 1954, that figure was 29 percent. In 1954, 44 percent of all farmland sales required financing. By 1979, that proportion had grown to 90 percent (Healy and Short, 1981:35-97).

Rapid expansion in export demand in the 1970s has not dramatically changed the middle sector that actually conducts trade for the U.S. The basic structure, both on the private side and the public side, was already in place. What has happened is that the distinctions between how agriculture is organized to participate in international trade and how it is organized to participate in domestic marketing have become less significant. The large supermarkets are important as fruit importers. The large processors are important as oilseed, coarse grain, and wheat exporters. From one perspective this means that variability emanating from international markets can resonate more rapidly through the agricultural system. That can be seen in the farm crisis of the early 1980s. But it had been seen before in the early 1970s and numerous times before that. What is different now is the velocity of that resonance, not its existence. For similar reasons, macroeconomic forces emanating from _outside_ agriculture entirely can resonate through the agricultural trading system very rapidly as well. In all this, a well-developed set of public policies are merged closely with the needs of various parts of the private sector to be insulated from risk and instability.

Ultimately, food security is a public interest question. And from that perspective it is appropriate to ask whether the complex linkages between the state, agriculture's middle sector and larger producers, and between agriculture generally and the rest of the economy can actually identify, protect, and arbitrate a public interest question such as food security. Where in the system is such a public interest institutionalized? That is a very complex question, but it is worthwhile to note that with the possible exceptions of the consumer outrage of 1973-75 and the structure of agriculture debate in 1979-80, the question has received no central attention in policy discussions since the rediscovery of international markets began in the early 1970s.

CONCLUSION AND QUANDARIES: SOME QUESTIONS THAT NEED ANSWERING

Agriculture's history and development in the United States are closely wedded to international markets and the relationship of agriculture to other parts of the American economy. In the twentieth century, a variety of policy instruments and program strategies have evolved in the public sector for managing agriculture and confronting certain problems faced by and associated with agriculture. Yet, today, the agricultural system seems more fragile in social, economic, and ecologic terms than ever. That does not mean that collapse is imminent or even eventual. What it does mean is that one is drawn to some elementary questions about the policy management of U.S. agriculture, and whether that management can identify and pursue a public interest issue such as food security in the light of the broader context in which agriculture is located and which this chapter has reviewed.

1. _Why the same old answers_? American agriculture

is in a continuing crisis of structural change in which family-operated farms continue to vanish, economic viability for producers and many parts of the financial, equipment, and supply system is difficult and episodic, and for which international markets have been and remain a mixed blessing. Why does this situation persist? Why does the policy system continue to be befuddled and uninnovative, limiting itself to reactive, often counterproductive, and frequently perverse forms of intervention and neglect? Peterson concludes a study of agricultural trade policy in the 1950s by noting that the agricultural policy of

> the Eisenhower administration, despite its goal of prosperity and balance through trade expansion and free market revival, remained one characterized by dumping, state trading, anticommunist embargoes, and shortsighted pursuit of commodity sales (Peterson, 1979:151).

Is it any different 25 years later? Will it be any different 25 years from now? With a few exceptions such as the debate surrounding the Brannan plan (Christenson, 1959; Matusow, 1967) and the short life of the Farm Security Administration, where is evidence of a serious policy discussion that suggests a fresh vision? Why is this the case and why is it likely to continue?

2. Does the left hand know what the right hand is doing? At the very same time that the policy system has attempted to balance the attractions of international markets and the pitfalls of overproduction through a collection of acreage control, price support and reserve policies, the publicly supported land-grant system turned out a continuing flow of mechanical, chemical, and biological technologies that contributed to increasing production and dislocating smaller farm operators and agricultural labor. Why did these contradictory strategies co-exist and draw sustenance ultimately from the same public treasury? If, as would seem to be the case, the broader agricultural policy system was uncoordinated in its response to problems associated with international trade, how can that same system pursue with any consistency a broad public interest issue such as food security?

3. Whose interests are promoted by the middle sector? In a recent interview, Representative James Jeffords of Vermont questioned whether it was even appropriate to speak of a crisis in farm-support policy:

> The people in the United States spend less on food as a percentage of their net disposable income than just about any other country in the world. So we'd better first of all find out whether we really have a problem. I'm not so sure we do (New York Times, August 7, 1983).

Despite the overall affluence of the United States, it would be at least as appropriate to ask whether American consumers pay more than they should. Some observers believe they do and they point to the inefficient and often uncompetitive food processing and distribution system. Why does the loose and sometimes inverse relationship between producer and consumer prices continue? The food price inflation of the mid-1970s generated a consumer backlash that led some to believe that consumers had captured the farm agenda (Paarlberg, 1980). Today, that backlash continues with regard to food safety and food quality, but in a period when farm prices are falling and food prices are rising, why is there so little concern? For example, in 1982 consumer food prices rose by 4 percent, the smallest rise in six years. In 1982 farm prices declined by 4 percent. The middle sector has captured the farm policy agenda, a middle sector that is the leading edge of America's linkage to international markets. Why is there so little interest in who that middle sector is and whether its interests are consistent with something called food security?

4. What happens when public programs get too expensive? The American economy is in a growth crisis that is unlikely to be reduced without significant reorganization and redirection of industry and education. Until that redirection is initiated and shows results, international economic relationships will be an arena of stress and mostly muted conflict. In this context, the rapidly rising costs of domestic commodity programs can hardly continue. The payment-in-kind program is a short-term (and costly) effort to reduce storage costs, but with a reduced reserve, what will happen to price stability for American consumers? If the U.S. economy generally and the public sector particularly again enters a period of deepened recession and fiscal stress, what will be the consequences for the government's role in managing agriculture? Who within the overall food system will be the most vulnerable to abandonment? The experience of the early 1980s suggests that low income consumers and smaller full-time farmers will face difficult times. What role will international agricultural markets have in a larger economic and fiscal recession and which interests will define what that role should be? Are those the same interests that are capable of protecting the food system's overall security? If so, on what terms?

5. Can incremental vision and short-term solutions be overcome? A set of impacts as potentially massive and long-term as those that would accompany climate change receives virtually no recognition. One major reason is that an issue such as climate change raises two related questions that the American policy system, the private sector, and the technology generating

community have been notably reluctant to confront.

The first question is: What is the longer-term future of the agricultural system? Can incremental vision even seriously comprehend that question? If not, does it mean that the U.S. is destined to see its agriculture and food system deteriorate without anything better than a patchwork of short-term measures to provide needed changes? This problem is especially serious if the U.S. continues to believe that the biological sustainability of its agricultural system is strictly a domestic problem, or that participation in international trade and the sustainability of U.S. agriculture are independent issues.

The second and closely related question is: Can the basic premises of American agriculture be reconsidered, especially if the conditions which accompanied the development of agriculture in this century change in some fundamental way? There is evidence that a reassessment of what American agriculture is can develop and find political expression, but the evidence is not altogether promising. The focus on organic farming and other supply side adjustments has not effectively encompassed recognition of the relationship of U.S. agriculture to a national and international demand system. What will it take to find political expression for a broader perspective on what is, after all, a broader issue: the basic assumptions which define the structure and development of American agriculture may not be consistent with the biological sustainability of that system.

6. Is there a better way of thinking about food security? Finally, along what axis should the food security issue be approached, particularly given the international market dimensions this chapter has reviewed? Three trade-offs characterize discussions in the policy arenas which are likely to confront the juxtaposition of international trade and food security issues: consumer/producer, foreign/domestic, and public/private. How adequate are any of the three?

The consumer/producer trade-off is a mainstay of most neoclassical economic analyses of food policy, and a very attractive way of describing political conflict lines that characterize the drafting of recent farm policy legislation. However, the trade-off is not compelling and may not even be relevant given the heterogeneity of those two categories and given the significant role of the middle sector. For example, as an aftermath of the 1980 grain embargo, the government concerned itself with compensating traders, not producers or consumers. Similarly, the urban poor may share more nutrition interests with poor farmers than they do with middle and upper class urban consumers. That alliance developed political expression and bureaucratic presence in the late 1970s, but more recently both expression and presence have weakened

(Auleta, 1983).

The foreign/domestic tradeoff was discussed earlier in terms of prospective competition between American and foreign consumers and producers. This trade-off is also popular among those who see agricultural exports as an American foreign policy resource as well as an American national interest. The problem with employing this trade-off as an axis for discussion is that many significant actors in the American food security picture cannot be neatly assigned to domestic or foreign categories (cf. Burbach and Flynn, 1980; Lappe and Collins, 1977:213-225). The international grain traders, food processors, and many of the large supermarket chains are examples. All are active as both exporters and importers and all are active in production and marketing operations inside and outside the U.S. Discussions of food security in terms of this tradeoff focus on international trade policy, but exclude comparable attention to internal trade policy.

A public/private tradeoff is proposed by that brand of political economy which considers public intervention in markets to be inefficient (cf. Gardner, 1981b) by definition. In this tradeoff, discussion of food security would advocate trade liberalization and reliance on forward contracting and futures markets as price-setting mechanisms. This also is not compelling for at least three reasons. First, it significantly underestimates the existence and significance of unequal and noncompetitive market power. Second, it assumes but rarely establishes that decisions made by those in the public sector are less informed and socially useful than decisions made in the private sector. Third, and perhaps most important, it ignores the role of the state, an orchestration of class and bureaucratic interests, in determining agriculture's fate. As recent publications suggest (e.g., Reich, 1983), improved public-private cooperation is necessary if the U.S. is to compete effectively as an international economic power generally.

WHAT NEXT?

An understanding of how international markets impact American food security should be historically based, politically sensitive, and socially astute. In analytic terms that means understanding the American agricultural system as a construction that systemically and systematically allocates degrees of risk and security. The vision of the U.S. feeding the world has to be understood as the complex hyperbole it is. The U.S. may indeed have a significant capacity to help feed the world, but where would American agriculture be without a world to be fed? And who among the hungry are we actually feeding? The vision of the U.S. feeding the world, within the symbolism of politics, also needs to be understood as a distraction from the issues of hunger and rural poverty within the United States itself. The hyperbole and the distraction are linked, an important and even essential characteristic of the contemporary dialogue on farm and food policy. A key to the growth and maintenance of that link is that agriculture is integrated into a national and international economic system. Food security will be more an outcome of that relationship than a product of forces unique to agriculture.

In practical terms, the issue of how international markets impact American food security cannot be separated from the broader need to establish a fresh vision of what the public interest in agriculture is and how that interest can be best institutionalized. Establishing such vision will not be easy. Public discussion that is informed and sensitive will be needed in at least three broad, related issue areas.

First, substance must be carefully separated from image in characterizing both international demand and need for American agricultural products and what the U.S. stakes are in the distribution and level of demand and need. Who pays, who is able to pay, and who benefits, for example, from significant shifts in the proportions, quantities, and accessibility of commercial and concessional exports? Are there other ways American resources might be employed to combat global hunger? How do different American approaches to international demand and need for agricultural exports impact hunger and rural poverty within as well as outside the U.S.?

Second, inefficiency and concentration (in terms of spatial distribution, market power, and productivity) in what this chapter has called the middle sector may not bode well for poorer consumers or producers in the U.S. This may well be the most significant consequence of the recent farm crisis. Nor is it clear that the middle sector's organization and performance are the best or only ways to protect or promote America's competitive prowess internationally. A middle sector that is more clearly consistent with the welfare of domestic producers and consumers is needed first of all. Although policy is not directly responsible for structuring or restructuring that sector, policy has provided and undoubtedly will continue to provide an environment in which some paths of evolution are considerably more likely than others. Consequently, a review will be needed of the relationships between structure, performance, and change in the middle sector and tax legislation, monetary policy, trade policy, welfare policy, land and water policies, and commodity policies. In effect, what will be needed is a serious discussion of the structure of the American _food_ system, not agriculture alone.

Third, pursuing an agricultural export policy that rests on a platform of an open international trading system is problematic when (a) protectionism exists in export markets, (b) forms of subsidization exist in the domestic producer market, and (c) domestic internal trade policy, of which those subsidies are part, may be shifting income away from producers to a middle sector which can, in turn, redistribute the risks involved in international trade. In these complex circumstances, understanding and more sensitive debate is needed on the following two issues:

(1) The sources and impacts of protectionism elsewhere. It serves little purpose to portray protectionism in foreign markets as "anti-American" or economic irrationality. We know surprisingly little about the roots of food policy in Japan, the European Economic Community, and the newly industrialized countries of the Third World.
(2) The impacts of domestic production subsidies on the risks involved in international trade participation. The impacts of implicit subsidies involved in commodity policies and tax measures on costs and returns in U.S. agriculture are becoming better understood, but the impacts of those same implicit subsidies on who gains and loses in international trade are less well understood.

Domestic farm policy, internal trade policy (traditionally summarized by the parity ratio, but more appropriately visualized through the distributive consequences of tax, land, and labor policies), and international trade policy need to be comprehended in a more coherent and balanced matter. That means

(1) striking a balance that does not simply reward existing market power and the ability to shelter risk in commodity and tax policies, and
(2) facing the very difficult question of "who governs" or at least "who decides."

The diffusion of responsibility among the Departments of Agriculture, Commerce, State, and Treasury, just to name a few, is too prone to special interests rather than public interest and it defeats the feasibility of Congressional oversight and public accountability.

Without the possibility of the public being able to confront food policy (as distinct from a large collection of other policies), the objective of even defining, let alone implementing, food security is likely to be elusive. This is especially the case when food security is not simply an outcome of domestic food supply-demand relationships, but rather an outcome of complex international trade relationships and domestic economic relationships.

NOTES

1. The distinction involves the efficiency with which plants absorb solar radiation and fix carbon in the photosynthesis process.

REFERENCES

Auleta, K. 1983. The Underclass. New York: Vintage Books.

Barnett, A. D. 1982. "China's Food Prospects and Import Needs." In China Among the Nations of the Pacific, ed. H. Brown, pp. 59-67. Boulder: Westview Press.

Batie, S.S., and Healy, R. G. 1980. "American Agriculture as a Strategic Resource." In The Future of American Agriculture as a Strategic Resource, eds. S. S. Batie and R. G. Healy, pp. 1-40. Washington, D.C.: Conservation Foundation.

______. 1983. "The Future of American Agriculture." Scientific American 248(February):45-53.

Bernard, Jr., H. W. 1980. The Greenhouse Effect. Cambridge: Ballinger Publishing Co.

Bosworth, B. P., and Lawrence, R. Z. 1982. Commodity Prices and the New Inflation. Washington, D.C.: Brookings Institution.

Breimyer, H. F. 1982. Preparing for the Contingency of Intense Pressure on U.S. Food-Producing Resources. NPA Report No. 192. Washington, D.C.: NPA.

Brown, L. 1975. The Politics and Responsibility of the North American Breadbasket. Worldwatch Paper 2. Washington, D.C.: Worldwatch Institute.

Burbach, R., and Flynn, P. 1980. Agribusiness in the Americas. New York: Monthly Review Press.

Burki, S. J., and Goering, T. J. 1977. A Perspective on the Foodgrain Situation in the Poorest Countries. World Bank Staff Working Paper No. 251. Washington, D.C.: IBRD. Agricultural and Rural Development Dept.

Caves, R. E., and Pugel, T. A. 1982. "New Evidence on Competition in the Grain Trade." Food Research Institute Studies 18:261-74.

Christenson, R. M. 1959. The Brannan Plan: Farm Politics and Policy. Ann Arbor: Univ. of Michigan Press.

Cochrane, W. W. 1979. The Development of American Agriculture: A Historical Analysis. Minneapolis: Univ. of Minnesota Press.

Crosson, P. R., and Brubaker, S. 1982. Resource and Environmental Effects of U.S. Agriculture.

Washington, D.C.: Resources for the Future.
Csaki, C. 1982. Long-term Prospects for Agricultural Development in the European CMEA Countries Including the Soviet Union. Research Report 82-25. Laxenburg, Austria: International Institute for Applied Systems Analysis.
Dawell, A. A., and Jesness, O. B. 1934. The American Farmer and the Export Market. Minneapolis: Univ. of Minnesota Press.
Destler, I. M. 1980. Making Foreign Economic Policy. Washington, D.C.: Brookings Institution.
Farnsworth, C. H. 1983. "U.S. Vow on Grain Delivery." New York Times, August 26, p. D4.
Foreign Agriculture. 1982a. "WUSATA: Young Cooperator Hits the Ground Running." 20.10(October):8-9.
______. 1982b. "SUSTA: A Firm Believer in Foreign Development." 20.10(October):10.
______. 1982c. "MIATCO Gears Up to Boost Agricultural Exports in 1983." 20.10(October):4-5.
Fornari, H. D. 1982. "Recent Developments in the American Grain Storage Industry." Agricultural History 56(January):264-71.
Gardner, B. 1981a. "Consequences of Farm Policies During the 1970s." In Food and Agricultural Policy for the 1980s, ed. G. Johnson, pp. 48-72. Washington, D.C.: American Enterprise Institute for Policy Research.
______. 1981b. The Governing of Agriculture. Lawrence: Regents Press of Kansas.
Grennes, T.; Johnson, P. R.; and Thursby, M. 1978. The Economics of World Grain Trade. New York: Praeger Publishing.
Groenewegen, J. R., and Cochrane, W. W. 1980. "A Proposal to Further Increase the Stability of the American Grain Sector." American Journal of Agricultural Economics 62(November):806-11.
Harlow, J. 1980. "International Middlemen in Grain and Oilseeds Markets." American Journal of Agricultural Economics 62(December):895-98.
Haynes, D. L. 1982. Effects of Climate Change on Agriculture Plant Pests. Environmental and Societal Consequences of a Possible Carbon Dioxide-Induced Climate Change, vol. 2, part 10. Washington, D.C.: Office of Energy Research. Carbon Dioxide Research Division. Department of Energy.
Healy, R. G., and Short, J. L. 1981. The Market for Rural Land: Trends, Issues, Policies. Washington, D.C.: Conservation Foundation.
Hightower, J. 1975. Eat Your Heart Out: Food Profiteering in America. New York: Crown Publishers.
Hitzhusen, F. J. 1982. The Political Economy of Soil Erosion: Some National Security Implications. Columbus: Mershon Center of the Ohio State Univ.
Houck, J. P. 1982. "Agreements and Policy in U.S.-Japanese Agricultural Trade." In U.S.-Japanese

Agricultural Trade Relations, eds. E. Castle and K. Hemmi, pp. 58-87. Baltimore: Johns Hopkins Univ. Press.

Johnson, J., and Clayton, K. 1982. "Organization and Well-being of the Farming Industry: Reflections on the Agriculture and Food Act of 1981." American Journal of Agricultural Economics 64(December): 947-56.

Josling, T. 1980. Developed-Country Agricultural Policies and Developing-Country Food Supplies: The Case of Wheat. Research Report 14. Washington, D.C.: International Food Policy Research Institute.

Kelejian, H. V., and Vavrichek, B. C. 1982. "Pollution, Climate Change and the Consequent Economic Cost Concerning Agriculture Production." In The Economics of Managing Chloroflurocarbons: Stratospheric Ozone and Climate Issues, eds. J. H. Cumberland, J. R. Hibbs, and I. Hoch, pp. 224-68. Washington, D.C.: Resources for the Future.

Kellogg, W. W., and Schware, R. 1981. Climate Change and Society: Consequences of Increasing Atmospheric Carbon Dioxide. Boulder: Westview Press.

______. 1982. "Society, Science and Climate Change." Foreign Affairs 60(Summer):1076-1109.

Koerselman, G. H. 1971. "Herbert Hoover and the Farm Crisis of the Twenties: A Study of the Commerce Department's Efforts to Solve the Agricultural Depression, 1921-1928." Ph.D. dissertation. Dekalb: Northern Illinois Univ.

Lappe, F. M., and Collins, J. 1977. Food First: Beyond the Myth of Scarcity. Boston: Houghton Mifflin Co.

Lewis, P. 1983. "Europe's Farm Policies Clash with American Export Goals." New York Times, February 22, p. 1.

Lyons, M. S., and Taylor, M. W. 1981. "Farm Politics in Transition: The House Agriculture Committee." Agricultural History 55(April):128-46.

Macleans. 1981. "Anticipating Another 'Grapes of Wrath.'" May 18. 94.20:16-18.

Martinson, O. B. 1978. "The American Grain Marketing System: An Organizational Analysis." Ph.D. dissertation. Madison: Univ. of Wisconsin.

Martinson, O. B., and Campbell, G. 1980. "Betwixt and Between: Farmers and the Marketing of Agricultural Inputs and Outputs." In The Rural Sociology of the Advanced Societies: Critical Perspectives, eds. F. Buttel and H. Newby, pp. 215-53. Monclair, NJ: Allanheld, Osmun and Co.

Matusow, A. J. 1967. Farm Policies and Politics in the Truman Years. Cambridge: Harvard Univ. Press.

McKinna, D. A. 1978. U.S. Government Sponsored Agricultural Export Market Development Programs. Agricultural Economics Extension 78-29. Ithaca, NY: Cornell Univ. New York State College of Agriculture and Life Sciences. Dept. of Agricultural Economics.

Morgan, D. 1979. Merchants of Grain. New York: Viking Press.
National Research Council. 1979. Carbon Dioxide and Climate: A Scientific Assessment. Washington, D.C.: National Academy of Sciences.
______. 1982. Carbon Dioxide and Climate: A Second Assessment. Carbon Dioxide/Climate Review Panel. Washington, D.C.: National Academy Press.
Nelson, B. 1983. "Trees that Once Saved Crops Fall Victim to Farm Economy." New York Times. March 6, p. E20.
Newman, J. E. 1982. Impacts of Rising Atmospheric Carbon Dioxide Levels on Agricultural Growing Seasons and Crop Water Use Efficiencies. Environmental and Societal Consequences of a Possible Carbon Dioxide-Induced Climate Change, Vol. 2, Part 8. Washington, D.C.: Office of Energy Research. Carbon Dioxide Research Division. Department of Energy.
New York Times. 1973. "A Roundtable: The Crisis in Farm-Support Policy." August 7, p. E5.
______. 1983. "Block Defends Grain Sales." August 22, p. D7.
Nourse, E. G. 1924. American Agriculture and the European Market. New York: McGraw-Hill.
O'Donnell, T. 1982. "Waiting for Ivan." Forbes. November 22.
Paarlberg, D. 1980. Farm and Food Policy: Issues of the 1980s. Lincoln: Univ. of Nebraska Press.
Perelman, M. 1977. Farming for Profit in a Hungry World. Montclair, NJ: Allanheld, Osmun and Co.
Peterson, T. H. 1979. Agricultural Exports, Farm Income, and the Eisenhower Administration. Lincoln: Univ. of Nebraska Press.
Polopolus, L. 1982. "Agricultural Economics Beyond the Farm Gate." American Journal of Agricultural Economics 64(December):803-10.
Raup, P. M. 1980. "Competition for Land and the Future of American Agriculture." In The Future of American Agriculture as a Strategic Resource, eds. S. S. Batie and R. G. Healy, pp. 41-77. Washington, D.C.: Conservation Foundation.
Reich, R. B. 1983. The Next American Frontier. New York: Times Books.
Robison, L. J., and Leatham, D. J. 1978. "Interest Rates Charged and Amounts Loaned by Major Farm Real Estate Lenders." Agricultural Economics Research 30.2(April):1-9.
Rockefeller Foundation. 1976. "Climate Change, Food Production and Interstate Conflict." Conference held at the Bellagio Study and Conference Center, June 4-8, 1975. New York: Rockefeller Foundation.
Sanderson, F. H. 1978. Japan's Food Prospects and Policies. Washington, D.C.: Brookings Institution.
______. 1982. "Managing Our Agricultural Interdependence." In U.S.-Japanese Trade Relations,

eds. E. Castle and K. Hemmi, pp. 393-426. Baltimore: Johns Hopkins Univ. Press.

Sarris, A. H., and Schmitz, A. 1981. "Toward a U.S. Agricultural Export Policy for the 1980s." American Journal of Agricultural Economics 63(December): 832-39.

Schertz, L. P., et al. 1979. Another Revolution in U.S. Farming? Washington, D.C.: USDA.

Schideler, J. H. 1957. Farm Crisis 1919-1923. Berkeley: Univ. of California Press.

Schneider, S. H. 1976. The Genesis Strategy: Climate and Global Survival. New York: Plenum Press.

Schuh, G. E. 1976. "The New Macroeconomics of Agriculture." American Journal of Agricultural Economics 58(December):802-11.

______. 1981. "U.S. Agriculture in an Interdependent World Economy: Policy Alternatives for the 1980s." In Food and Agriculture Policy for the 1980s, ed. D. G. Johnson, pp. 157-82. Washington, D.C.: American Enterprise Institute for Policy Research.

Soth, L. 1981. "The Grain Export Boom: Should It Be Tamed?" Foreign Affairs 59(Spring):895-912.

Strickland, R. P. 1982. "Alternative Indicators of Farm Operator's Earnings." Agricultural Economics Research 34.3(July):28-33.

Successful Farming. 1982. "Japanese are Buying U.S. Grain Elevators." 80(January):19-21.

Talbot, R. B., and Kihl, Y. W. 1982. "The Politics of Domestic and Foreign Policy Linkages in U.S.-Japanese Agricultural Policy Making." In U.S.-Japanese Agricultural Trade Relations, eds. E. Castle and K. Hemmi, pp. 275-338. Baltimore: Johns Hopkins Univ. Press.

Tucker, H. A. 1982. Effects of Climate Change on Animal Agriculture. Environmental and Societal Consequences of a Possible Carbon Dioxide-Induced Climate Change, vol. 2, part 11. Washington, D.C.: Office of Energy Research. Carbon Dioxide Research Division. Department of Energy.

USDA. 1981a. "A Time to Choose." Summary Report on the Structure of Agriculture. Washington, D.C.

______. 1981b. National Agricultural Land Study. Final Report. Washington, D.C.

______. 1982a. "Agricultural Exports by Cooperatives." Research Report No. 26. Agricultural Cooperative Service.

______. 1982b. Agricultural Outlook. Economic Research Service. December.

______. 1982c. "Farm Real Estate Debt." Statistical Bulletin 31 (October). Farm Credit Administration. Economic Analysis Division.

______. 1983a. Foreign Agricultural Trade of the United States (January/February). Economic Research Service.

______. 1983b. Outlook for U.S. Agricultural Exports.

(February 17). Economic Research Service. Foreign Agricultural Service.

______. 1983c. Agricultural Outlook (June).

United States Senate. 1982. Committee on Governmental Affairs. Subcommittee on Energy, Nuclear Proliferation, and Government Processes. "Growing Agricultural Trade Protectionism in Europe." Hearings held October 16, 1981. Washington, D.C.: Government Printing Office.

Wadeken, K. 1982. "Soviet Agriculture's Dependence on the West." Foreign Affairs 60(Spring):882-903.

Wall Street Journal. 1982. "Falling Grain Exports Bruise the Farm Belt; Damage Could Spread." February 1, p. 1.

Winsberg, M. D. 1982. "Agricultural Specialization in the United States since World War II." Agricultural History 56(October):692-701.

Yamada, S. 1982. "The Problem of Food Security in Japan." In Food Security: Theory, Policy and Perspectives from Asia and the Pacific Rim, eds. A. H. Chisholm and R. Tyers, pp. 217-37. Lexington, MA: Lexington Books.

Yoshioka, Y. 1982. "The Personal View of a Japanese Negotiator." In U.S.-Japanese Agricultural Trade Relations, eds. E. Castle and K. Hemmi, pp. 341-67. Baltimore: Johns Hopkins Univ. Press.

Overview

18
Food Security in the U.S.— An Overview

Sylvan H. Wittwer

INTRODUCTION

Some unique features characterize the food and agricultural industry of the United States. It is by far the largest, since one out of every five jobs is in the food and agricultural business. Agricultural products rank first among all U.S. exports, and U.S. agricultural exports are almost equal to those of all other nations combined. Food costs are among the lowest of any country in the world. The agricultural business consists of a vast infra-structure of supporting and allied industries including food processing, storage, transportation, credit and employment, fertilizers, and chemical and food companies.

No one questions the historical productivity of U.S. farms. Contributing factors have been abundant resources of land, water, and a favorable climate with a history of resourceful farmers and technological innovations. This nation has supplied not only its own needs, but a substantial portion of the agricultural products used elsewhere in the world. A mere three percent of the work force provides not only this abundance of food for domestic consumption but a substantial surplus for foreign trade that has offset much of the nation's expenditure for imported oil. Even the most severe critics of the American agricultural and food producing system recognize that it has been, and remains, one of the most successful sectors of the U.S. economy (Office of Technology Assessment, 1981; Wittwer, 1980a; Wittwer, 1983e).

SUSTAINABILITY OF U.S. AGRICULTURAL PRODUCTION

The sustainability of this successful record has now come under scrutiny by many groups within and external to the agricultural establishment itself. The critics are many and they take on a variety of forms and issues. Although the criticisms are heavy, the viable alternative options for production systems are light.

Achieving an adequacy and security of food supply is both a humanistic goal and a mark of progress. There was never a greater outpouring of reports from conferences, workshops, commissions, symposia, and individually written books and articles on issues of the adequacy, the security, the sustainability, the safety, the health aspects, the strategic values, and the dependability of our food supply.

Food security has been defined here as having three dimensions--availability, accessibility, and adequacy. I will add a fourth, that of dependability. Of equal importance to that of production itself is dependability of supply (Wittwer, 1980b; Wittwer, 1982b; Wittwer, 1983b). The variations in productivity conditioned by drought and pestilence must be recognized as the single variable which accounts for most problems in achieving food security.

The current expectations and needs for agriculture and the agricultural sciences are twofold. First, food must be produced to satisfy the most basic of all human requirements and second, it must be done in a resource sustainable manner (Altieri, et al.; Dundon, 1982; Edens and Haynes, 1982; Edens and Koenig, 1980).

Many issues have been raised here concerning the sustainability of contemporary agriculture and food producing systems in the U.S., and bearing upon our nation's food security, and that of the world. Further, concern has been expressed as to the social, economic, and health costs of current food producing systems. Emphasis has been given to the devastations of soil erosion, heralded as the most serious ecological problem in the world (Pimentel, Ch. 5), resulting in depletion of groundwater (fossil) resources, ecosystem degradation, farm land conversion to other uses, loss of soil organic matter, soil compaction, salinity problems and acid precipitation (acid rain), desertification, and in beneficial soil biota and the loss of plant and animal species.

The point is made repeatedly that if agricultural research is to contribute to the long-term strength of agriculture the focus has to change. According to the participants here, and others, technologies must be developed that build a sustainable agriculture without soil and water depletion and loss of plant and animal species. Doing the wrong kind of research and adopting the wrong technologies have been problems (Jackson and Bender, Ch. 2; English, Ch. 3; Pimentel, Ch. 5; Friedland, Ch. 8; Baumel and Hayenga, Ch. 12; Freudenberger, Ch. 16; Busch and Lacy, 1983; Hadwiger, 1982; Hildreth, 1982; Johnson, 1983; National Research Council, 1972; Presidential Commission on World Hunger, 1980).

To achieve these idealistic goals is no small expectation or assignment for the agricultural research establishment. Serious questions must be raised as to the sustainability of current and projected expansion of our food exports to offset the cost of imported oil, very little of which is used for food production. There is the question of continued low cost foodstuffs--the expectation of all U.S. citizens--being provided on a sustained basis. The question is how much are we willing to pay directly or indirectly in higher food prices, and the issue is how long can we remain competitive.

Toxic chemicals in the environment, some of them pesticides and fertilizers, now used and which will continue to be used for food production, have been declared environmental threats and hazards to human health and well being. Debates are continuing on issues of food safety, the deleterious effects on fish and wildlife habitats, endangered species, and carcinogenicity. There are issues of environmental degradation, concerns for animal welfare, impacts on the safety of farmers, agricultural workers and consumers (Coye, Ch. 9), adverse effects of production and processing technologies, the extrusion of the smaller family farms in agriculture (commonly referred to as the demise of the family farm), erosion of rural communities, the increased concentration of production and wealth, inadequate resource conservation, and exploitation of fragile lands (Bonnen, 1983). Although some have tried no one has yet clarified or defined an environmentally acceptable set of agricultural production technologies that will meet the needs and expectations of a public imbued with a cheap food policy and even approach the level of food exports of recent years considered essential to maintain a reasonable balance of payments in international trade. A further problem is now encountered with an urban public soured on paying for malfunctioning farm programs to support farmers who are wealthier than the typical tax payer (Bonnen, 1983).

RESOURCE AND TECHNOLOGY INPUTS

The timing of this workshop was on the eve of an all-time record in crop production (September 1982). The nation's farmers were hounded with agricultural and food surpluses, overproduction, and low prices. Accordingly, technologies to improve production were severely criticized here. That situation has changed dramatically within one year. During the past decade the nation and the world have gone through a full cycle of food security, moving from shortages, high prices, and inadequate production to surpluses, low prices, and overproduction. That cycle is now repeating itself. If the past is any indication of the future, it will be repeated again and again.

Climate is the determining factor in the stability

and magnitude of our food production capability (National Academy of Sciences, 1976; Slater and Levin, 1981). It was neither discussed nor highlighted in this workshop as a resource input. Climate should be considered both as a resource to be used wisely and as a hazard to be dealt with in food producing systems, with major impacts on food security.

How can we commit to aggressively expanded food production programs, agricultural exports, and biomass production for renewable energy resources, while at the same time achieve nondegradation of soils, reduce overdraft of groundwaters, and avoid modifications and pollutions of the environment through increased use of irrigation waters, fertilizers, and pesticides? This stands as the supreme challenge for future agricultural science and technology. To reduce or eliminate some of the inputs such as fertilizers, pesticides, and water, and still maintain productivity and food security, poses an even greater challenge. Little help in finding a solution to these basic dilemmas of the nation's current food needs at home and abroad and providing food security with sustainable food production technologies came out of the workshop. As Koppel (Ch. 17) has suggested, the basic assumptions which define the structure and development of U.S. agriculture may not be consistent with the biological, physical, and economic sustainability of the system.

A reflection of resource inputs and those of technology are depicted in Figure 18.1, giving the composite index of crop yields in the U.S. since 1910. The "golden age" of agricultural productivity in the U.S. was in the decades of the '50s and '60s. Climatic perturbations are noted in the dust bowl periods of the '30s and three major droughts and heat waves in 1974, 1980, and 1983. In 1983 production of corn and sorghum was reduced to less than half of that for the record crops in 1982, and soybean production dropped to two thirds of the previous year. Coupled with the Payment in Kind (PIK) program of 83 million acres taken out of production, this was the greatest shortfall in production performance the U.S. has experienced in the last 50 years (Wittwer, 1983d).

Significant changes have occurred in many aspects of American agriculture during the past decade. The percentage of income expended for food, however, has not changed. It is consistently the lowest in the world. Costs, inputs, and prices have fluctuated wildly. Great degrees of uncertainty characterize the U.S. food production system. Federal expenditures for agricultural research have scarcely kept pace with inflation. Productivity, in general, has increased, but not consistently (Johnson and Wittwer, 1983; Wittwer, 1982c).

The decade of the '70s witnessed dramatic changes in world food economy. A series of world events between

1973 and 1975 sent farm commodity prices soaring to new heights. Suddenly the world seemed to have run out of food, and shortages, some real and others articulated, were experienced. The demand for U.S. exports jumped dramatically. With this so-called "world food crisis" modern-day Malthusians proclaimed that population growth had outstripped food production (LeVeen, Ch. 4). The "lifeboat" ethic and the principle of triage took sway. Thus, history is filled with prophetic accounts of imminent world hunger, famine, and starvation that would overtake the world. These dire prophecies have not come to pass. Some economists in the mid-1970s even proclaimed "a new era" for American agriculture and predicted that previous excess production capabilities would be replaced by shortages and rising prices. "All out" food production coupled with attractive price incentives were proclaimed and became bywords. Cash cropping as contrasted with mixed crop-livestock farming systems was initiated in much of the grain belt of the United States, to meet what was considered ever expanding export markets. Added to problems of soil erosion, this dramatically changed much of the agricultural landscape of the heartland of the U.S.

RESOURCE INPUTS

The determinants of agricultural production are three--new technologies, resource inputs, and economic incentives (Wittwer, 1980c).

Land, as a resource, comes first for establishing food security. More than 98 percent of our food supply is land dependent. Preservation and protection of our land resources should be paramount. Concerns with respect to the land resource relate to soil erosion. This has been the subject of many reports, symposia, and speeches (Brown, 1981; Larson, et al., 1981; Larson, et al., 1983; Schultz, 1982). Corrective action is difficult because most soil conservation practices do not pay for themselves (Batie and Healy, 1980; Batie and Healy, 1983). Other concerns relate to irreversible farm land conversion to other uses, compaction, loss of organic matter, salinity, desertification, soil subsidence from overdrafts of groundwater, and toxic substances. Inseparable from the land resource is that of water and irrigation (Napier, et al., 1983), the more subtle effects of acid rain (Connecticut Agricultural Experiment Station, 1983; National Acid Precipitation Assessment Program, 1982; Walker, 1982), and changes in the soil biota (Sampson, Ch. 1). Small farms with limited land resources can be economically viable units (USDA, 1982). A challenge for the future is to promote the production of small farm units.

Another challenge will be to focus on technologies that build a sustainable agriculture without soil or water depletion and with minimal fossil energy requirements. One approach is conservation tillage which, if coupled with allelopathic properties of some crop residues (Putnam, 1983), could achieve a sustainable soil tillage system which would reduce soil erosion to near zero and simultaneously conserve water, energy, and soil organic matter (Phillips, et al., 1980). Other unexplored areas for greatly enhancing the productive capacity of land would be at the microbiological level, including nitrogen fixers (Brown, et al., 1975), mycorrhizae, and disease-suppressing soil and root colonizing bacteria (Schroth and Hancock, 1982). Industrial institutions are now alerted to the possibilities of applying microbiological methods and products to agriculture (Campbell, et al., 1983; Kenny, et al., 1982; Sharp and Evans, 1982).

During 1983, the U.S. had less land under cultivation than at any time within the last 100 years. The "Payment in Kind" program resulted in a drop of 83 million acres from the 1982 level. This should have provided an unprecedented opportunity to plant soil-improving crops and to reverse the phenomenon of soil erosion on at least 25 percent of the crop land area of the nation (Wittwer, 1983d). Accurate assessments of changes in land resources in the U.S. leave much to be desired. The ability to accurately classify and qualify changes in land use, either by ground observations or from satellites, needs dramatic improvement (National Aeronautics and Space Administration, 1983). The widely-quoted figure of 2 million acres of prime agricultural land in the U.S. being taken annually out of production may be in error by 50 to 100 percent. Global estimates for clearing of closed canopy forests range from 7 to 30 million acres per year.

Water is America's most precious and wasted resource. Agriculture is by far the largest consumer; it requires, through crop irrigation, 80 to 85 percent of the fresh water resources of the nation, and wastes about half of it. As Sampson points out (Ch. 1), irrigated agriculture in the U.S. produces 27 percent of the crops harvested in only 12 percent of the cultivated acres (English, Ch. 3). There are many useful technologies for greater efficiency of water use in irrigated agriculture. These include lining and covering canals, computerized programming for releasing irrigation water by installing wires and gates, closer scheduling of applications, and irrigation at night (English, Ch. 3). The current overdraft of groundwater resources in the U.S. is largely irreversible use of a nonrenewable resource which will deplete supplies for the future and may result in severe soil subsidence. Water for expanded supplemental crop irrigation in the subhumid areas is the one important option for enhancing both the magnitude and stability of crop production. The future value of water for food production will not be limited to desert areas. There is just not enough rain at the right time. The issue is not whether to irrigate, but how. The new technologies of drip irrigation, coupled with conservation tillage and plastic mulching, could have significant impacts on reducing water requirements for crop production and increasing efficiency in the use of fertilizers (Stark, et al., 1983). Increasing competition for water use by industry, energy extraction, and recreation, coupled with current depletions of groundwater reserves, will result in rising costs and concerted water conservation technologies for the most crucial of all resource inputs into the food system during the coming decades. Water, not land, energy, or fertilizer, will become the most critical of all resources for food production (English, Ch. 3; Larson, et al., 1981; Napier, et al., 1983; USDA, 1983; Wittwer, 1983b).

Energy. Food production results in a renewable energy resource, but food is more than energy. Agriculture, through the production of green plants, is the only major industry that processes solar energy. Green plants are biological suntraps. The aim of agriculture is to adjust species of crops to locations, and with planting designs, cropping systems, and cultural practices to maximize through leaves the biological harvest of sunlight to produce things useful for people (Wittwer, 1982a). Residues from the harvest of crops and forestry products, if properly managed, could provide a substantial renewable energy resource for the nation (Pimentel, Ch. 5; Pimentel, et al., 1981).

While the heavy dependence of agriculture on fossil fuels has been emphasized (Pimentel, Ch. 5), water will be even more crucial in the future, because only three percent of the nation's fossil energy budget now goes into food production. The big energy culprits in the U.S. food system come after the food leaves the farm gate. They include food processing, transportation, storage, and marketing. There is also a close dependence of crop irrigation on energy inputs. As Pimentel (Ch. 5) rightly points out, irrigated corn takes three times the total energy inputs compared with a crop grown under normal rainfall.

The largest energy, as well as industrial, input into food production is nitrogen fertilizer. Considerable progress has been made and will likely continue to be made in the development of technologies which will reduce energy requirements for chemical fixation of nitrogen and increase biological nitrogen fixation under field conditions (Geissbuhler, et al., 1982).

A final agricultural production opportunity resides with photosynthesis, the most important biological energy fixation process on earth. Solar energy is clean, makes no noise, is widely available, requires no fuels, is renewable, is nonpolluting and cannot be embargoed by other nations (Wittwer, 1982a). Few research efforts on photosynthesis have focused on crop productivity and increased biomass production. This is an important option for the future, and it was scarcely emphasized in the workshop.

CROP PROTECTION

Closely allied to alternative food production systems, widely applauded in the workshop but not identified as to essential components, is the popularized concept of integrated pest management. We have known for 20 years that we must seek new approaches to pest control (Carson, 1962). Integrated pest management systems (the use of natural enemies (parasites), resistant varieties, cultural practices, and chemicals) will become essential for the survival (increased production, yield stability, improved habitability, and reduced costs) of most food producing systems. Next to drought, pestilence is most feared by farmers. There is an ever-increasing concern in the use of pesticides as to the safety of farm workers, pesticide applicators, people in proximal communities, and all consumers of agricultural products (Friedland, Ch. 8; Coye, Ch. 9).

A matter of major concern is resistance among pests to chemical pesticides. There are now approximately 430 insects resistant to insecticides, 100 diseases resistant to fungicides and bacteriocides, and 36 weeds resistant to herbicides. Chemicals for pest control in the future should be applied in concert with computer programming for both volume and amount, and cost reductions for precise targeting of treatments; also, safer, more powerful, and more selective materials should be used (Campbell, et al., 1983). Novel approaches to pest control and pesticide resistance should be sought (Geissbuhler, et al., 1982).

Integrated pest management in the broad sense and even with individual crops is still only a concept. No system has yet been widely accepted by growers that does not use chemicals. Possible exceptions are those involving organic farming. Integration of disciplines at the institutional level has not yet occurred. It now appears that integrated pest management, because of institutional constraints and funding strategies, will not, in reality, happen in this generation (Wittwer, 1980b; Wittwer, 1983b).

GENETIC RESOURCES

Plants provide directly or indirectly up to 95 percent of the world's total food supply. Worldwide, of the 350,000 species of plants, only about 0.1 percent (fewer than 300) are important current sources of food. Globally, 21 crops essentially stand between human life and starvation. In the approximate order of importance they are rice, wheat, corn, potato, barley, sweet potato, cassava, soybean, oat, sorghum, millet, sugar cane, sugar beet, rye, peanut, field bean, chickpea, pigeon pea, cowpea, banana, and coconut. New technologies, resource inputs, and economic incentives must focus on these crops (Wittwer, 1983b).

The genetic vulnerability of almost all major food crops now grown in the U.S. has been forcibly emphasized (Butler, Ch. 7). Single crop variety concentrations for soybeans (16 percent), cotton (8 percent), wheat (6 percent), rice (30 percent), barley (20 percent), and corn (37 percent) are appalling. While accessions of genetic stocks for the major food crops are extensive--65,000 for rice, 26,000 for wheat, 13,000 for maize, more than 14,200 for sorghum, 10,500 for soybeans, 12,000 for potatoes, and 11,500 for cowpeas--the so-called world collections are sadly deficient in wild species (Myers, 1983). The erosion of food crop materials in many parts of the world is, according to some, facing a crisis. That there is a crisis in the current erosion of genetic food crop materials, however, is not accepted by Duvick (Ch. 6), who believes there is ample available or in reserve. Meanwhile, there is no clear indication that there exists any focus, national or international, which would create incentives to increase the genetic diversity of our currently accepted major food crops (Butler, Ch. 7).

One partial solution suggested here would be perennial cereal grains (Jackson and Bender, Ch. 2). It is pointed out that our country's grass lands contain species of perennial grains and prairie grasses that could be turned to agriculture and food uses. This is a questionable approach. Perennial grains are a good idea, but where are they? A perennial corn has been described but it is a long way from something that would feed people. There are perennial grasses and legumes, even cucurbits, that will survive the winters of temperate zone agriculture, but few, if any, provide a good food resource if grown in the temperate zones characteristic of the U.S. and other food surplus nations. Many crops now grown in the U.S. as annuals (tomato, potato) are perennials in the tropics. The perennial food crops we now grow have severe problems with weed competition.

ATMOSPHERIC COMPONENTS

Another development, assumed by some to be of crisis proportions, is the rising level of atmospheric carbon dioxide. Climate changes are projected that would influence the availability of energy resources, dislocate agriculture, melt the Antarctic ice caps, raise sea levels, and sink coastal cities. Such projections of climate change have captured the attention and imagination of scientists as well as the public (Clark, 1982; National Academy of Sciences, 1983a; Seidel and Keyes, 1983). Elevated levels of atmospheric carbon dioxide also have biological effects which should be further evaluated (Lemon, 1983). Although most of these appear positive and may have already occurred, they are seldom mentioned.

We know a great deal about how crops respond in controlled environments to increases in atmospheric carbon dioxide up to double and triple present levels. All major crops in short-term experiments and those cultured long-term in commercial greenhouses have produced higher yields and grown more rapidly when given increased carbon dioxide under otherwise identical conditions of sunlight, soil fertility, water supply, and temperature, and in the absence of devastating pests. This perspective is absent in almost all writings on the carbon dioxide issue. Elevated levels of atmospheric carbon dioxide may also alleviate water, high and low temperature, low light, and air pollutant stresses. The current and projected rising levels of atmospheric carbon dioxide (2 ppm/year) could have a major positive effect on total U.S. and global food production during the next several decades, with significant impacts on food security. Important potential benefits would be better (more efficient) use of water by plants and an extension of boundaries for crop production into arid and semi-arid lands, with implications for reductions in desertification (Lemon, 1983; Wittwer, 1983f).

There is also mounting evidence of the effects of air pollutants on food production; these include sulfur and nitrogen compounds, and ozone, all of which now have confirmed effects on food crop production (Heck, et al., 1982). It will be increasingly important for U.S. agriculture to identify accurately the sources of air pollutants, monitor changes, assess their effects on food crops, and seek means of reducing their adverse effects on agricultural productivity.

One of the environmental issues alluded to in this volume is acid rain (Connecticut Agricultural Experiment Station, 1983; National Acid Precipitation Assessment Program, 1982; Walker, 1982). The effects are subtle and neither the short nor long-term effects are known. No adverse effects on food crop production in the U.S. have yet been identified. Understanding and monitoring changes in atmospheric constituents and atmospheric depositions and their effects on plants, animals, and fish is an important challenge for food research in the future.

ASSESSMENT OF RESOURCES

Mention should be made of the remarkable progress in thematic mapping and multispectral imagery which now make it possible to inventory accurately changes in land cover, cropping patterns, and plant species, water and energy availability, and the biogeochemical cycles that affect food production and the future biological productivity of the earth. Remote sensing and observations from satellites will become increasingly important for monitoring changes in resource inputs into food systems and for long-term weather forecasting (National Aeronautics and Space Administration, 1983). The development of predictive models of natural and man-made physical, chemical, and biological changes concerned with food production will be of great significance. Space technology now allows worldwide agricultural crop forecasting and early warning of impending disasters resulting from severe climatic variations and pestilence. Food security may be directly linked to such disasters and alleviation of them.

FOOD RESOURCES--CROPS

It has been emphasized that approximately 21 crops stand between people and starvation. For the establishment of food security, their production must have top priority. Food habits of people are not changed easily or quickly. What people consume is based on tradition and more on appearance, taste, color, and texture than nutritional values. High lysine corn, Triticale, and Amaranth are classical examples of how little progress has been made in dietary change over the past 20 years, even with nutritionally superior cereal grains. New crops are not readily accepted as replacements for the old. School lunch programs for the youth still offer the best mechanism for introducing new food items and changes in diets. The American public has an enormous number of food products from which to choose (Gussow, Ch. 10).

One striking change in the food habits of people of most nations during the past two decades has been the increased consumption of meat (Barr, 1981). For such countries as Japan and Israel, there have been several-fold increases. Primary increases in meat consumption in the U.S. have involved pork and poultry. The goals of most nations, including the U.S., are for even higher meat consumption rates in the future.

Grain exports from the U.S. and other exporting nations to both developing and industrialized countries and to those with centrally controlled economies are used primarily to feed ever-expanding livestock populations and not starving or malnourished people. Food security and needs for the future must include the dimension of population growth as well as the desire for peoples of all nations for improved diets. Improved diets usually mean more meat, milk, and eggs, coupled with increased consumption of fruit and vegetables.

FOOD RESOURCES--ANIMALS

Food animals, although not addressed here in great detail, will continue to play an important role in national and world food security. They are important contributors of food, energy, and byproducts. They convert indigestible forages (cellulose) and other indigestible products to human food. They constitute a global food reserve that can be easily transported, is mobile, and is more evenly distributed, and exceeds the present grain reserves. Animal products are also the chief source of vitamin B12. Animals also serve as a worldwide buffer for stabilizing grain prices and supplies. In the United States, they provide 3/4 of the protein, 1/3 of the energy, and most of the calcium and phosphorus for the diet (Pond, et al., 1980).

Flocks and herds have been with people from the beginning. They graze off the land. Forages provide a feed supply for approximately 2.5 billion ruminant animals useful to man. They are the primary feed for livestock in developing countries. Because forages may reduce soil erosion, provide biomass, and enrich the soil, their production is an essential component of appropriate, long-term land use. Animals are natural protein factories that harvest vast food resources, otherwise of little value, and convert them to milk, eggs, meat, and other useful products. They are living storehouses of mobile food. The opportunities through improved reproductive efficiency and environmental control are truly great for rapid progress in genetic improvement and for disease control (Seidel, 1981). Industrialization of the poultry and hog industries in the U.S. has proceeded at a rapid pace (Wittwer, 1983e). A counterpart now exists in aquaculture.

OPTIMIZING OUTPUT PER UNIT RESOURCE INPUT

Food production for the future will see almost all increases resulting from improvement in yields (output per unit land area per unit time) and from growing additional crops during a given year on the same land. There really are no other viable options. The technologies that will make this possible must be developed now. These include crop varieties that are pest resistant and climatically resilient. Appropriate or selective mechanization will play a dominant role

(Martin, 1983; Rasmussen, 1982). The goal will be to increase the productivity of both land and labor. There will be genetically engineered vaccines for disease control in food animals, new highly potent pesticides (some already exist), mono-clonal antibodies, animal growth hormones, and interferons for improved performance of livestock and crop productivity. Explants from super plant selections will be clonally propagated through tissue culture, which will serve as the pathway for field applications of genetically engineered plants (National Academy of Sciences, 1983b,c).

CLIMATE AND WEATHER

The most determinant and disrupting variable in achieving food security--meaning adequacy, availability, and accessibility--is that of climate and weather. Greater resistance to climatic stress must hold a prime spot in any program for agricultural research of the future (Boyer, 1982). The topic of climate and food security has not gone unnoticed (National Academy of Sciences, 1976; Slater and Levin, 1981), but there is little to show for implementation of recommendations for correction. Climatic stresses are the major causes of instabilities in both food production and marketing. Shortfalls and surpluses and frequent price disruptions result. Improved resistance to environmental and climatic stresses--drought, heat, cold, problem soils, salinity, and air pollutants--will be achieved through genetic improvement of crops and breeds of livestock, chemical treatments, and better management (Boyer, 1982; National Academy of Sciences, 1976; Slater and Levin, 1981). The effects of unfavorable climate on agricultural productivity are as pervasive as those of unfavorable soils. Productivity will need to be enhanced in the now marginal growing areas, resulting in greater stability of supplies and improvements in quality. If there is to be climatic change from the rising level of atmospheric carbon dioxide, coupled with the reality now of interannual variations in climate which cause current disruptions in food supplies and markets, then there are valid reasons for seeking means of alleviation of climatic stresses. Highest priority should be given to research on alleviation of climatic and environmental stresses for both crops and livestock.

EXTERNALITIES OF NEW FOOD TECHNOLOGIES

Much concern was expressed in the workshop as to social consequences or costs created by new technologies, including mechanization, and other products of agricultural research. Technological change is never neutral. An externality is either an external cost or benefit which is not priced on the market. The recipient of a new technology may not be held accountable for all the cost or he may not receive all the benefits (Center for Rural Affairs, 1982). Moreover, the market is often priced in such a way that the benefits, costs, or both are passed on to the consumer rather than to the producer or the user.

Some major externalities have been generated during the past several decades, some of which have been highly beneficial to societies, both at home and abroad. Others have been costly. Greater appreciation of and more effective anticipation and estimates of costs and benefits to all people from the introduction of new technologies as products of agricultural research must be an important part of future programming of research investments (Knorr and Clancy, Ch. 11; Meyers, Ch. 15; Bonnen, 1981; Buttel, et al., 1983; Hadwiger, 1982; Hightower and Demarko, 1972; Johnson, 1983; National Academy of Sciences, 1982; Office of Technology Assessment, 1982; Paarlberg, 1981; USDA, 1983). Belief that only good can come from technological change and that all externalities are positive is no longer acceptable, as was pointed out in the workshop. Indeed, much of the current criticism of agricultural research is directed at failures to consider all costs and benefits from adoption of the new technological innovations. The concept of "growing two ears of corn or two blades of grass where only one grew before" no longer applies. One must consider, among other things, the resource inputs--their availability, costs, renewability, and other impacts on society and the environment (Bonnen, 1981).

Both positive and negative externalities can be identified. These are associated with new food producing technologies such as mechanization, genetic improvement, the use of fertilizers and pesticides, antibiotics, confinement housing, and industrialization of poultry and hog production.

The following are generally viewed by American society as negative externalities, although benefits as well as costs may be associated with some of them:

(1) Migration of 33 million people from farms to cities and to urban and industrial employment between 1930 and 1974.
(2) Food safety and human health issues, and costs for chemical use (Gussow, Ch. 10; and Coye, Ch. 9).
(3) Environmental degradation from the use of agricultural chemicals.
(4) Exploitation of geological endowments (soil, water, minerals) and nonrenewable energy resources.
(5) Displacement of farm workers by mechanization.

(6) Animal welfare and animal rights vs. animals for food.
(7) Destruction of fish and wildlife habitats (National Academy of Sciences, 1982)
(8) Inadequate funding of research to ensure future food security (Ruttan, 1982).

Conversely, there are externalities of new agricultural technologies that may be characterized as positive, or predominantly so:

(1) Enhanced productivity of crops and livestock.
(2) Reasonable food prices for everyone--domestic and foreign.
(3) Lowest expenditure of personal income for food in the world; approximately 16 percent during the past decade.
(4) Enormous international trade and exporting earnings--45 billion for 1981.
(5) Greater productivity per farmer than anywhere in the world (each farmer produces food for himself and for 80 additional people).
(6) Release of land resources for other uses from a tripling of crop yields during the last 40 years.
(7) Food security, abundance, and variety.

Agriculture in the U.S. and in other industrialized countries has substituted machines for human labor to make the greatest use of the short periods of good weather available for certain farming operations (planting, tillage, pest control, harvesting). The goals of mechanization in many agriculturally developing nations, however, must be viewed differently. For those nations, the emphasis on increased production and convenience is not applicable.

CONCLUSIONS

The agricultural progress witnessed in the United States since the Second World War, and technologies which emerged therefrom, now has parallels in other nations, where a high degree of food security, in some instances, has been achieved. Grain production in India's Punjab has tripled and food supplies in India have not been critical during the past decade. India is now exporting rather than importing grain. Other nations, including Pakistan, Turkey, Colombia, Mexico, and Thailand, have made great progress as a result of "the green revolution" in new seeds, irrigation, and fertilizer. Indonesia and the Philippines have achieved near self-sufficiency within a decade in rice production. Both Japan and Taiwan are each confronted with more than one million metric tons of surplus rice resulting from over production for which there is no market. In India today there is a "white revolution" in milk production. It had its origin in the Gugarat state northwest of Bombay. Millions of farmers, some landless workers, deliver milk twice a day to hundreds of cooperatives. A daily cash income is provided and standards of living for farmers are raised in a system which is labor intensive at the production level and labor sparing for marketing. Its spread is termed "operation flood."

Food production problems, however, remain. Malnutrition is prevalent and hunger persists in most of Africa, as well as some countries in Latin America; and in most small nations in the semi-arid tropics where population growth is rapid, political instability exists, resources are limited, and there are no migration options.

The United States, Canada, Argentina, Brazil, and Australia are the major grain and oil exporting nations of the world, upon which many countries rely for vast quantities of food. They are the bread baskets of last resort. World grain trade approximated 230 million metric tons for 1982 or about 14 percent of the total production. World grain stocks exceeded 15 percent of utilization. More than 60 percent of the grain for international trade in 1980 had its origin in the United States. The U.S. cannot indefinitely serve as the bread basket of the world. Food production and its delivery, along with resource inputs, will become increasingly expensive. Ultimately, food will need to be produced closer to the people who consume it.

Increased food production will go far to alleviate hunger and malnutrition, and to establish food security. It, alone, however, is not enough. Distribution and delivery of food and enough income to buy food are the critical problems (Christenson, et al., Ch. 13). Today, food surpluses, malnutrition, and hunger exist side by side in many nations. Only poor people go hungry. The challenge, as to food security for the future, is to get food to people who need it. People are malnourished and go hungry because of ineffective distribution, lack of purchasing power, and, in some cases, lack of information. Government decisions may also prohibit ready access to available food.

Improved local production would solve some food distribution and malnutrition problems. It should improve availability, accessibility, and adequacy of food supplies. China has fewer land resources than the United States, yet it feeds more than a billion people--and quite well at that. China's achievement and experience suggest that one answer to both worldwide food problems and food security--availability, accessibility, and adequacy--is that production must be where the people are. This means improving the chances for feeding the hungry and poor by bringing the means of food production closer. In addition, local growing, harvesting, processing, and sale of food alleviates shortages. Among the challenges of the future will be to promote the productivity of the small farm units which constitute most of the world. They can become economically viable units (Wittwer, 1983a).

Although great differences of opinion were expressed in the workshop as to the types of research which should be conducted by both publicly and privately supported institutions in the U.S., there was general agreement that more research of the right kind was needed (Buttel, et al., 1983; Rockefeller Foundation and Office of Science and Technology Policy, 1982). The U.S. and the world continue to undervalue investments in agricultural research and fail to increase investments to economically justified levels (Ruttan, 1982). The world is not giving high priority to its future ability, through research and technology, to feed itself and to achieve food security (Meyers, Ch. 15).

Maximum yield or productivity trials could be a useful approach (Wortman and Cummings, Jr, 1978). They could represent the ultimate in use of new technology, resource inputs, and economic incentives. They would be as important for practice as for theory. Records of production for many food commodities of 20 years ago are averages for today. Record yields should not be regarded as abnormal occurrences, and a study of the comparative productivity of previously successful agricultural systems could be most rewarding. Scientists should be encouraged to constantly test the limits of available technology, as well as their ability to combine technology and resource components in a systems approach to achieve maximum production within constraints of resource inputs (Cardwell, 1982). This is what successful farmers must do. The operational approach for maximizing yields must be one which integrates the total of inputs. These include climate and weather, appropriate cultivars, soil nutrients, soil structure, soil moisture, temperature, and sunlight. There are cropping systems, pest management systems, water management systems, soil management systems, and tillage systems. Finally, much of the role of science for future food security will focus on biological improvements. The extensive use of commercial hybrid crop varieties important for feeding people is still confined to corn, sorghum, millet, sugar beets, and coconuts.

Development of human resources for agricultural improvement and food security will be the great challenge for the future. The talents of some of the very best minds among young scientists are being denied to agricultural and food research. This cannot be rationalized with true national or international interest if food security is important. There must be a reaffirmation and recognition by American and international academicians that the pursuit of food research and technology is a respectable and modern science. There is no limit to the scientific creativity of the human mind. Scientific and biological limits are far from being reached; rather, people have only begun to explore the possibilities in food production and food security. The resource base can change with time and technology. Science is extending the limits and opening the frontiers of the world.

Figure 18.1

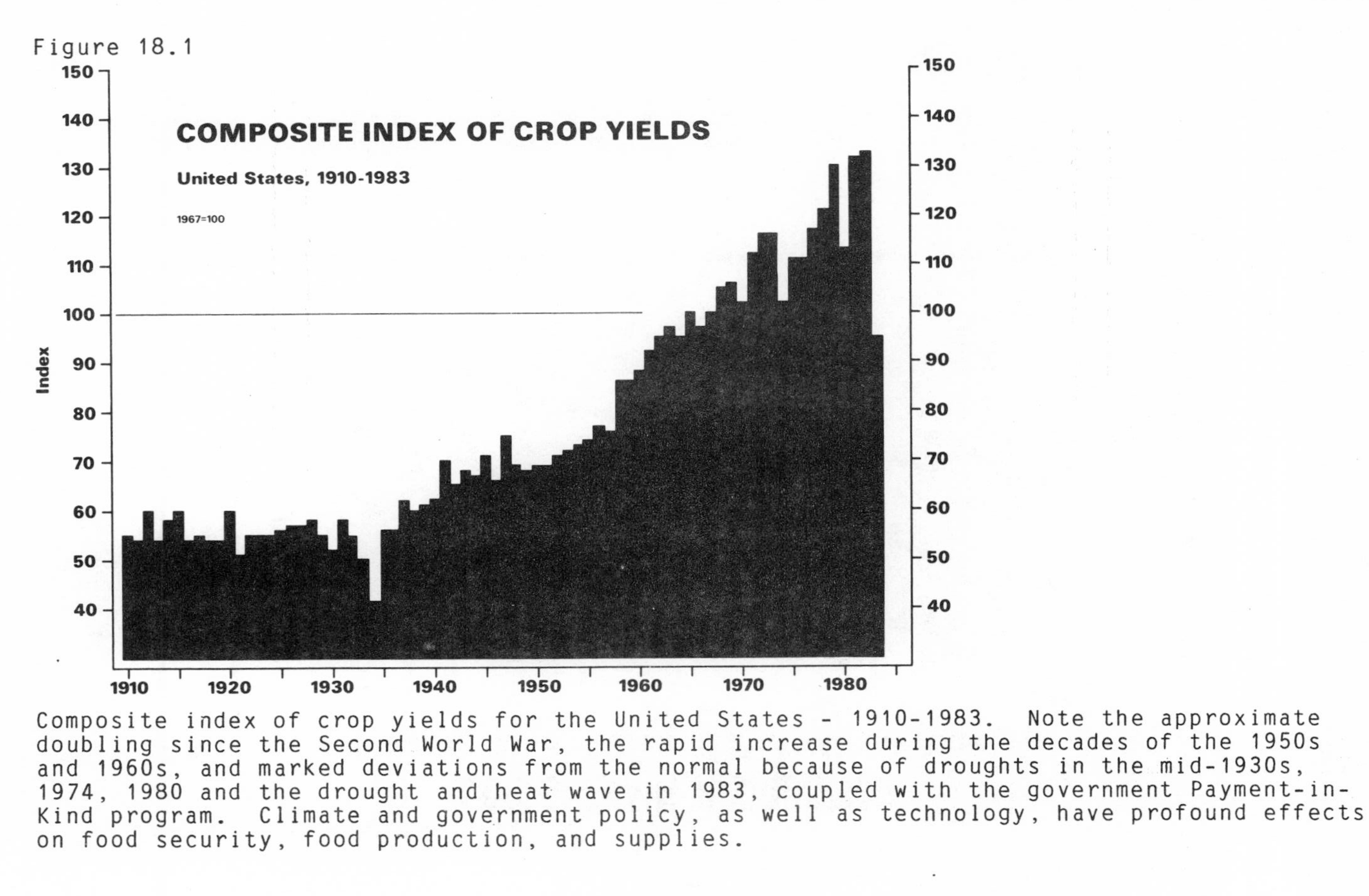

Composite index of crop yields for the United States - 1910-1983. Note the approximate doubling since the Second World War, the rapid increase during the decades of the 1950s and 1960s, and marked deviations from the normal because of droughts in the mid-1930s, 1974, 1980 and the drought and heat wave in 1983, coupled with the government Payment-in-Kind program. Climate and government policy, as well as technology, have profound effects on food security, food production, and supplies.

Figure 18.2

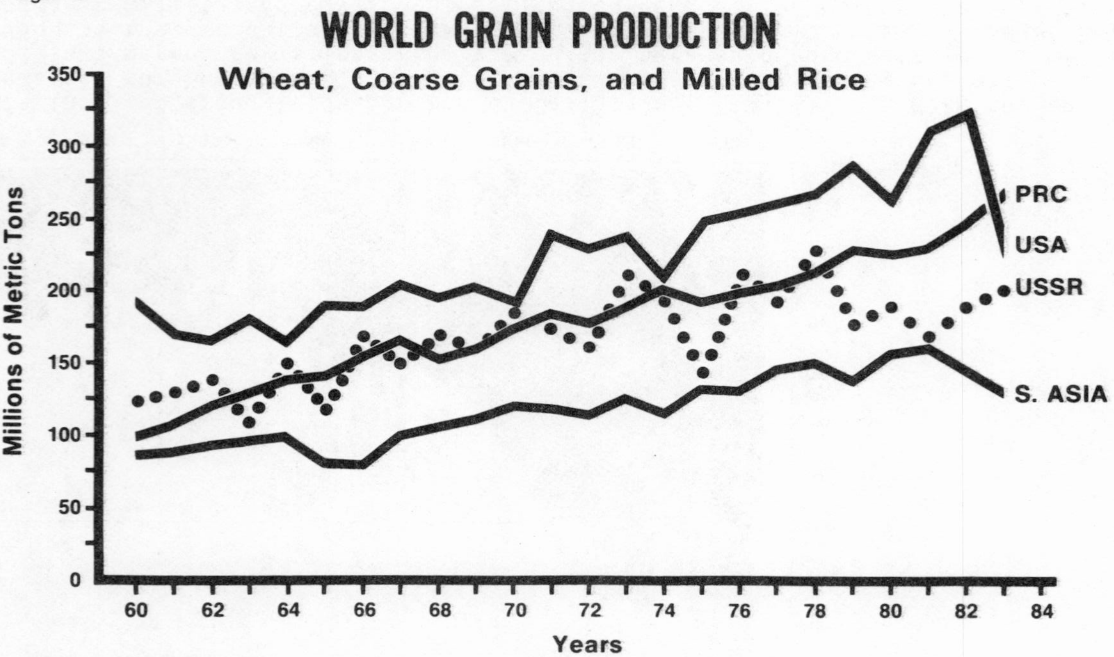

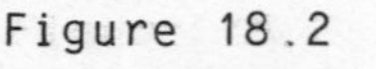

World grain production (1960-1983) as reflected by the main producing areas. The general trend is upward, but with many climatic and policy induced perturbations especially in the USA and the USSR. Adverse climates seldom strike all grain producing areas during the same year. Note the consistent upward trend in the People's Republic of China (PRC). This can largely be ascribed to over 50 percent irrigation of cultivated crops.

REFERENCES

Altieri, M. A.; Letourneau, D. K.; and Davis, J. R. 1983. "Developing Sustainable Agroecosystems." Bioscience 33(1):45-49.

Barr, T. N. 1981. "The World Food Situation and Global Grain Prospects." Science 214(4525):1087-94.

Batie, S. S., and Healy, R. G., eds. 1980. The Future of American Agriculture as a Strategic Resource. Washington, D.C.: The Conservation Foundation.

______. 1983. "The Future of American Agriculture." Scientific American 248(2):45-53.

Bonnen, J. T. 1981. "Agriculture's System of Developing Institutions: Reflections on the U.S. Experience." Paper presented at the 1981 Symposium on Rural Economics: Quebec Agriculture and Food Economy and its Development in the 1980's, October 19-22, at the University of Laval, Quebec, Canada.

______. 1983. "Historical Sources of U.S. Agricultural Productivity: Implications for R and D Policy and Social Science Research." American Journal of Agricultural Economics 65(5):958-966.

Boyer, J. S. 1982. "Plant Productivity and Environment." Science 218:443-48.

Brown, L. R. 1981. "World Population Growth, Soil Erosion and Food Security." Science 214(4524): 995-1002.

Brown, A. W. A.; Byerly, T. C.; Gibbs, M.; and San Pietro, A., eds. 1975. "Crop Productivity--Research Imperatives." Proceedings of an International Conference at Michigan State University, East Lansing, Michigan, and Charles F. Kettering Foundation, Yellow Springs, Ohio.

Busch, L., and Lacy, W. B. 1983. Science, Agriculture, and the Politics of Research. Boulder: Westview Press.

Buttel, F. H.; Kenney, M.; Kloppenburg Jr., J.; and Cowan, J. T. 1983. "Problems and Prospects of Agricultural Research: the Winrock Report." The Rural Sociologist 3(2):67-75.

Campbell, W. C.; Fisher, M. H.; Stapley, E. O.; Albers-Schonberg, G.; and Jacob, T. A. 1983. "Ivermectin: A Potent New Antiparasitic Agent." Science 221:823-28.
Cardwell, V. B. 1982. "Fifty Years of Minnesota Corn Production: Sources of Yield Increase." Agronomy Journal 74:984-90.
Carson, R. 1962. Silent Spring. Cambridge: Riverside Press.
Center for Rural Affairs. 1982. "The Path Not Taken. A Case Study of Agricultural Research Decision-Making at the Animal Science Department of the University of Nebraska." Walthill, Nebraska.
Clark, W. C., ed. 1982. Carbon Dioxide Review: 1982 New York: Clarendon Press, Oxford Univ. Press.
Connecticut Agricultural Experiment Station. 1983. Acid Rain, Sources and Effects in Connecticut. Report of the Acid Rain Task Force. New Haven.
Dundon, S. 1982. "Hidden Obstacles to Creativity in Agricultural Sciences." Mimeographed.
Edens, T. C., and Haynes, D. L. 1982. "Closed System Agriculture: Resource Constraints, Management Options, and Design Alternatives." Annual Review Phytopathology 20:363-95.
Edens, T. C., and Koenig, H. E. 1980. "Agroecosystem Management in a Resource-limited World." Bioscience 30:697-701.
Geissbuhler, H.; Brenneisen, P.; and Fisher, H. P. 1982. "Frontiers in Crop Production: Chemical Research Objectives." Science 217:505-10.
Hadwiger, D. F. 1982. The Politics of Agricultural Research. Lincoln: Univ. of Nebraska Press.
Heck, W. W.; Taylor, O. C.; Adams, R.; Bingham, G.; Miller, J.; Preston, E.; and Weinstein, L. 1982. "Assessment of Crop Loss from Ozone." Journal of the Air Pollution Association 32(4):353-61.
Hightower, J., and Demarko, S. 1972. "Hard Tomatoes, Hard Times." Agribusiness Accountability Project. Washington, D.C.
Hildreth, R. J. 1982. "The Agricultural Research Establishment in Transition." Proceedings. Academy of Political Science 34(3):235-47.
Johnson, G. L. 1983. "Ethical Dilemmas Posed by Recent and Prospective Developments with Respect to Agricultural Research." Paper presented at the Symposium on Societal Implications of Scientific Advances in Agricultural Sciences. Annual Meetings of the American Association for the Advancement of Science, May 27. Detroit, Michigan.
Johnson, G. L., and Wittwer, S. H. 1983. "Perspective on the Role of Technology in Determining Future Supplies of Food, Fiber and Forest Products in the U.S." East Lansing: Michigan Agricultural Experiment Station; Washington, D.C.: Resources for the Future.
Kenny, M.; Buttel, F. H.; Cowan, J. T.; and Kloppenburg, Jr., J. K. 1982. "Genetic Engineering and

Agriculture: Exploring the Impacts of Biotechnology on Industrial Structure, Industry-University Relationships, and the Social Organization of U.S. Agriculture." Cornell Rural Sociology Bulletin(125). Ithaca: Cornell Univ.

Larson, W. E.; Walsh, L. M.; Stewart, B. A.; and Boelter, D. H., eds. 1981. Soil and Water Resources: Research Priorities for the Nation. Madison: Soil Science Society of America.

Larson, W. E.; Pierce, F. J.; and Dowdy, R. H. 1983. "The Threat of Soil Erosion to Long-term Crop Production." Science 219:458-65.

Lemon, E. R., ed. 1983. "CO_2 and Plants. The Response of Plants to Rising Levels of Atmospheric Carbon Dioxide." AAAS Selected Symposium 84. Boulder: Westview Press.

Martin, P. L. 1983. "Labor-intensive Agriculture." Scientific American 249(4):54-59.

Mayer, A., and Mayer, J. 1973. "Agriculture: the Island Empire." Daedalus 103 (Summer):83-95.

Myers, N. 1983. A Wealth of Wild Species. Storehouse for Human Welfare. Boulder: Westview Press.

Napier, T. L.; Scott, D.; Easter, K. W.; and Supalla, R., eds. 1983. Water Resources Research: Problems and Potentials for Agriculture and Rural Communities. Ankeny, Iowa: Soil Conservation Society of America.

National Academy of Sciences. 1976. Climate and Food. Climatic Fluctuation and U.S. Agricultural Production. Washington, D.C.

______. 1982. Impacts of Emerging Agricultural Trends on Fish and Wildlife Habitat. National Research Council. Washington, D.C.: National Academy Press.

______. 1983a. Changing Climate. National Research Council. Washington, D.C.: National Academy Press.

______. 1983b. "The Molecular and Genetic Technology of Plants." In Frontiers in Science and Technology, pp. 45-62. Washington, D.C.

______. 1983c. Report of the Briefing Panel on Agricultural Research. Committee on Science, Engineering, and Public Policy. Washington, D.C.: National Academy Press.

National Acid Precipitation Assessment Program. 1982. Annual Report to the President and Congress. Washington, D.C.

National Aeronautics and Space Administration. 1983. "Land-Related Global Habitability Science Issues." Land-Related Global Habitability Science Working Group. NASA Office of Space Science and Applications, Washington, D.C.

National Research Council. 1972. National Technical Information Science, PB213-338. Washington, D.C.

Office of Technology Assessment. 1981. An Assessment of the United States Food and Agricultural Research System. Congress of the United States, Washington, D.C.

______. 1982. Impacts of Technology on U.S. Cropland and Rangeland Productivity. Congress of the United States, Washington, D.C.

Paarlberg, D. 1981. "The Land Grant Colleges and the Structure Issue." American Journal of Agricultural Economics 63 (February):129-34.

Phillips, R. E.; Blevins, R. L.; Thomas, G. W.; Frye, W. W.; and Phillips, S. 1980. "No-tillage Agriculture." Science 208:1108-13.

Pimentel, D.; Moran, M. A.; Fast, S; Weber, G.; Bukantis, R.; Balliett, L.; Boveng, P.; Cleveland, C.; Hindman, S.; and Young, M. 1981. "Biomass Energy from Crop and Forest Residues." Science 212(4499): 1110-15.

Pond, W. G.; Merkel, R. A.; McGilliard, L. D.; and Rhodes, V. J., eds. 1980. Animal Agriculture--Research to Meet Human Needs in the 21st Century. Boulder: Westview Press.

Presidential Commission on World Hunger. 1980. Overcoming World Hunger: The Challenge Ahead. Washington, D.C.

Putnam, A. R. 1983. "Allelopathic Chemicals." Chemical Engineering News 61 (April 4):34-45.

Rasmussen, W. D. 1982. "The Mechanization of Agriculture." Scientific American 247(3):77-89.

Rockefeller Foundation and Office of Science and Technology Policy, the White House. 1982. Science for Agriculture. New York: Rockefeller Foundation.

Ruttan, V. W. 1982. Agricultural Research Policy. Minneapolis: Univ. of Minnesota Press.

Schroth, M. N., and Hancock, J. G. 1982 "Disease-suppressive Soil and Root-colonizing Bacteria." Science 216(4553):1376-81.

Schultz, T. W. 1982. "The Dynamics of Soil Erosion in the United States." Agricultural Economics Paper 82:8. Chicago: Univ. of Chicago.

Seidel, Jr., G. F. 1981. "Superovulation and Embryo Transfer in Cattle." Science 211:351-58.

Seidel, S., and Keyes, D. 1983. Can We Delay a Greenhouse Warming? The Effectiveness and Feasibility of Options to Slow a Build-up of Carbon Dioxide in the Atmosphere. Washington, D.C.: U.S. Environmental Protection Agency.

Sharp, W. R., and Evans, D. A. "Plant Tissue Culture: the Foundation for Genetic Engineering in Higher Plants." Paper presented at the North Atlantic Region of the American Society of Agricultural Engineers Annual Meetings, August 9, at Burlington, Vermont.

Slater, L. E., and Levin, S. K., eds. 1981. "Climatic Impact on Food Supplies." AAAS Selected Symposium 62. Boulder: Westview Press.

Stark, J. C.; Jarrell, W. M.; Letey, J.; and Valoras, N. 1983. "Nitrogen Use Efficiency of Trickle Irrigated Tomatoes Receiving Continuous Injections of N."

Agronomy Journal 75:672-76.
USDA. 1982. Research for Small Farms. Proceedings of the Special Symposium, H. K. Kerr, Jr. and K. Krudson, eds. Agricultural Research Service. Misc. Publ. No. 1422. Washington, D.C.
______. 1983. Agricultural Research Service Program Plan. Agricultural Research Service Miscellaneous Publication No. 1429. Washington, D.C.
Walker, J. T. 1982. Characterization of Rain in the Georgia Piedmont and Effects of Acidified Water on Crops and Ornamental Plants. Georgia Agricultural Experiment Station Research Bulletin No. 283.
Wilkes, G. 1983. "Current Status of Crop Plant Germplasm." CRC Critical Reviews in Plant Science 1(2):133-81. Boca Raton: CRC Press.
Wittwer, S. H. 1980a. "Agriculture: America's Number 1 Industry." In Reflections of America, Commemorating the Statistical Abstract Centennial. Washington, D.C.: U.S. Department of Commerce.
______. 1980b. "Future Trends in Agricultural Technology and Management." In Long-Range Environmental Outlook. Washington, D.C.: National Research Council, National Academy of Sciences.
______. 1980c. "Food Production Prospects: Technology and Resource Options." In The Politics of Food, ed. D. Gale Johnson. Chicago: Chicago Council of Foreign Relations.
______. 1982a. "Solar Energy and Agriculture." Experimentia 38:10-12.
______. 1982b. "U.S. Agriculture in the Context of the World Food Situation." In Science, Technology, and the Issues of the Eighties: Policy Outlook, eds. A. H. Teich and Ray Thornton, pp. 191-214. Boulder: Westview Press.
______. 1982c. "The Michigan Challenge: Adjustments in Agricultural Research to Meet a Balanced Economy." Proceedings of the Agricultural Research Institute, pp. 135-60, October 4-6, Fort Worth, Texas and Bethesda, Maryland.
______. 1983a. "Nutrition, Agriculture and World Health." Food and Nutrition News 55(1) (January/February). Chicago: National Livestock and Meat Board.
______. 1983b. "The New Agriculture: A View from the 21st Century." In Agriculture in the Twenty-First Century, pp. 337-67, Philip Morris Operations Complex, Richmond, Virginia, April 11-13. New York: Wiley.
______. 1983c. "Role of Science and Technology in Animal Production: Creating a Vision for the Future." Special lecture at the Fifth International Conference on Animal Production, August 14, Tokyo.
______. 1983d. "Decisions about Agricultural Uses of Land." Paper presented at the Western Hemisphere Nutrition Congress VII, August 10, at Miami Beach,

Florida. Published in the Proceedings of the conference.

______. 1983e. "Food, Our Most Important Renewable Resource." An essay for Change. New York: Copper-Hewitt Museum.

______. 1983f. "Rising Atmospheric CO_2 and Crop Productivity." HortScience 18:667-73.

Wortman, S., and Cummings, Jr., R. W. 1978. To Feed This World. Baltimore: Johns Hopkins Univ. Press.

Index